THEORETICAL PHYSICS

Previous Proceedings in the Series of MRST Conferences

Year		Held in	Publisher	ISBN
2001	23rd	London, Ont. Canada	AIP Conf. Proceedings Vol. 601	0-7354-0045-8
2000	22nd	Rochester, New York, USA	AIP Conf. Proceedings Vol. 541	1-56396-966-1
1999	21st	Ottawa, Ont. Canada	AIP Conf. Proceedings Vol. 488	1-56396-902-5
1998	20th	Montreal, Que. Canada	AIP Conf. Proceedings Vol. 452	1-56396-845-2

Other Related Titles from AIP Conference Proceedings

624 Cosmology and Elementary Particle Physics: Coral Gables Conference on Cosmology and Elementary Particle Physics
Edited by B. N. Kursunoglu, S. L. Mintz, and A. Perlmutter, July 2002, 0-7354-0073-3

616 Experimental Cosmology at Millimetre Wavelengths: 2K1BC Workshop
Edited by Marco DePetris and Massimo Gervasi, May 2002, 0-7354-0062-8

609 Astrophysical Polarized Backgrounds: Workshop on Astrophysical Polarized Backgrounds
Edited by Stefano Cecchini, Stefano Cortiglioni, Robert Sault, and Carla Sbarra, March 2002, 0-7354-0055-5

607 String Theory: 10th Tohwa University International Symposium on String Theory
Edited by H. Aoki and T. Tada, February 2002, 0-7354-0051-2

589 New Developments in Fundamental Interaction Theories: 37th Karpacz Winterschool of Theoretical Physics
Edited by Jerzy Lukierski and Jakub Rembieliński, October 2001, 0-7354-0029-6

562 Particles and Fields: Ninth Mexican School
Edited by Gerardo Herrera Corral and Lukas Nellen, April 2001, 1-56396-998-X

539 Symmetries in Subatomic Physics: 3rd International Symposium
Edited by X.-H. Guo, A. W. Thomas, and A. G. Williams, October 2000, 1-56396-964-5

493 General Relativity and Relativistic Astrophysics: Eighth Canadian Conference
Edited by C. P. Burgess and R. C. Myers, November 1999, 1-56396-905-X

484 Trends in Theoretical Physics II
Edited by Horacio Falomir, Ricardo E. Gamboa Saraví, and Fidel A. Schaposnik, July 1999, 1-56396-894-0

To learn more about these titles, or the AIP Conference Proceedings Series, please visit the webpage **http://proceedings.aip.org/proceedings**

THEORETICAL PHYSICS

MRST 2002: A Tribute to George Leibbrandt

Waterloo, Ontario, Canada   15–17 May 2002

EDITORS
Victor Elias
Richard J. Epp
Robert C. Myers
Perimeter Institute for Theoretical Physics
Waterloo, Ontario, Canada

SPONSORING ORGANIZATIONS
Perimeter Institute for Theoretical Physics
The Institute of Particle Physics/
  L'Institut de la Physique des Particules

Melville, New York, 2002
AIP CONFERENCE PROCEEDINGS ■ VOLUME 646

Editors:

Victor Elias
Richard J. Epp
Robert C. Myers

Perimeter Institute for Theoretical Physics
35 King Street North
Waterloo, Ontario N2J 2W9
CANADA

E-mail: velias@perimeterinstitute.ca
 repp@perimeterinstitute.ca
 rmyers@perimeterinstitute.ca

Authorization to photocopy items for internal or personal use, beyond the free copying permitted under the 1978 U.S. Copyright Law (see statement below), is granted by the American Institute of Physics for users registered with the Copyright Clearance Center (CCC) Transactional Reporting Service, provided that the base fee of $18.00 per copy is paid directly to CCC, 222 Rosewood Drive, Danvers, MA 01923. For those organizations that have been granted a photocopy license by CCC, a separate system of payment has been arranged. The fee code for users of the Transactional Reporting Service is: 0-7354-0101-2/02/$19.00.

© 2002 American Institute of Physics

Individual readers of this volume and nonprofit libraries, acting for them, are permitted to make fair use of the material in it, such as copying an article for use in teaching or research. Permission is granted to quote from this volume in scientific work with the customary acknowledgment of the source. To reprint a figure, table, or other excerpt requires the consent of one of the original authors and notification to AIP. Republication or systematic or multiple reproduction of any material in this volume is permitted only under license from AIP. Address inquiries to Office of Rights and Permissions, Suite 1NO1, 2 Huntington Quadrangle, Melville, N.Y. 11747-4502; phone: 516-576-2268; fax: 516-576-2450; e-mail: rights@aip.org.

L.C. Catalog Card No. 2002114216
ISBN 0-7354-0101-2
ISSN 0094-243X
Printed in the United States of America

CONTENTS

Preface..ix
George Leibbrandt (1937-2001) x
    M. C. C. Leibbrandt
Opening Remarks...xvii
    H. Burton

PLENARY TALKS

Strings in Flat Space and pp Waves from $\mathcal{N}=4$ Super Yang Mills 3
    D. Berenstein, *J. Maldacena*, and H. Nastase
The Cosmic Microwave Background and Inflation, Then & Now..............15
    J. R. Bond, C. Contaldi, D. Pogosyan, B. Mason, S. Myers, T. Pearson,
    U.-L. Pen, S. Prunet, T. Readhead, and J. Sievers
Black Holes in High-energy Collisions...................................34
    S. B. Giddings
Direct Evidence for Neutrino Flavor Transformation from
Neutral-current Interactions in SNO....................................43
    A. B. McDonald (representing the SNO Collaboration)
The Ground State of Quantum Gravity with Positive
Cosmological Constant..59
    L. Smolin

SUPERSTRINGS

String Bits and the Myers Effect 77
    P. J. Silva
Fixed-topology Solutions in the Myers Effect...........................83
    G. G. Alexanian, A. P. Balachandran, and *P. J. Silva*
A New Non-commutative Field Theory 89
    K. Savvidy

TOPICS IN QUANTUM FIELD THEORY

Trace Anomaly and Quantization of Maxwell's Theory on
Non-commutative Spaces...99
    S. I. Kruglov
Variational Two Fermion Wave Equations in QED.........................105
    A. G. Terekidi and J. W. Darewych
Breaking of Supersymmetry in a U(1) Model with Stueckelberg Fields........111
    S. V. Kuzmin and D. G. C. McKeon

Italicized name indicates the author who presented the paper.

Self Energy of Chiral Leptons in a Background Hypermagnetic Field 117
    J. Cannellos, E. J. Ferrer, and V. de la Incera
Magnetic Behaviour of SO(5) Superconductors 123
    R. MacKenzie
The Cosmological Constant Problem and Nonlocal Quantum Gravity 130
    J. W. Moffat

GENERAL RELATIVITY

Critical Behaviour in Spherically Symmetric Scalar Field Collapse in Any Dimension 139
    M. Birukou, V. Husain, *G. Kunstatter*, E. Vaz, M. Olivier, and B. Preston
Spacetime Ambiguities 147
    K. Lake
Imposing a 4-Dimensional Background on General Relativity 152
    E. M. Schaefer

MATRIX MODELS

Two-Dimensional Quantum Gravity as String Bit Hamiltonian Models 161
    B. Durhuus and *C.-W. H. Lee*
On Quantization of Matrix Models 167
    A. Starodubtsev
Large N Matrix Models and Noncommutative Fisher Information 173
    A. Agarwal, *L. Akant*, G. S. Krishnaswami, and S. G. Rajeev

PARTICLE PHENOMENOLOGY

Single Top Production and Extra Dimensions 179
    A. Datta
Constraints on the Vector Quark Model from Rare B Decays 183
    M. R. Ahmady, M. Nagashima, and A. Sugamoto
Chiral Lagrangian Treatment of the Isosinglet Scalar Mesons in 1-2 GeV Region 189
    A. H. Fariborz

BRANE WORLDS

Quest for a Self-tuning Brane-world Solution to the Cosmological Constant Problem 197
    J. M. Cline and H. Firouzjahi
Dynamical Stability of the AdS Soliton in Randall-Sundrum Model 203
    C. P. Burgess, J. M. Cline, N. R. Constable, and *H. Firouzjahi*

Italicized name indicates the author who presented the paper.

Order ρ^2 Corrections to Cosmology with Two Branes 209
    J. Vinet and J. M. Cline

dS/CFT CORRESPONDENCE

Conserved Quantities, Entropy and the dS/CFT Correspondence 217
    R. B. Mann and A. M. Ghezelbash
Static and Dynamic Vortices in de Sitter Spacetimes 223
    A. M. Ghezelbash and R. B. Mann
A Short Tale from de Sitter Space 229
    F. Leblond
Non-Abelian Monopoles and Dyons in a Modified Einstein-Yang-Mills-Higgs System 235
    J. S. Rozowsky

DEVELOPMENTS IN QUANTUM THEORY

Realism in the Realized Popper's Experiment 243
    G. Hunter
The Dirac Equation in Classical Statistical Mechanics 249
    G. N. Ord

Conference Schedule 255
List of Participants 259
Author Index 261

Italicized name indicates the author who presented the paper.

PREFACE

On May 15-17 2002, 73 physicists from over 25 universities and research institutes in Canada and the United States gathered together in Waterloo for the 24th annual Montréal-Rochester-Syracuse-Toronto (MRST) conference. MRST-02 was hosted by Perimeter Institute for Theoretical Physics to honour the memory of George Leibbrandt, an outstanding Canadian theorist. He played a vital role in all aspects of the founding of the Institute and served on its Board of Directors. It is a most fitting tribute to George that MRST has extended its orbit beyond its four founding universities to include Perimeter Institute. Traditionally, MRST meetings have provided an informal forum where researchers, primarily from eastern Canada and the northeastern United States, discuss recent developments in theoretical high energy physics. Indeed this tradition continued in Waterloo with talks ranging from formal topics in field theory, brane worlds and superstrings, to phenomenological issues. Another tradition that continued is the active participation of many graduate students and postdoctoral researchers. However, in keeping with the wider range of research activity at Perimeter Institute, the scope of MRST-02 reflected a broader range of topics in theoretical physics, including talks on general relativity, quantum gravity, quantum computation and high T_c superconductivity.

A number of stimulating plenary speakers contributed to MRST-02. Steve Giddings (from UC Santa Barbara) spoke on the possible detection of microscopic black holes at the LHC. Juan Maldacena (from Princeton's IAS) spoke on the recently discovered equivalence between certain strongly coupled gauge theories and quantum gravity with a negative cosmological constant. Perimeter's own Lee Smolin spoke on recent developments in the understanding of quantum gravity with a positive cosmological constant. Finally, showcasing the world-class physics that is already being done in the local region, two plenary speakers gave stimulating experimental talks. Dick Bond discussed the latest understanding of the cosmological parameters obtained from new data from the BOOMERANG experiment. Art MacDonald spoke on the recent exciting results from the SNO neutrino detector.

In hosting the first conference at Perimeter Institute, we received help from a number of people. In particular, we must thank Janet Fesnoux for her aid in anticipating and carrying out a myriad of organizational tasks. Rita Schwander helped in manning the registration table and overseeing the catering. John McCormick and Dmitry Brovkovitch gave technical support setting up the computer facilities and audio-visual systems. Further, we need to thank Steve Roth and Julie Blake for allowing use of the Waterloo Stage Theatre to host the conference on short notice and at minimal cost. Also, we very much appreciate the secretarial support provided by Audrey Kager at University of Western Ontario. Finally we have to thank our gang of "roadies", Frederic Leblond and David Winters, for their efforts in ensuring the smooth running of the conference.

We would also like to thank The Institute of Particle Physics (Canada) for generous financial assistance, as well as all MRST-02 participants for contributing to the lively, friendly and stimulating exchange of ideas.

Victor Elias, Richard J. Epp and Robert C. Myers
August 2002, Waterloo

George Leibbrandt
(1937 - 2001)

George Leibbrandt (October 23, 1937 - April 3, 2001)

Martha Cecily Coventry Leibbrandt

The George Leibbrandt I met in the fall of 1960 was a shy, modest guy, with a keen mind and a great sense of humour—someone I fell in love with in no time at all. His sterling qualities were immediately apparent, but, even so, I underestimated him. He was utterly fearless and a true and loyal friend. Perhaps because of his close bond with his mother, a remarkable lady, he was always considerate and supportive of women. Courteous to a fault and a good listener, he was quick to size up people and situations, kept his own counsel and remained his own person. Theoretical physics was clearly the love of his life, but I soon learned that ancient history, the Classics, Greek mythology, the Mayan civilization, languages, music, theatre, to name a few, were vigorous contenders. Almost everything in life held value and fascination for him, and that made him a wonderful companion.

That is not to say that I was entirely content with the way things turned out. In my innocence, in the early days of our relationship, I thought that once established and the pressure off, George would engage in his professorial duties—lecturing, conducting research, meeting with students—then relax. I had no idea I had married someone who was happiest working the better part of every day, seven days a week. Yet it surprised me how he kept to this schedule while packing so much more into his daily life. He would be off, with the least encouragement, to Stratford or the Shaw Festival, or to watch a hockey game, especially if his favourite Maple Leafs were playing. Soccer matches on television, especially when Germany was on the field, got his competitive juices flowing. He was a vocal armchair quarterback during football season, a staunch supporter of tennis stars like Boris Becker and Steffi Graf, and an avid Olympics watcher. Whether playing his zither or watching a meteor shower, or sitting with his feet in a mountain stream, or surprising me with a gala evening at the Vienna State Opera, George was in his element.

About twelve years ago, we became interested in ballroom dancing and, as our proficiency increased, so did our involvement. George approached dancing as he did everything else—rigorously. He kept lists of things to remember, wanted to know every detail for proper execution of the steps and made sure we practised regularly. Even while working in Cambridge, Vienna and Geneva, George found a practice site and kept up the routine.

In the early 50s, George emigrated from Germany knowing very little English. Through dint of perseverance he became not only fluent, but skilled in the language, delighting in its subtleties. He became a good writer. Graduate students and post-docs alike will remember how much emphasis George placed on a well-written paper. Sentences and paragraphs were to flow logically and naturally; grammar and spelling to be above reproach. He spent hours vexing over every turn of phrase. Like Churchill, he made lists of substitute choices above words in his working copy.

His penchant for language and his natural wit made him irrepressible. I can well recall on one occasion when I was trying to get myself pulled together for some function or other—George was always ready ahead of time—he came to the door of the bathroom as I was putting on my "face", gave me a mischievous grin and said, "Ah, beauty in preparation!" I could have strangled him. When we arrived home after a working day I seldom took time to do more than doff my coat before rushing to the stove to throw something together for supper. George was no cook, but he did his part, peeling vegetables, or setting the table. All of which I was grateful for, except that at such a time he couldn't resist the temptation to ask, "And what culinary delights have you in store for us tonight?"

George could be maddening and funny, but when it came to physics he was in earnest. It's my belief he found majesty and beauty and confirmation of the existence of a Supreme Being in the laws of physics. He hungered to know more. He took on the thorniest problems as a personal challenge. Weeks and months could be spent working in his home office on a problem that refused to be solved. Meanwhile, pages and pages of formulae piled up in his wastebasket. One night, after such a lengthy period of discouragement, I woke to find his side of the bed empty and the light from his study indicating where he was. Some time later he returned. "I've got it," was all he said, but I knew this was one of those sweet moments of victory, something to savour, something to spur him on.

In the late 60s, George began spending a period away each summer, doing research, first at Harvard and then overseas. I told him years later that such periods of "development time" meant we were married only half as long as in actual fact, because he was away so much. Admittedly, in the early years we usually went away for longer stints as a family, going to London (Imperial College), Trieste (The International Centre for Theoretical Physics) and finally to Hamburg (DESY). Then I had to wait until the children were grown before joining him for a time each year at CERN.

Travelling was hard on the kids, especially as they dealt with immersion in two languages other than their own in fairly rapid succession. But we had no choice. George did his best to turn it to their advantage, making opportunities for them to see the sights and absorb the culture. Despite his busy schedule, he was a proud and caring father. In Italy, sometimes, he would conduct his "Uncle Daddy's Father's School", a light-hearted way for him to see that the children maintained their edge in arithmetic and other subjects. No bedtime was complete until George related another episode of "Grumbading the Whale", a story he made up as he went along about bad guys and good guys, some stupid and wise pirates and, of course, lots of marine life. It was full of adventure, suspense and some very funny moments. The youngsters loved it.

In George's filing cabinet I recently found all the file folders he had maintained, each with a child's name attached, containing a clutch of mementos—crayon drawings, letters and cards. His children were central to his life, even when he was away from them. He followed their progress with pride and interest. His overriding concern, however, and one he stressed, was that they take full advantage of their God-given talents and act on every opportunity that came their way.

This attitude of a parent was present in his relationship with his students. He prepared his lectures with care and tried his best to make the subject matter meaningful to them. In return he expected them to work hard. A student who struggled against odds to finish creditably won his unqualified respect. On the other hand, he had little patience with those looking for "a free ride". Along with the required material, he tried to provide a view of where the field was heading, bringing back to his students news and tidbits from conferences he had attended. He frequently larded his lectures with additional references, sometimes from ancient history or the Classics, which he hoped would give them a more liberal, well-rounded education. His witticisms and wry sense of humour were probably lost on a certain proportion, but fortunately, a goodly number of students appreciated what he was trying to do. Letters from them at the time of his death spoke of his infectious enthusiasm for learning and his skill as a teacher, but most of all of his friendship. One student wrote: "Words cannot express how kind and generous George was to me. His support has deeply touched me and his love and passion for his work has inspired me."

Whenever George was slated to give a talk to colleagues, he moved into high gear, which meant weeks of marshalling his data and preparing his overheads. George had a "thing" about overheads. He had suffered through too many talks where overheads were almost unintelligible with too much data on each, and inscribed in a way that only the presenter could decipher. As a presentation loomed, George brought out his arsenal of coloured pens and set to work on what for me were works of art—very few lines of formulae per page, writ large and with certain features set off by different coloured inks. His aim was to present his material as clearly and logically as possible. For him, one of the greatest compliments he every received was being told his talk had been "pedagogically excellent."

George had many high moments in his life, but I am sure it would be no exaggeration to say that one of the most thrilling happened on his birthday, October 23, 2000. That was the day of the public announcement officially launching Perimeter Institute for Theoretical Physics in Waterloo, Ontario. During almost two years as a director on its board, he had become deeply involved in most aspects of its development.

George brought a number of attributes to this new role. Over thirty years he had formed a bond of mutual trust and respect with people at various institutions around the world whose own knowledge, experience and foresight could, and indeed proved to be, invaluable in planning such an international centre. Drawing upon over three decades of experience as a physicist working abroad, he had also come to some conclusions himself which he could offer as advice. And finally, George had a vision for the Institute that was completely objective. He did not have a single, self-serving bone in his body.

The Institute for George was a dream-come-true. He was on fire. As time went on, and the project remained "unofficial", I became the recipient of news on progress, problems and possibilities. Especially over the dinner table he would review the status so far. He mulled over every detail of mandate, governance and scope, so that PI would take its place in the world as a viable, important centre for research and learning. Oddly enough, I was particularly surprised at how much he got involved with the design of the building. As a member of the building committee, he wanted the eventual structure to complement the physicists in everything they did. He envisioned it as light, spacious and airy, with places where the occupants could either get together in congenial, comfortable surroundings or be removed, uninterrupted and quiet. His notes and memos at the time indicate he worried over everything from the materials on the exterior, walls and floors, the type of windows—he wanted them to open—the dimensions of the foyer and offices, the access to green space, even to the availability of washrooms and showers. He abhorred "computerized structures", like some he had seen elsewhere "where light switches come 'on' and blinds open and close automatically". He wanted something more user-friendly, flexible and homelike.

When I see how far the Perimeter Institute has come towards fulfillment of George's dream—seeing colleagues, students and visiting scientists gather and productive work underway, seeing the site prepared and, more recently, the sod turned, I am so grateful—grateful that I had George, whom his colleagues wished to honour by naming MRST 2002 in his memory, and grateful that George had that time, brief though it was, to make his own unique contribution to theoretical physics in Canada.

Perimeter Institute's temporary home, dubbed "Spacetime Square"

OPENING REMARKS

On behalf of Perimeter Institute, I would like to take this opportunity to welcome you to both MRST 2002 and Perimeter Institute. Perimeter is motivated to play an increasingly active role in the surrounding theoretical physics community and we are correspondingly delighted to be able to host our first conference even before our new facility is completed. MRST is a particularly appropriate event to be the inaugural conference hosted by Perimeter, as it is historically an occasion for substantial numbers of graduate students and postdocs to present their work—and Perimeter, as you may know, is a research institute with a strong youth-oriented mandate, determined to serve as a free-wheeling environment where innovative young scientists can boldly pursue their ideas.

This conference could not have occurred without the dedication and commitment of several people. Richard Epp, Perimeter's Program Coordinator, worked tirelessly with Rob Myers, our renowned string theorist, to ensure the conference's success. Victor Elias, from the University of Western Ontario, was a font of wisdom and support. John McCormick was responsible for electronically transforming this theatre to an environment which could support a physics conference, while Janet Fesnoux provided invaluable logistic support and spun her usual magic to ensure that all participants were duly welcomed.

All that is left for me to do is to therefore officially welcome you all to MRST 2002 and wish everyone a most successful conference. I look forward to seeing you all at Perimeter in the months and years to come!

Howard Burton
Executive Director
Perimeter Institute for Theoretical Physics

PLENARY TALKS

Strings in flat space and pp waves from $\mathcal{N} = 4$ Super Yang Mills[1]

David Berenstein*, Juan Maldacena* and Horatiu Nastase*

*Institute for Advanced Study, Princeton, New Jersey 08540, USA

Abstract. We explain how the string spectrum in flat space and pp-waves arises from the large N limit, at fixed g_{YM}^2, of U(N) $\mathcal{N} = 4$ super Yang Mills. We reproduce the spectrum by summing a subset of the planar Feynman diagrams. We give a heuristic argument for why we can neglect other diagrams.

INTRODUCTION

The fact that large N gauge theories have a string theory description was believed for a long time [1]. These strings live in more than four dimensions [2]. One of the surprising aspects of the AdS/CFT correspondence [3, 4, 5, 6] is the fact that for $\mathcal{N} = 4$ super Yang Mills these strings move in ten dimensions and are the usual strings of type IIB string theory. The radius of curvature of the ten dimensional space goes as $R/l_s \sim (g_{YM}^2 N)^{1/4}$. The spectrum of strings on $AdS_5 \times S^5$ corresponds to the spectrum of single trace operators in the Yang Mills theory. The perturbative string spectrum is not known exactly for general values of the 't Hooft coupling, but it is certainly known for large values of the 't Hooft coupling where we have the string spectrum in flat space. In this paper we will explain how to reproduce this spectrum from the gauge theory point of view. In fact we will be able to do slightly better than reproducing the flat space spectrum. We will reproduce the spectrum on a pp-wave. These pp-waves incorporate, in a precise sense, the first correction to the flat space result for certain states.

The basic idea is the following. We consider chiral primary operators such as $Tr[Z^J]$ with large J. This state corresponds to a graviton with large momentum p^+. Then we consider replacing some of the Zs in this operator by other fields, such as ϕ, one of the other transverse scalars. The position of ϕ inside the operator will matter since we are in the planar limit. When we include interactions ϕ can start shifting position inside the operator. This motion of ϕ among the Zs is described by a field in 1+1 dimensions. We then identify this field with the field corresponding to one of the transverse scalars of a string in light cone gauge. This can be shown by summing a subset of the Yang Mills Feynman diagrams. We will present a heuristic argument for why other diagrams are not important.

---

[1] The present article is a condensed version of an earlier publication appearing in *J. High Energy Phys.* JHEP 04 (2002) 013.

Since these results amount to a "derivation" of the string spectrum at large 't Hooft coupling from the gauge theory, it is quite plausible that by thinking along the lines sketched in this paper one could find the string theory for other cases, most interestingly cases where the string dual is not known (such as pure non-supersymmetric Yang Mills).

This paper is organized as follows. In section two we will describe a limit of $AdS_5 \times S^5$ that gives a plane wave. In section three we describe the spectrum of string theory on a plane wave. In section 4 we describe the computation of the spectrum from the $\mathcal{N}=4$ Yang Mills point of view.

PP waves as limits of $AdS \times S$

In this section we show how pp wave geometries arise as a limit of $AdS_p \times S^q$.[2] Let us first consider the case of $AdS_5 \times S^5$. The idea is to consider the trajectory of a particle that is moving very fast along the S^5 and to focus on the geometry that this particle sees. We start with the $AdS_5 \times S^5$ metric written as

$$ds^2 = R^2 \left[-dt^2 \cosh^2\rho + d\rho^2 + \sinh^2\rho d\Omega_3^2 + d\psi^2 \cos^2\theta + d\theta^2 + \sin^2\theta d\Omega_3'^2 \right] \quad (1)$$

We want to consider a particle moving along the ψ direction and sitting at $\rho = 0$ and $\theta = 0$. We will focus on the geometry near this trajectory. We can do this systematically by introducing coordinates $\tilde{x}^{\pm} = \frac{t \pm \psi}{2}$ and then performing the rescaling

$$x^+ = \tilde{x}^+, \quad x^- = R^2 \tilde{x}^-, \quad \rho = \frac{r}{R}, \quad \theta = \frac{y}{R}, \quad R \to \infty \quad (2)$$

In this limit the metric (1) becomes

$$ds^2 = -4dx^+dx^- - (\vec{r}^{\,2} + \vec{y}^{\,2})(dx^+)^2 + d\vec{y}^{\,2} + d\vec{r}^{\,2} \quad (3)$$

where $\vec{y}$ and $\vec{r}$ parameterize points on R^4. We can also see that only the components of F with a plus index survive the limit. We see that this metric is of the form of a plane wave metric[3]

$$ds^2 = -4dx^+dx^- - \mu^2 \vec{z}^{\,2} dx^{+2} + d\vec{z}^{\,2}$$
$$F_{+1234} = F_{+5678} = \text{const} \times \mu \quad (4)$$

where $\vec{z}$ parametrizes a point in R^8. The mass parameter μ can be introduced by rescaling (2) $x^- \to x^-/\mu$ and $x^+ \to \mu x^+$. These solutions where studied in [8].

It will be convenient for us to understand how the energy and angular momentum along ψ scale in the limit (2). The energy in global coordinates in AdS is given by $E = i\partial_t$ and the angular momentum by $J = -i\partial_\psi$. This angular momentum generator

---

[2] While this paper was being written the paper [7] appeared which contains the same point as this section.
[3] The constant in front of F depends on the normalizations of F and can be computed once a normalization is chosen.

can be thought of as the generator that rotates the 12 plane of R^6. In terms of the dual CFT these are the energy and R-charge of a state of the field theory on $S^3 \times R$ where the S^3 has unit radius. Alternatively, we can say that $E = \Delta$ is the conformal dimension of an operator on R^4. We find that

$$\begin{aligned} 2p^- = -p_+ &= i\partial_{x^+} = i\partial_{\tilde{x}^+} = i(\partial_t + \partial_\psi) = \Delta - J \\ 2p^+ = -p_- &= -\frac{\tilde{p}_-}{R^2} = \frac{1}{R^2}i\partial_{\tilde{x}^-} = \frac{1}{R^2}i(\partial_t - \partial_\psi) = \frac{\Delta + J}{R^2} \end{aligned} \quad (5)$$

Notice that $p^\pm$ are non-negative due to the BPS condition $\Delta \geq |J|$. Configurations with fixed non zero p^+ in the limit (2) correspond to states in AdS with large angular momentum $J \sim R^2 \sim N^{1/2}$. When we perform the rescalings (2) we take the $N \to \infty$ limit keeping the string coupling g fixed and we focus on operators with $J \sim N^{1/2}$ and $\Delta - J$ fixed.

From this point of view it is clear that the full supersymmetry algebra of the metric (1) is a contraction of that of $AdS_5 \times S^5$ [8]. This algebra implies that $p^\pm \geq 0$.

This limit is a particular case of Penrose's limit [9][4], see also [10, 11]. In other $AdS_d \times S^p$ geometries we can take similar limits. The only minor difference as compared to the above computation is that in general the radius of AdS_d and the sphere are not the same. Performing the limit for $AdS_7 \times S^4$ or $AdS_4 \times S^7$ we get the same geometry, the maximally supersymmetric plane wave metric discussed in [12, 13]. For the $AdS_3 \times S^3$ geometries that arise in the D1-D5 system the two radii are equal and the computation is identical to the one we did above for $AdS_5 \times S^5$.

In general the geometry could depend on other parameters besides the radius parameter R. It is clear that in such cases we could also define other interesting limits by rescaling these other parameters as well. For example one could consider the geometry that arises by considering D3 branes on A_{k-1} singularities [14]. These correspond to geometries of the form $AdS_5 \times S^5/Z_k$ [15]. The Z_k quotient leaves an S^1 fixed in the S^5 if we parameterize this S^1 by the ψ direction and we perform the above scaling limit we find the same geometry that we had above except that now $\vec{y}$ in (3) parameterizes an A_{k-1} singularity. It seems possible to deform a bit the singularity and scale the deformation parameter with R in such a way to retain a finite deformation in the limit. We will not study these limits in detail below but they are of clear physical interest.

Strings on pp-waves

It has been known for a while that strings on pp-wave NS backgrounds are exactly solvable [16]. The same is true for pp-waves on RR backgrounds. In fact, after we started thinking about this the paper by Metsaev [17] came out, so we will refer the reader to it for the details. The basic reason that strings on pp-waves are tractable is that the action dramatically simplifies in light cone gauge.

---

[4] We thank G. Horowitz for suggesting that plane waves could be obtained this way.

We start with the metric (4) and we choose light cone gauge $x^+ = \tau$ where τ is the worldsheet time. Then we see that the action for the eight transverse directions becomes just the action for eight massive bosons. Similarly the coupling to the RR background gives a mass for the eight transverse fermions.

So in light cone gauge we have eight massive bosons and fermions. It turns out that 16 of the 32 supersymmetries of the background are linearly realized in light cone gauge (just as in flat space). These sixteen supersymmetries commute with the light cone hamiltonian and so they imply that the bosons and fermions have the same mass, see [17].

After the usual gauge fixing (see [18, 17]) the light cone action becomes

$$S = \frac{1}{2\pi\alpha'} \int dt \int_0^{2\pi\alpha' p^+} d\sigma \left[\frac{1}{2}\dot{z}^2 - \frac{1}{2}z'^2 - \frac{1}{2}\mu^2 z^2 + i\bar{S}(\not{\partial} + \mu I)S \right] \qquad (6)$$

where $I = \Gamma^{1234}$ and S is a Majorana spinor on the worldsheet and a positive chirality SO(8) spinor under rotations in the eight transverse directions. We quantize this action by expanding all fields in Fourier modes on the circle labelled by σ. For each Fourier mode we get a harmonic oscillator (bosonic or fermionic depending on the field). Then the light cone Hamiltonian is

$$2p^- = -p_+ = H_{lc} = \sum_{n=-\infty}^{+\infty} N_n \sqrt{\mu^2 + \frac{n^2}{(\alpha' p^+)^2}} \qquad (7)$$

Here n is the label of the fourier mode, $n > 0$ label left movers and $n < 0$ right movers. N_n denotes the total occupation number of that mode, including bosons and fermions. Note that the ground state energy of bosonic oscillators is cancelled by that of the fermionic oscillators.

In addition we have the condition that the total momentum on the string vanishes

$$P = \sum_{n=-\infty}^{\infty} n N_n = 0 \qquad (8)$$

Note that for $n = 0$ we also have harmonic oscillators (as opposed to the situation in flat space). When only the $n = 0$ modes are excited we reproduce the spectrum of massless supergravity modes propagating on the plane wave geometry. A particle propagating on a plane wave geometry with fixed p^+ feels as if it was on a gravitational potential well, it cannot escape to infinity if its energy, p^-, is finite. Similarly a massless particle with zero p^+ can go to $r = \infty$ and back in finite x^- time (inversely proportional to μ). This is reminiscent to what happens for particles in AdS. In the limit that μ is very small, or in other words if

$$\mu \alpha' p^+ \ll 1 \qquad (9)$$

we recover the flat space spectrum. Indeed we see from (3) that the metric reduces to the flat space metric if we set μ to zero.

It is also interesting to consider the opposite limit, where

$$\mu \alpha' p^+ \gg 1 \qquad (10)$$

In this limit all the low lying string oscillator modes have almost the same energy. This limit (10) corresponds to a highly curved background with RR fields. In fact we will later see that the appearance of a large number of light modes is expected from the Yang-Mills theory.

It is useful to rewrite (7) in terms of the variables that are natural from the $AdS_5 \times S^5$ point of view. We find that the contribution to $\Delta - J = 2p^-$ of each oscillator is its frequency which can be written as

$$(\Delta - J)_n = w_n = \sqrt{1 + \frac{4\pi g N n^2}{J^2}} \qquad (11)$$

using (5) and the fact that the AdS radius is given by $R^4 = 4\pi g N \alpha'^2$. Notice that N/J^2 remains fixed in the $N \to \infty$ limit that we are taking.

It is interesting to note that in the plane wave (4) we can also have giant gravitons as we have in $AdS_5 \times S^5$. These giants are D3 branes classically sitting at fixed x^- and wrapping the S^3 of the first four directions or the S^3 of the second four directions with a size

$$r^2 = 2\pi g p^+ \mu \alpha'^2 \qquad (12)$$

where p^+ is the momentum carried by the giant graviton. This result follows in a straightforward fashion from the results in [19]. Its p^- eigenvalue is zero. We see that the description of these states in terms of D-branes is correct when their size is much bigger than the string scale. In terms of the Yang-Mills variables this happens when $\frac{J^2}{N} \gg \frac{1}{g}$

Strings from $\mathcal{N} = 4$ Super Yang Mills

We are interested in the limit $N \to \infty$ where g_{YM}^2 is kept fixed and small, $g_{YM}^2 \ll 1$. We want to consider states which carry parametrically large R charge $J \sim \sqrt{N}$.[5] This R charge generator, J, is the SO(2) generator rotating two of the six scalar fields. We want to find the spectrum of states with $\Delta - J$ finite in this limit. We are interested in single trace states of the Yang Mills theory on $S^3 \times R$, or equivalently, the spectrum of dimensions of single trace operators of the theory on R^4. We will often go back and forth between the states and the corresponding operators.

Let us first start by understanding the operator with lowest value of $\Delta - J = 0$. There is a unique single trace operator with $\Delta - J = 0$, namely $Tr[Z^J]$, where $Z \equiv \phi^5 + i\phi^6$ and the trace is over the N color indices. We are taking J to be the SO(2) generator rotating the plane 56. At weak coupling the dimension of this operator is J since each Z field has dimension one. This operator is a chiral primary and hence its dimension is protected by supersymmetry. It is associated to the vacuum state in light cone gauge, which is the unique state with zero light cone hamiltonian. In other words we have the

---

[5] For reasons that we will discuss later we also need that $J/N^{1/2} \ll 1/g_{YM}$. This latter condition comes from demanding that (12) is smaller than the string scale and it ensures that the states we consider are strings and not D-brane "giant gravitons" [19].

correspondence

$$\frac{1}{\sqrt{J}N^{J/2}}Tr[Z^J] \longleftrightarrow |0,p_+\rangle_{l.c.} \qquad (13)$$

We have normalized the operator as follows. When we compute $\langle Tr[\bar{Z}^J](x)Tr[Z^J](0)\rangle$ we have J possibilities for the contraction of the first $\bar{Z}$ but then planarity implies that we contract the second $\bar{Z}$ with a Z that is next to the first one we contracted and so on. Each of these contraction gives a factor of N. Normalizing this two point function to one we get the normalization factor in (13).[6]

Now we can consider other operators that we can build in the free theory. We can add other fields, or we can add derivatives of fields like $\partial_{(i_1}\cdots\partial_{i_n)}\phi^r$, where we only take the traceless combinations since the traces can be eliminated via the equations of motion. The order in which these operators are inserted in the trace is important. All operators are all "words" constructed by these fields up to the cyclic symmetry, these were discussed and counted in [2]. We will find it convenient to divide all fields, and derivatives of fields, that appear in the free theory according to their $\Delta - J$ eigenvalue. There is only one mode that has $\Delta - J = 0$, which is the mode used in (13). There are eight bosonic and eight fermionic modes with $\Delta - J = 1$. They arise as follows. First we have the four scalars in the directions not rotated by J, i.e. ϕ^i, $i = 1,2,3,4$. Then we have derivatives of the field Z, $D_i Z = \partial_i Z + [A_i, Z]$, where $i = 1,2,3,4$ are four directions in R^4. Finally there are eight fermionic operators $\chi^a_{J=\frac{1}{2}}$ which are the eight components with $J = \frac{1}{2}$ of the sixteen component gaugino χ (the other eight components have $J = -\frac{1}{2}$). These eight components transform in the positive chirality spinor representation of $SO(4) \times SO(4)$.[7] We will focus first on operators built out of these fields and then we will discuss what happens when we include other fields, with $\Delta - J > 1$, such as $\bar{Z}$.

The state (13) describes a particular mode of ten dimensional supergravity in a particular wavefunction [5]. Let us now discuss how to generate all other massless supergravity modes. On the string theory side we construct all these states by applying the zero momentum oscillators a_0^i, $i = 1,\ldots,8$ and S_0^b, $b = 1,\ldots 8$ on the light cone vacuum $|0,p_+\rangle_{l.c.}$. Since the modes on the string are massive all these zero momentum oscillators are harmonic oscillators, they all have the same light cone energy. So the total light cone energy is equal to the total number of oscillators that are acting on the light cone ground state. We know that in $AdS_5 \times S^5$ all gravity modes are in the same supermultiplet as the state of the form (13) [22]. The same is clearly true in the limit that we are considering. More precisely, the action of all supersymmetries and bosonic symmetries

---

[6] In general in the free theory any contraction of a single trace operator with its complex conjugate one will give us a factor of N^n, where n is the number of fields appearing in the operator. If the number of fields is very large it is possible that non-planar contractions dominate over planar ones [20, 21]. In our case, due to the way we scale J this does not occur in the free theory.

[7] The first SO(4) corresponds to rotations in R^4, the space where the Yang Mills theory is defined, the second $SO(4) \subset SO(6)$ corresponds to rotations of the first four scalar fields, this is the subgroup of $SO(6)$ that commutes with the $SO(2)$, generated by J, that we singled out to perform the analysis. By positive chirality in $SO(4) \times SO(4)$ we mean that it has positive chirality under both $SO(4)$s or negative under both $SO(4)$. Combining the spinor indices into $SO(8)$, $SO(4) \times SO(4) \subset SO(8)$ it has positive chirality under $SO(8)$. Note that $SO(8)$ is not a symmetry of the background.

of the plane wave background (which are intimately related to the $AdS_5 \times S^5$ symmetries) generate all other ten dimensional massless modes with given p_+. For example, by acting by some of the rotations of S^5 that do not commute with the SO(2) symmetry that we singled out we create states of the form

$$\frac{1}{\sqrt{J}} \sum_l \frac{1}{\sqrt{J} N^{J/2+1/2}} Tr[Z^l \phi^r Z^{J-l}] = \frac{1}{N^{J/2+1/2}} Tr[\phi^r Z^J] \qquad (14)$$

where ϕ^r, $r = 1, 2, 3, 4$ is one of the scalars neutral under J. In (14) we used the cyclicity of the trace. Note that we have normalized the states appropriately in the planar limit. We can act any number of times by these generators and we get operators roughly of the form $\sum Tr[\cdots z \phi^r z \cdots z \phi^k]$. where the sum is over all the possible orderings of the ϕs. We can repeat this discussion with the other $\Delta - J = 1$ fields. Each time we insert a new operator we sum over all possible locations where we can insert it. Here we are neglecting possible extra terms that we need when two $\Delta - J = 1$ fields are at the same position, these are subleading in a $1/J$ expansion and can be neglected in the large J limit that we are considering. In other words, when we act with the symmetries that do not leave Z invariant we will change one of the Zs in (13) to a field with $\Delta - J = 1$, when we act again with one of the symmetries we can change one of the Zs that was left unchanged in the first step or we can act on the field that was already changed in the first step. This second possibility is of lower order in a $1/J$ expansion and we neglect it. We will always work in a "dilute gas" approximation where most of the fields in the operator are Z's and there are a few other fields sprinkled in the operator.

For example, a state with two excitations will be of the form

$$\sim \frac{1}{N^{J/2+1}} \frac{1}{\sqrt{J}} \sum_{l=1}^{J} Tr[\phi^r Z^l \psi^b_{J=\frac{1}{2}} Z^{J-l}] \qquad (15)$$

where we used the cyclicity of the trace to put the ϕ^r operator at the beginning of the expression. We associate (15) to the string state $a_0^{\dagger k} S_0^{\dagger b} |0, p_+\rangle_{l.c.}$. Note that for planar diagrams it is very important to keep track of the position of the operators. For example, two operators of the form $Tr[\phi^1 Z^l \phi^2 Z^{J-l}]$ with different values of l are orthogonal to each other in the planar limit (in the free theory).

The conclusion is that there is a precise correspondence between the supergravity modes and the operators. This is of course well known [4, 5, 6]. Indeed, we see from (7) that their $\Delta - J = 2p^-$ is indeed what we compute at weak coupling, as we expect from the BPS argument.

In order to understand non-supergravity modes in the bulk it is clear that what we need to understand the Yang Mills description of the states obtained by the action of the string oscillators which have $n \neq 0$. Let us consider first one of the string oscillators which creates a bosonic mode along one of the four directions that came from the S^5, let's say $a_n^{\dagger 8}$. We already understood that the action of $a_0^{\dagger 8}$ corresponds to insertions of an operator ϕ^4 on all possible positions along the "string of Z's". By a "string of Zs" we just mean a sequence of Z fields one next to the other such as we have in (13). We propose that $a_n^{\dagger 8}$ corresponds to the insertion of the same field ϕ^4 but now with a position dependent phase

$$\frac{1}{\sqrt{J}}\sum_{l=1}^{J}\frac{1}{\sqrt{J}N^{J/2+1/2}}Tr[Z^l\phi^4 Z^{J-l}]e^{\frac{2\pi i n l}{J}} \qquad (16)$$

In fact the state (16) vanishes by cyclicity of the trace. This corresponds to the fact that we have the constraint that the total momentum along the string should vanish (8), so that we cannot insert only one $a_n^{\dagger\,i}$ oscillator. So we should insert more than one oscillator so that the total momentum is zero. For example we can consider the string state obtained by acting with the $a_n^{\dagger\,8}$ and $a_{-n}^{\dagger\,7}$, which has zero total momentum along the string. We propose that this state should be identified with

$$a_n^{\dagger\,8}a_{-n}^{\dagger\,7}|0,p_+\rangle_{l.c.} \longleftrightarrow \frac{1}{\sqrt{J}}\sum_{l=1}^{J}\frac{1}{N^{J/2+1}}Tr[\phi^3 Z^l \phi^4 Z^{J-l}]e^{\frac{2\pi i n l}{J}} \qquad (17)$$

where we used the cyclicity of the trace to simplify the expression. The general rule is pretty clear, for each oscillator mode along the string we associate one of the $\Delta - J = 1$ fields of the Yang-Mills theory and we sum over the insertion of this field at all possible positions with a phase proportional to the momentum. States whose total momentum is not zero along the string lead to operators that are automatically zero by cyclicity of the trace. In this way we enforce the $L_0 - \bar{L}_0 = 0$ constraint (8) on the string spectrum.

In summary, each string oscillator corresponds to the insertion of a $\Delta - J = 1$ field, summing over all positions with an n dependent phase, according to the rule

$$\begin{aligned} a^{\dagger i} &\longrightarrow D_i Z \quad \text{for } i=1,\cdots,4 \\ a^{\dagger j} &\longrightarrow \phi^{j-4} \quad \text{for } j=5,\cdots,8 \\ S^a &\longrightarrow \chi^a_{J=\frac{1}{2}} \end{aligned} \qquad (18)$$

In order to show that this identification makes sense we want to compute the conformal dimension, or more precisely $\Delta - J$, of these operators at large 't Hooft coupling and show that it matches (7). First note that if we set $\frac{gN}{J^2} \sim 0$ in (11) we find that all modes, independently of n have the same energy, namely one. This is what we find at weak 't Hooft coupling where all operators of the form (17) have the same energy, independently of n. Expanding the string theory result (11) we find that the first correction is of the form

$$(\Delta - J)_n = w_n = 1 + \frac{2\pi g N n^2}{J^2} + \cdots \qquad (19)$$

This looks like a first order correction in the 't Hooft coupling and we can wonder if we can reproduce it by a a simple perturbative computation. Manipulations with non BPS operators suggest that anomalous dimensions grow like $g^2 N$ and that they disappear from the spectrum of the theory at strong coupling. However, this line of reasoning assumes that we keep the dimension of the operator in the free field theory (J in this case) fixed as we take the large N limit. In our case the states we begin with are almost BPS; there are cancellations which depend on the free field theory dimension (J) which render the result finite even in the infinite 't Hooft coupling limit. The interesting diagrams arise from the following interaction vertex

$$\sim g_{YM}^2 Tr([Z,\phi^j][\bar{Z},\phi^j]) \tag{20}$$

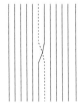

FIGURE 1. Diagrams that exchange the position of ϕ. They have "momentum", n, dependent contributions.

This vertex leads to diagrams, such as shown in 1 which move the position of the ϕ^j operator along the "string" of Z's. In the free theory, once a ϕ^j operator is inserted at one position along the string it will stay there, states with ϕ^j's at different positions are orthogonal to each other in the planar limit (up to the cyclicity of the trace). We can think of the string of Zs in (13) as defining a lattice, when we insert an operator ϕ^1 at different positions along the string of Zs we are exciting an oscillator $b_l^\dagger$ at the site l on the lattice, $l = 1, \cdots J$. The interaction term (20) can take an excitation from one site in the lattice to the neighboring site. So we see that the effects of (20) will be sensitive to the momentum n. In fact one can precisely reproduce (19) from (20) including the precise numerical coefficient. The details of the computation are in [23].

Encouraged by the success of this comparison we want to reproduce the full square root[8] in (11). At first sight this seems a daunting computation since it involves an infinite number of corrections. These corrections nevertheless can be obtained from exponentiating (20) and taking into account that in (20) there are terms involving two creation operators $b^\dagger$ and two annihilation operators b. In other words we have $\phi \sim b + b^\dagger$. As we explained above, we can view ϕ's at different positions as different operators. So we introduce an operator $b_l^\dagger$ which introduces a ϕ operator at the site l along the string of Zs. Then the free hamiltonian plus the interaction term (20) can be thought of as

$$H \sim \sum_l b_l^\dagger b_l + \frac{g_{YM}^2 N}{(2\pi)^2}[(b_l + b_l^\dagger) - (b_{l+1} + b_{l+1}^\dagger)]^2 \tag{21}$$

In [23] we give more details on the derivation of (21). In the large N and J limit it is clear that (21) reduces to the continuum Hamiltonian

$$H = \int_0^L d\sigma \, \frac{1}{2}[\dot\phi^2 + \phi'^2 + \phi^2], \quad L = J\sqrt{\frac{\pi}{gN}} \sim p^+ \tag{22}$$

which is the correct expression for $H = p^- = \Delta - J$ for strings in the light cone gauge.

---

[8] Square roots of the 't Hooft coupling are ubiquitous in the AdS computations.

In summary, the "string of Z's" becomes the physical string and that each Z carries one unit of J which is one unit of p^+. Locality along the worldsheet of the string comes from the fact that planar diagrams allow only contractions of neighboring operators. So the Yang Mills theory gives a string bit model (see [24]) where each bit is a Z operator. Each bit carries one unit of J which through (22) is one unit of p^+.

The reader might, correctly, be thinking that all this seems too good to be true. In fact, we have neglected many other diagrams and many other operators which, at weak 't Hooft coupling also have small $\Delta - J$. In particular, we considered operators which arise by inserting the fields with $\Delta - J = 1$ but we did not consider the possibility of inserting fields corresponding to $\Delta - J = 2, 3, \ldots$, such as $\bar{Z}$, $\partial_k \phi^r$, $\partial_{(l} \partial_{k)} Z$, etc.. The diagrams of the type we considered above would give rise to other 1+1 dimensional fields for each of these modes. These are present at weak 't Hooft coupling but they should not be present at strong coupling, since we do not see them in the string spectrum. We believe that what happens is that these fields get a large mass in the $N \to \infty$ limit. In other words, the operators get a large conformal dimension. In [23], we discuss the computation of the first correction to the energy (the conformal weight) of the of the state that results from inserting $\bar{Z}$ with some "momentum" n. In contrast to our previous computation for $\Delta - J = 1$ fields we find that besides an effective kinetic term as in (19) there is an n independent contribution that goes as gN with no extra powers of $1/J^2$. This is an indication that these excitations become very massive in the large gN limit. In addition, we can compute the decay amplitude of $\bar{Z}$ into a pair of ϕ insertions. This is also very large, of order gN.

Though we have not done a similar computation for other fields with $\Delta - J > 1$, we believe that the same will be true for the other fields. In general we expect to find many terms in the effective Lagrangian with coefficients that are of order gN with no inverse powers of J to suppress them. In other words, the lagrangian of Yang-Mills on S^3 acting on a state which contains a large number of Zs gives a lagrangian on a discretized spatial circle with an infinite number of KK modes. The coefficients of this effective lagrangian are factors of gN, so all fields will generically get very large masses.

The only fields that will not get a large mass are those whose mass is protected for some reason. The fields with $\Delta - J = 1$ correspond to Goldstone bosons and fermions of the symmetries broken by the state (13). Note that despite the fact that they morally are Goldstone bosons and fermions, their mass is non-zero, due to the fact that the symmetries that are broken do not commute with p^-, the light cone Hamiltonian. The point is that their masses are determined, and hence protected, by the (super)symmetry algebra.

Having described how the single string Hilbert space arises it is natural to ask whether we can incorporate properly the string interactions. Clearly string interactions come when we include non-planar diagrams [1]. There are non-planar diagrams coming from the cubic vertex which are proportional to $g_{YM}/N^{1/2}$. These go to zero in the large N limit. There are also non-planar contributions that come from iterating the three point vertex or from the quartic vertex in the action. These are of order $g_{YM}^2 \sim g$ compared to planar diagrams so that we get the right dependence on the string coupling g. In the discussion in this paragraph we have ignored the fact that J also becomes large in the limit we are considering. If we naively compute the factors of J that would appear we would seem to get a divergent contribution for the non-planar diagrams in this limit. Once we take into account that the cubic and quartic vertices contain commutators then the powers of J get reduced. From the gravity side we expect that some string interactions should become strong when $\frac{J}{N^{1/2}} \sim \frac{1}{g_{YM}}$. In other words, at these values of J we expect to find D-brane states in the gravity side, which means that the usual single trace description of operators is not valid any more, see discussion around (12). We have not been able to successfully reproduce this bound from the gauge theory side.

Some of the arguments used in this section look very reminiscent of the DLCQ description of matrix strings [25, 26]. It would be interesting to see if one can establish a connection between them. Notice that the DLCQ description of ten dimensional IIB theory is in terms of the M2 brane field theory. Since here we are extracting also a light cone description of IIB string theory we expect that there should be a direct connection.

It would also be nice to see if using any of these ideas we can get a better handle on other large N Yang Mills theories, particularly non-supersymmetric ones. The mechanism by which strings appear in this paper is somewhat reminiscent of [27].

REFERENCES

1. 't Hooft, G., *Nucl. Phys.* **B72**, 461 (1974).
2. Polyakov, A.M., *Gauge fields and space-time*, arXiv:hep-th/0110196.
3. Maldacena, J., *Adv. Theor. Math. Phys.* **2**, 231 (1998).
4. Gubser, S.S., Klebanov, I.R., and Polyakov, A.M., *Phys. Lett.* **B428**, 105 (1998).
5. Witten, E., *Adv. Theor. Math. Phys.* **2**, 253 (1998).
6. Aharony, O., Gubser, S.S., Maldacena, J., Ooguri, H., and Oz, Y., *Phys. Rept.* **323**, 183 (2000).
7. Blau, M., Figueroa-O'Farrill, J., Hull, C., and Papadopoulos, G., *Class. Quant. Grav.* **19**, L87 (2002).
8. Blau, M., Figueroa-O'Farrill, J., Hull, C., and Papadopoulos, G., *JHEP* **0201**, 047 (2001).
9. Penrose, R., *Any spacetime has a plane wave as a limit*, in *Differential geometry and relativity*, Reidel, Dordrecht, 1976, pp. 271-275.
10. Sfetsos, K., *Phys. Lett.* **B324**, 335 (1994);
    Sfetsos, K., and Tseytlin, A.A., *Nucl. Phys.* **B427**, 245 (1994).
11. Gueven, R., *Phys. Lett.* **B191**, 275 (1987); *Phys. Lett.* **B482**, 255 (2000).
12. Kowalski-Glikman, J., *Phys. Lett.* **B134**, 194 (1984).
13. Figueroa-O'Farrill, J., and Papadopoulos, G., *JHEP* **0108**, 036 (2001).
14. Douglas, M.R., and Moore, G.W., *D-branes, Quivers, and ALE Instantons*, arXiv:hep-th/9603167.
15. Kachru, S., and Silverstein, E., *Phys. Rev. Lett.* **80**, 4855 (1998).
16. See for example:
    Amati, D., and Klimcik, C., *Phys. Lett.* **B210**, 92 (1988);
    Horowitz, G.T., and Steif, A.R., *Phys. Rev. Lett.* **64**, 260 (1990);
    de Vega, H.J., and Sanchez, N., *Phys. Lett.* **B244**, 215 (1990); *Phys. Rev. Lett.* **65**, 1517 (1990);
    Jofre, O., and Nunez, C., *Phys. Rev.* **D50**, 5232 (1994).
17. Metsaev, R.R., *Type IIB Green-Schwarz superstring in plane wave Ramond-Ramond background*, arXiv:hep-th/0112044.
18. Polchinski, J., *String Theory* Cambridge University Press, Cambridge (UK), 1998.
19. McGreevy, J., Susskind, L., and Toumbas, N., *JHEP* **0006**, 008 (2000).
20. Balasubramanian, V., Berkooz, M., Naqvi, A., and Strassler, M.J., *JHEP* **0204**, 034 (2002).
21. Corley, S., Jevicki A., and Ramgoolam, S., *Adv. Theor. Math. Phys.* **5**, 808 (2001).
22. Gunaydin, M., and Marcus, N., *Class. Quant. Grav.* **2**, L11 (1985);
    Kim, H.J., Romans, L.J., and van Nieuwenhuizen, P., *Phys. Rev.* **D32**, 389 (1985).
23. Berenstein, D., Maldacena, J.M., and Nastase, H., *JHEP* **0204**, 013 (2002).
24. Giles, R., and Thorn, C.B., *Phys. Rev.* **D16**, 366 (1977);
    Thorn, C.B., *Reformulating string theory with the 1/N expansion*, arXiv:hep-th/9405069.

25. Motl, L., *Proposals on nonperturbative superstring interactions,* arXiv:hep-th/9701025.
26. Dijkgraaf, R., Verlinde, E., and Verlinde, H., *Nucl. Phys.* **B500**, 43 (1997).
27. Klebanov, I.R., and Susskind, L., *Nucl. Phys.* **B309**, 175 (1988).

The Cosmic Microwave Background & Inflation, Then & Now

J. Richard Bond[1], Carlo Contaldi[1], Dmitry Pogosyan[2], Brian Mason[3,4], Steve Myers[4], Tim Pearson[3], Ue-Li Pen[1], Simon Prunet[5,1], Tony Readhead[3], Jonathan Sievers[3]

1. CIAR Cosmology Program, Canadian Institute for Theoretical Astrophysics,
Toronto, Ontario, Canada
2. Physics Department, University of Alberta, Edmonton, Alberta, Canada
3. Astronomy Department, California Institute of Technology, Pasadena, California, USA
4. National Radio Astronomy Observatory, Socorro, New Mexico, USA
5. Institut d'Astrophysique de Paris, Paris, France

Abstract. The most recent results from the Boomerang, Maxima, DASI, CBI and VSA CMB experiments significantly increase the case for accelerated expansion in the early universe (the inflationary paradigm) and at the current epoch (dark energy dominance). This is especially so when combined with data on high redshift supernovae (SN1) and large scale structure (LSS), encoding information from local cluster abundances, galaxy clustering, and gravitational lensing. There are "7 pillars of Inflation" that can be shown with the CMB probe, and at least 5, and possibly 6, of these have already been demonstrated in the CMB data: (1) the effects of a large scale gravitational potential, demonstrated with COBE/DMR in 1992-96; (2) acoustic peaks/dips in the angular power spectrum of the radiation, which tell about the geometry of the Universe, with the large first peak convincingly shown with Boomerang and Maxima data in 2000, a multiple peak/dip pattern shown in data from Boomerang and DASI (2nd, 3rd peaks, first and 2nd dips in 2001) and from CBI (2nd, 3rd, 4th, 5th peaks, 3rd, 4th dips at 1-sigma in 2002); (3) damping due to shear viscosity and the width of the region over which hydrogen recombination occurred when the universe was 400000 years old (CBI 2002); (4) the primary anisotropies should have a Gaussian distribution (be maximally random) in almost all inflationary models, the best data on this coming from Boomerang; (5) secondary anisotropies associated with nonlinear phenomena subsequent to 400000 years, which must be there and may have been detected by CBI and another experiment, BIMA. Showing the 5 "pillars" involves detailed confrontation of the experimental data with theory; e.g., (5) compares the CBI data with predictions from two of the largest cosmological hydrodynamics simulations ever done. DASI, Boomerang and CBI in 2002, AMiBA in 2003, and many other experiments have the sensitivity to demonstrate the next pillar, (6) polarization, which must be there at the $\sim 7\%$ level. A broad-band DASI detection consistent with inflation models was just reported. A 7th pillar, anisotropies induced by gravity wave quantum noise, could be too small to detect. A minimal inflation parameter set, $\{\omega_b, \omega_{cdm}, \Omega_{tot}, \Omega_Q, w_Q, n_s, \tau_C, \sigma_8\}$, is used to illustrate the power of the current data. After marginalizing over the other cosmic and experimental variables, we find the current CMB+LSS+SN1 data give $\Omega_{tot} = 1.00^{+.07}_{-.03}$, consistent with (non-baroque) inflation theory. Restricting to $\Omega_{tot} = 1$, we find a nearly scale invariant spectrum, $n_s = 0.97^{+.08}_{-.05}$. The CDM density, $\omega_{cdm} = \Omega_{cdm}h^2 = .12^{+.01}_{-.01}$, and baryon density, $\omega_b \equiv \Omega_b h^2 = .022^{+.003}_{-.002}$, are in the expected range. (The Big Bang nucleosynthesis estimate is 0.019 ± 0.002.) Substantial dark (unclustered) energy is inferred, $\Omega_Q \approx 0.68 \pm 0.05$, and CMB+LSS Ω_Q values are compatible with the independent SN1 estimates. The dark energy equation of state, crudely parameterized by a quintessence-field pressure-to-density ratio w_Q, is not well determined by CMB+LSS ($w_Q < -0.4$ at 95% CL), but when combined with SN1 the resulting $w_Q < -0.7$ limit is quite consistent with the $w_Q = -1$ cosmological constant case.

A SYNOPSIS OF CMB EXPERIMENTS

We are in the midst of a remarkable outpouring of results from the CMB that has seen major announcements in each of the last three years, with no sign of abatement in the pace as more experiments are scheduled to release analyses of their results. This paper is an update of [1] to take into account how the new data have improved the case for primordial acceleration, and for acceleration occurring now. The simplest inflation models are strongly preferred by the data. This does not mean inflation is proved, it just fits the available information better than ever. It also does not mean that competitor theories are ruled out, but they would have to look awfully like inflation for them to work. As a result many competitors have now fallen into extreme disfavour as the data have improved.

The CMB Spectrum: The CMB is a nearly perfect blackbody of $2.725 \pm 0.002\,K$ [2], with a $3.372 \pm 0.007\,mK$ dipole associated with the 300 km s^{-1} flow of the earth in the CMB, and a rich pattern of higher multipole anisotropies at tens of μK arising from fluctuations at photon decoupling and later. Spectral distortions from the blackbody have been detected in the COBE FIRAS and DIRBE data. These are associated with starbursting galaxies due to stellar and accretion disk radiation downshifted into the infrared by dust then redshifted into the submillimetre; they have energy about twice all that in optical light, about a tenth of a percent of that in the CMB. The spectrally well-defined Sunyaev-Zeldovich (SZ) distortion associated with Compton-upscatting of CMB photons from hot gas has not been observed in the spectrum. The FIRAS 95% CL upper limit of 6.0×10^{-5} of the energy in the CMB is compatible with the $\lesssim 10^{-5}$ expected from clusters, groups and filaments in structure formation models, and places strong constraints on the allowed amount of earlier energy injection, e.g., ruling out mostly hydrodynamic models of LSS. The SZ effect has been well observed at high resolution with very high signal-to-noise along lines-of-sight through dozens of clusters. The SZ effect in random fields may have been observed with the CBI and BIMA, again at high resolution, although multifrequency observations to differentiate the signal from the CMB primary and radio source contributions will be needed to show this.

The Era of Upper Limits: The story of the experimental quest for spatial anisotropies in the CMB temperature is a heroic one. The original 1965 Penzias and Wilson discovery paper quoted angular anisotropies below 10%, but by the late sixties 10^{-3} limits were reached, by Partridge and Wilkinson and by Conklin and Bracewell. As calculations of baryon-dominated adiabatic and isocurvature models improved in the 70s and early 80s, the theoretical expectation was that the experimentalists just had to get to 10^{-4}, as they did, e.g., Boynton and Partridge in 73. The only signal found was the dipole, hinted at by Conklin and Bracewell in 73, but found definitively in Berkeley and Princeton balloon experiments in the late 70s, along with upper limits on the quadrupole. Throughout the 1980s, the upper limits kept coming down, punctuated by a few experiments widely used by theorists to constrain models: the small angle 84 Uson and Wilkinson and 87 OVRO limits, the large angle 81 Melchiorri limit, early (87) limits from the large angle Tenerife experiment, the small angle RATAN-600 limits, the 7°-beam Relict-1 satellite limit of 87, and Lubin and Meinhold's 89 half-degree South Pole limit, marking a first assault on the peak.

Primordial fluctuations from which structure would have grown in the Universe can be one of two modes: adiabatic scalar perturbations associated with gravitational curvature variations or isocurvature scalar perturbations, with no initial curvature variation, but variations in the relative amounts of matter of different types, e.g., in the number of photons per baryon, or per dark matter particle, i.e., in the entropy per particle. Both modes could be present at once, and in addition there may also be tensor perturbations associated with primordial gravitational radiation which can leave an imprint on the CMB. The statistical distribution of the primordial fluctuations determines the statistics of the radiation pattern; e.g., the distribution could be generically non-Gaussian, needing an infinity of N-point correlation functions to characterize it. The special case when only the 3D 2-point correlation function is needed is that of a Gaussian random field; if dependent upon spatial separations only, not absolute positions or orientations, it is homogeneous and isotropic. The radiation pattern is then a 2D Gaussian field fully specified by its correlation function, whose spherical transform is the angular power spectrum, $\mathscr{C}_\ell$, where ℓ is the multipole number. If the 3D correlation does not depend upon multiplication by a scale factor, it is scale invariant. This does not translate into scale invariance in the 2D radiation correlation, whose features reflect the physical transport processes of the radiation through photon decoupling.

The upper limit experiments were in fact highly useful in ruling out broad ranges of theoretical possibilities. In particular adiabatic baryon-dominated models were ruled out. In the early 80s, universes dominated by dark matter relics of the hot Big Bang lowered theoretical predictions by about an order of magnitude over those of the baryon-only models. In the 82 to mid-90s period, many groups developed codes to solve the perturbed Boltzmann–Einstein equations when such collisionless relic dark matter was present. Armed with these pre-COBE computations, plus the LSS information of the time, a number of otherwise interesting models fell victim to the data: scale invariant isocurvature cold dark matter models in 86, large regions of parameter space for isocurvature baryon models in 87, inflation models with radically broken scale invariance leading to enhanced power on large scales in 87-89, CDM models with a decaying ($\sim$ keV) neutrino if its lifetime was too long ($\gtrsim$ 10yr) in 87 and 91. Also in this period there were some limited constraints on "standard" CDM models, restricting Ω_{tot}, Ω_B, and the amplitude parameter σ_8. (σ_8^2 is a bandpower for density fluctuations on a scale associated with rare clusters of galaxies, $8h^{-1}$ Mpc, where $h = H_0/(100 \text{ km s}^{-1} \text{ Mpc}^{-1})$.)

DMR and Post-DMR Experiments to April 1999: The familiar motley pattern of anisotropies associated with $2 \leq \ell \lesssim 20$ multipoles at the $30\mu K$ level revealed by COBE at $7°$ resolution was shortly followed by detections, and a few upper limits (UL), at higher ℓ in 19 other ground-based (gb) or balloon-borne (bb) experiments — most with many fewer resolution elements than the 600 or so for COBE. Some predated in design and even in data delivery the 1992 COBE announcement. We have the intermediate angle SP91 (gb), the large angle FIRS (bb), both with strong hints of detection before COBE, then, post-COBE, more Tenerife (gb), MAX (bb), MSAM (bb), white-dish (gb, UL), argo (bb), SP94 (gb), SK93-95 (gb), Python (gb), BAM (bb), CAT (gb), OVRO-22 (gb), SuZIE (gb, UL), QMAP (bb), VIPER (gb) and Python V (gb). A list valid to April 1999 with associated bandpowers is given in [3], and are referred here as 4.99 data. They showed evidence for a first peak [3], although it was not well localized. A strong first peak, followed by a sequence of smaller peaks diminished by damping in the $\mathscr{C}_\ell$

spectrum was a long-standing prediction of adiabatic models. For restricted parameter sets, good constraints were given on n_s, and on Ω_{tot} and Ω_Λ when LSS was added [4].

TOCO, BOOMERANG & MAXIMA: The picture dramatically improved over the 3 years since April 1999. In summer 99, the ground-based TOCO experiment in Chile [5], and in November 99 the North American balloon test flight of Boomerang [6], gave results that greatly improved first-peak localization, pointing to $\Omega_{tot} \sim 1$. Then in April 2000 dramatic results from the first CMB long duration balloon (LDB) flight, Boomerang [7, 8], were announced, followed in May 2000 by results from the night flight of Maxima [10]. Boomerang's best resolution was $10.7' \pm 1.4'$, about 40 times better than that of COBE, with tens of thousands of resolution elements. (The corresponding Gaussian beam filtering scale in multipole space is $\ell_s \sim 800$.) Maxima had a similar resolution but covered an order of magnitude less sky. In April 2001, the Boomerang analysis was improved and much more of the data were included, delivering information on the spectrum up to $\ell \sim 1000$ [11, 12]. Maxima also increased its ℓ range [13].

Boomerang carried a 1.2m telescope with 16 bolometers cooled to 300 mK in the focal plane aloft from McMurdo Bay in Antarctica in late December 1998, circled the Pole for 10.6 days and landed just 50 km from the launch site, only slightly damaged. Maps at 90, 150 and 220 GHz showed the same basic spatial features and the intensities were shown to fall precisely on the CMB blackbody curve. The fourth frequency channel at 400 GHz is dust-dominated. Fig. 1 shows a 150 GHz map derived using four of the six bolometers at 150 GHz. There were 10 bolometers at the other frequencies. Although Boomerang altogether probed 1800 square degrees, the April 2000 analysis used only one channel and 440 sq. deg., and the April 2001 analyses used 4 channels and the region in the ellipse covering 800 sq. deg. That is the Boomerang data used in this paper. In [14], the coverage is extended to 1200 sq deg, 2.9% of the sky.

Maxima covered a 124 square degree region of sky in the Northern Hemisphere. Though Maxima was not an LDB, it did well because its bolometers were cooled even more than Boomerang's, to 100 mK, leading to higher sensitivity per unit observing time, it had a star camera so the pointing was well determined, and, further, all frequency channels were used in its analysis.

DASI, CBI & VSA: DASI (the Degree Angular Scale Interferometer), located at the South Pole, has 13 dishes of size 0.2m. Instead of bolometers, it uses HEMTs, operating at 10 frequency channels spanning the band $26 - 36$GHz. An interferometer baseline directly translates into a Fourier mode on the sky. The dish spacing and operating frequency dictate the ℓ range. In DASI's case, the range covered is $125 \lesssim \ell \lesssim 900$. 32 independent maps were constructed, each of size $3.4°$, the field-of-view (fov). The total area covered was 288 sq. deg. DASI's spectacular results were also announced in April 2001, unveiling a spectrum close to that reported by Boomerang at the same time. The two results together reinforced each other and lent considerable confidence to the emerging $\mathscr{C}_\ell$ spectrum in the $\ell < 1000$ regime.

CBI (the Cosmic Background Imager), based at 16000 feet on a high plateau in Chile, is the sister experiment to DASI. It has 13 0.9m dishes operating in the same HEMT channels as DASI. The instrument measures 78 baselines simultaneously. The larger dishes by a factor of 4 and longer baselines imply higher resolution by about the same factor: the CBI results reported in May 2002 go to ℓ of 3500, a huge increase

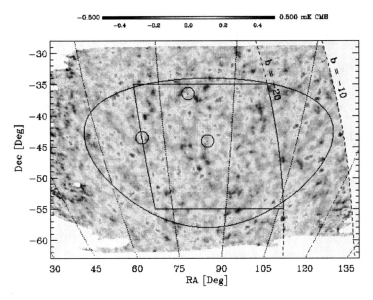

FIGURE 1. The Boomerang 150 GHz bolometer map is shown in the top figure. Of the entire 1800 square degrees covered, only the interior 800 sq. degs. (within the ellipse) were used in [11] and this analysis. In the April 2000 analysis only one channel within the rectangle was used. In the Ruhl et al., 2002 analysis, essentially the entire region is covered. The three circles show regions cut out of the analysis because they contain quasars with emission at 150 GHz. The resolution is $10.7' \pm 1.4'$ fwhm.

over Boomerang, Maxima and DASI. Only the analyses of data from the year 2000 observing campaign were reported. During 2000, CBI covered three deep fields of diameter roughly 0.75° [16], and three mosaic regions, each of size roughly 13 square degrees [17]. In analyzing such high resolution data at 30 GHz, great attention must be paid to the contamination by point sources, but we are confident that this is handled well [16]. Data from 2001 roughly doubles the amount, increases the area covered, and its analysis is currently underway.

Fig. 2 shows one of the CBI mosaic regions of the sky and Fig. 3 one of the deep regions.

The VSA (Very Small Array) in Tenerife, also an interferometer, operating at 30 GHz, covered the ℓ range of DASI, and confirmed the spectrum emerging from the Boomerang, Maxima and DASI data in that region. The VSA is now observing at longer baselines to increase its ℓ range.

The Optimal Spectrum, circa Summer 2002: The power spectrum shown in Fig. 4 combines all of the data in a way that takes all of the uncertainties in each experiment (calibration and beam) into account. The point at small ℓ is dominated by DMR, at $900 \lesssim \ell \lesssim 2000$ by the CBI mosaic data, with that beyond 2000 by the CBI deep data. In between, Boomerang drives the small error bars, DASI and CBI set the calibration and beam of Boomerang and give spectra totally compatible with Boomerang. Both VSA and Maxima are in agreement with this data as well. Further, although

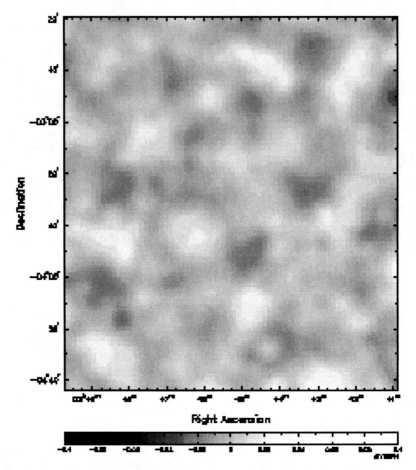

FIGURE 2. The inner 2.5 degrees of one of the three 13 square degree CBI mosaic fields is shown. The mosaic image is a standard radio astronomy map, with point sources removed. The regions are smaller than Boomerang covered, but the resolution is a factor of at least three better. The mass subtended by the CBI resolution scale ($\sim 4'$) easily encompasses the mass that collapses later in the universe to generate clusters of galaxies.

the errors from the experiments before April 2000 are larger, the quite heterogeneous 4.99+TOCO+Boomerang-NA mix of CMB data is very consistent with what the newer experiments show. It is an amazing concordance of data. Accompanying this story is a convergence with decreasing errors over time on the values of the cosmological parameters given in Table 1.

Primary CMB Processes and Soundwave Maps at Decoupling: Boomerang, Maxima, DASI, CBI and VSA were designed to measure the *primary* anisotropies of the CMB, those which can be calculated using linear perturbation theory. What we see in Figs. 1, 2 are, basically, images of soundwave patterns that existed about 400,000 years

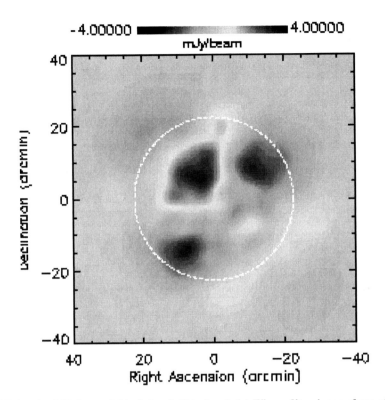

FIGURE 3. An 0.75 degree field-of-view (within the circle) Wiener-filtered map of one of the three deep CBI fields is shown. Apart from the primary anisotropy signal, the deep images contain power at high ℓ that might be from the SZ effect in collapsed clusters at redshift of order 1. We know that the predicted SZ signal is near to that seen, and is very sensitive to σ_8 ($\mathscr{C}_\ell^{(SZ)} \propto \sigma_8^7$). $\sigma_8 \approx 1$ is needed for the SZ effect as determined by hydro simulations to agree with the data. The primary CMB data prefer values between 0.8 and 0.9, and so although we do expect the SZ signal to be lurking within the CBI signal, it may not be quite as large as the extra power seen in these deep maps.

after the Big Bang, when the photons were freed from the plasma. The visually evident structure on degree scales is even more apparent in the power spectra of the maps, which show a dominant (first acoustic) peak, with less prominent subsequent ones detected at varying levels of statistical significance.

The images are actually a projected mixture of dominant and subdominant physical processes through the photon decoupling "surface", a fuzzy wall at redshift $z_{dec} \approx 1050$, when the Universe passed from optically thick to thin to Thomson scattering over a comoving distance ≈ 19 Mpc. Prior to this, acoustic wave patterns in the tightly-coupled photon-baryon fluid on scales below the comoving "sound crossing distance" at decoupling, $\lesssim 150$ Mpc (i.e., $\lesssim 150$ kpc physical), were viscously damped, strongly so on scales below the ≈ 10 Mpc damping scale. After, photons freely-streamed along geodesics to us, mapping (through the angular diameter distance relation) the post-decoupling spatial structures in the temperature to the angular patterns we observe now

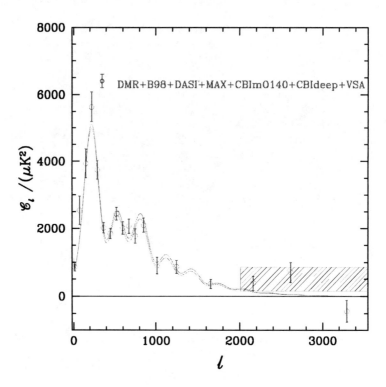

FIGURE 4. The $\mathscr{C}_\ell$ are defined in terms of CMB temperature anisotropy multipoles by $\mathscr{C}_\ell \equiv \ell(\ell+1)\langle|(\Delta T/T)_{\ell m}|^2\rangle/(2\pi)$. The optimal $\mathscr{C}_\ell$ spectrum corresponds to a maximum-likelihood fit to the power in bands marginalized over beam and calibration uncertainties of the various experiments, This one uses "all-data" (DMR + Boomerang + Maxima + DASI + CBI mosaic + CBI deep + VSA + TOCO + Boomerang-NA + the 4.99 data). A $\Delta\ell = 75$ binning was chosen up to ~ 800, going over to the CBI deep binning at large ℓ. The $\ell > 2000$ excess found with the deep CBI data is denoted by the light blue hatched region (95% confidence limit). Two best fit models to "all-data" are shown. They are both ΛCDM models; e.g., the upper curve (magenta) has parameters $\{\Omega_{\text{tot}}, \Omega_\Lambda, \Omega_b h^2, \Omega_{\text{cdm}} h^2, n_s, \tau_C\}=\{1.0, 0.7, 0.02, 0.14, 0.975, 0\}$.

as the primary anisotropies. The maps are images projected through the fuzzy decoupling surface of the acoustic waves (photon bunching), the electron flow (Doppler effect) and the gravitational potential peaks and troughs ("naive" Sachs-Wolfe effect) back then. Free-streaming along our (linearly perturbed) past light cone leaves the pattern largely unaffected, except that temporal evolution in the gravitational potential wells as the photons propagate through them leaves a further ΔT imprint, called the integrated Sachs-Wolfe effect. Intense theoretical work over three decades has put accurate calculations of this linear cosmological radiative transfer on a firm footing, and there are speedy, publicly available and widely used codes for evaluation of anisotropies in a variety of cosmological scenarios, "CMBfast" and "CAMB" [18]. Extensions to more cosmological models have been added by a variety of researchers.

Of course there are a number of nonlinear effects that are also present in the maps.

These *secondary* anisotropies include weak-lensing by intervening mass, Thomson-scattering by the nonlinear flowing gas once it became "reionized" at $z \sim 10$, the thermal and kinematic SZ effects, and the red-shifted emission from dusty galaxies. They all leave non-Gaussian imprints on the CMB sky.

The Immediate Future: Results are in and the analysis is underway for the bolometer single dish ACBAR experiment based at the South Pole (with about $5'$ resolution, allowing coverage to $\ell \sim 2000$). Results from two balloon flights, Archeops and Tophat, are also expected soon. MINT, a Princeton interferometer, also has results soon to be released.

In June 2001, NASA launched the all-sky HEMT-based MAP satellite, with $12'$ resolution. We expect spectacularly accurate results covering the ℓ range to about 600 to be announced in Jan. 2003, with higher ℓ and smaller errors expected as the observing period increases. Eventually four years of data are expected.

Further downstream, in 2007, ESA will launch the bolometer+HEMT-based Planck satellite, with $5'$ resolution.

Targeting Polarization of the Primary CMB Signal: The polarization dependence of Compton scattering induces a well defined polarization signal emerging from photon decoupling. Given the total $\mathscr{C}_\ell$ of Fig. 4, we can forecast what the strength of that signal and its cross-correlation with the total anisotropy will be, and which ℓ range gives the maximum signal: $\sim 5\mu K$ over $\ell \sim 400 - 1600$ is a target for the "E-mode" that scalar fluctuations give.

A great race was on to first detect the E-mode. Experiments range from many degrees to subarcminute scales. DASI has analyzed 271 days of polarization data on 2 deep fields ($3.4°\ fov$) and just announced a 5σ detection at a level consistent with inflation-based models [19]. Boomerang will fly again, in December 2002, with polarization-sensitive bolometers of the kind that will also be used on Planck; as well, MAXIMA will fly again as the polarization-targeting MAXIPOL; the detectors on CBI have been reconfigured to target polarization and the CBI is beginning to take data; MAP can also measure polarization with its HEMTs. Other experiments, operating or planned include AMiBA, COMPASS, CUPMAP, PIQUE and its sequel, POLAR, Polarbear, Polatron, QUEST, Sport/BaRSport, BICEP, among others. The amplitude of the DASI E detection is 0.8 ± 0.3 of the forecasted amplitude from the total anisotropy data; the cross-correlation of the polarization with the total anisotropy has also been detected, with amplitude 0.9 ± 0.4 of the forecast. These detections used a broad-band shape covering the ℓ range $\sim 250 - 750$ derived from the theoretical forecasts. The forecasts indicate solid detections with enough well-determined bandpowers for use in cosmological parameter studies are soon likely.

We cannot yet forecast the strength of the "B-mode" signal induced by gravity waves, since there is as yet no evidence for or against them in the data. However, the amplitude would be very small indeed even at $\ell \sim 100$. Nonetheless there are experiments such as BICEP (Caltech) being planned to go after polarization in these low ℓ ranges.

Targeting Secondary Anisotropies: SZ anisotropies have been probed by single dishes, the OVRO and BIMA mm arrays, and the Ryle interferometer. Detections of individual clusters are now routine. The power at $\ell > 2000$ seen in the CBI deep data (Fig. 4) [16] and in the BIMA data at $\ell \sim 6000$ [24] may be due to the SZ effect in ambient fields, e.g., [23]. A number of planned HEMT-based interferometers are being built with this ambient effect as a target: CARMA (OVRO+BIMA together), the SZA (Chicago, to be incorporated in CARMA), AMI (Britain, including the Ryle telescope), and AMiBA (Taiwan). Bolometer-based experiments will also be used to probe the SZ effect, including the CSO (Caltech submm observatory, a 10m dish) with BOLOCAM on Mauna Kea, ACBAR at the South Pole and the LMT (large mm telescope, with a 50 metre dish) in Mexico. As well, APEX, a 12 m German single dish, Kobyama, a 10 m Japanese single dish, and the 100m Green Bank telescope can all be brought to bear down on the SZ sky. Very large bolometer arrays with thousands of elements are planned: the South Pole Telescope (SPT, Chicago) and the Atacama Cosmology Telescope (ACT, Princeton), with resolution below $2'$, will be powerful probes of the SZ effect as well as of primary anisotropies.

Anisotropies from dust emission from high redshift galaxies are being targeted by the JCMT with the SCUBA bolometer array, the OVRO mm interferometer, the CSO, the SMA (submm array) on Mauna Kea, the LMT, the ambitious US/ESO ALMA mm array in Chile, the LDB BLAST, and ESA's Herschel satellite. About 50% of the submm background has so far been identified with sources that SCUBA has found.

The CMB Analysis Pipeline: Analyzing Boomerang and other single-dish experiments involves a pipeline, reviewed in [25], that takes the timestream in each of the bolometer channels coming from the balloon plus information on where it is pointing and turns it into spatial maps for each frequency characterized by average temperature fluctuation values in each pixel (Fig. 1) and a pixel-pixel correlation matrix characterizing the noise. From this, various statistical quantities are derived, in particular the temperature power spectrum as a function of multipole, grouped into bands, and band-band error matrices which together determine the full likelihood distribution of the bandpowers [20, 3]. Fundamental to the first step is the extraction of the sky signal from the noise, using the only information we have, the pointing matrix mapping a bit in time onto a pixel position on the sky. In the April 2001 analysis of Boomerang, and subsequent work, powerful use of Monte Carlo simulations was made to evaluate the power spectrum and other statistical indicators in maps with many more pixels than was possible with conventional matrix methods for estimating power spectra [11].

For interferometer experiments, the basic data are visibilities as a function of baseline and frequency, with contributions from random detector noise as well as from the sky signals. As mentioned above, a baseline is a direct probe of a given angular wavenumber vector on the sky, hence suggests we should make "generalized maps" in "momentum space" (i.e., Fourier transform space) rather than in position space, as for Boomerang. A major advance was made in [21] to deal with the large volume of interferometer data that we got with CBI, especially for the mosaics with their large number of overlapping fields. We "optimally" compressed the $> \mathcal{O}(10^5)$ visibility measurements of each field into a $< \mathcal{O}(10^4)$ coarse grained lattice in momentum space, and used the information in that "generalized pixel" basis to estimate the power spectrum and statistical distribution of the signals.

There is generally another step in between the maps and the final power spectra, namely separating the multifrequency spatial maps into the physical components on the sky: the primary CMB, the thermal and kinematic Sunyaev-Zeldovich effects, the dust, synchrotron and bremsstrahlung Galactic signals, the extragalactic radio and submillimetre sources. The strong agreement among the Boomerang maps indicates that to first order we can ignore this step, but it has to be taken into account as the precision increases.

Because of the 1 cm observing wavelength of CBI and its resolution, the contribution from extragalactic radio sources is significant. We project out of the data sets known point sources when estimating the primary anisotropy spectrum by using a number of constraint matrices. The positions are obtained from the (1.4 GHz) NVSS catalog. When projecting out the sources we use large amplitudes which effectively marginalize over all affected modes. This insures robustness with respect to errors in the assumed fluxes of the sources. The residual contribution of sources below our known-source cutoff is treated as a white noise background with an amplitude (and error) estimated as well from the NVSS database [16].

The CMB Statistical Distributions are Nearly Gaussian: The primary CMB fluctuations are quite Gaussian, according to COBE, Maxima, and now Boomerang and CBI analyses. Analysis of data like that in the Fig. 1 map show a one-point distribution of temperature anisotropy values that is well fit by a Gaussian distribution. Higher order (concentration) statistics (3,4-point functions, etc.) tell us of non-Gaussian aspects, necessarily expected from the Galactic foreground and extragalactic source signals, but possible even in the early Universe fluctuations. For example, though non-Gaussianity occurs only in the more baroque inflation models of quantum noise, it is a necessary outcome of defect-driven models of structure formation. (Peaks compatible with Fig. 4 do not appear in non-baroque defect models, which now appear highly unlikely.) There is currently no evidence for a breakdown of the Gaussianity in the 150 GHz maps as long as one does not include regions near the Galactic plane in the analysis. We have also found that the one-point distribution of the CBI data is also compatible with a Gaussian. However, since we know non-Gaussianity is necessarily there at some level, more exploration is needed.

Though great strides have been made in the analysis of the Boomerang-style and CBI-style experiments, there is intense effort worldwide developing fast and accurate algorithms to deal with the looming megapixel datasets of LDBs and the satellites. Dealing more effectively with the various component signals and the statistical distribution of the errors resulting from the component separation is a high priority.

COSMIC PARAMETER ESTIMATION

Parameters of Structure Formation: Following [1], we adopt a restricted set of 8 cosmological parameters, augmenting the basic 7 used in [8, 9, 11, 22], $\{\Omega_\Lambda, \Omega_k, \omega_b, \omega_{cdm}, n_s, \tau_C, \sigma_8\}$, by one. The vacuum or dark energy encoded in the cosmological constant Ω_Λ is reinterpreted as Ω_Q, the energy in a scalar field Q which dominates at late times, which, unlike Λ, could have complex dynamics associated with it. Q is now often termed a quintessence field. One popular phenomenology is to add one more parameter, $w_Q = p_Q/\rho_Q$, where p_Q and ρ_Q are the pressure and density of the Q-field, related to its kinetic and potential energy by $\rho_Q = \dot{Q}^2/2 + (\nabla Q)^2/2 + V(Q)$, $p_Q = \dot{Q}^2/2 - (\nabla Q)^2/6 - V(Q)$. Thus $w_Q = -1$ for the cosmological constant. Spatial fluctuations of Q are expected to leave a direct imprint on the CMB for small ℓ. This will depend in detail upon the specific model for Q. We ignore this complication here, but caution that using DMR data which is sensitive to low ℓ behaviour in conjunction with the rest will give somewhat misleading results. To be self consistent, a model must be complete: e.g., even the ludicrous models with constant w_Q would have necessary fluctuations to take into account. As well, as long as w_Q is not exactly -1, it will vary with time, but the data will have to improve for there to be sensitivity to this, and for now we can just interpret w_Q as an appropriate time-average of the equation of state. The curvature energy $\Omega_k \equiv 1 - \Omega_{tot}$ also can dominate at late times, as well as affecting the geometry.

We use only 2 parameters to characterize the early universe primordial power spectrum of gravitational potential fluctuations Φ, one giving the overall power spectrum amplitude $\mathscr{P}_\Phi(k_n)$, and one defining the shape, a spectral tilt $n_s(k_n) \equiv 1 + d\ln\mathscr{P}_\Phi/d\ln k$, at some (comoving) normalization wavenumber k_n. We really need another 2, $\mathscr{P}_{GW}(k_n)$ and $n_t(k_n)$, associated with the gravitational wave component. In inflation, the amplitude ratio is related to n_t to lowest order, with $\mathcal{O}(n_s - n_t)$ corrections at higher order, e.g., [26]. There are also useful limiting cases for the $n_s - n_t$ relation. However, as one allows the baroqueness of the inflation models to increase, one can entertain essentially any power spectrum (fully k-dependent $n_s(k)$ and $n_t(k)$) if one is artful enough in designing inflaton potential surfaces. Actually $n_t(k)$ does not have as much freedom as $n_s(k)$ in inflation. For example, it is very difficult to get $n_t(k)$ to be positive. As well, one can have more types of modes present, e.g., scalar isocurvature modes ($\mathscr{P}_{is}(k_n), n_{is}(k)$) in addition to, or in place of, the scalar curvature modes ($\mathscr{P}_\Phi(k_n), n_s(k)$). However, our philosophy is consider minimal models first, then see how progressive relaxation of the constraints on the inflation models, at the expense of increasing baroqueness, causes the parameter errors to open up. For example, with COBE-DMR and Boomerang, we can probe the GW contribution, but the data are not powerful enough to determine much. Planck can in principle probe the gravity wave contribution reasonably well.

We use another 2 parameters to characterize the transport of the radiation through the era of photon decoupling, which is sensitive to the physical density of the various species of particles present then, $\omega_j \equiv \Omega_j h^2$. We really need 4: ω_b for the baryons, ω_{cdm} for the cold dark matter, ω_{hdm} for the hot dark matter (massive but light neutrinos), and ω_{er} for the relativistic particles present at that time (photons, very light neutrinos, and possibly weakly interacting products of late time particle decays). For simplicity, though, we restrict ourselves to the conventional 3 species of relativistic neutrinos plus photons, with ω_{er} therefore fixed by the CMB temperature and the relationship between the neutrino and photon temperatures determined by the extra photon entropy accompanying e^+e^- annihilation. Of particular importance for the pattern of the radiation is the (comoving) distance sound can have influenced by recombination (at redshift $z_{dec} = a_{dec}^{-1} - 1$),

$$r_s = 6000/\sqrt{3}\ \text{Mpc} \int_0^{\sqrt{a_{dec}}} (\omega_m + \omega_{er} a^{-1})^{-1/2}(1 + \omega_b a/(4\omega_\gamma/3))^{-1/2}\ d\sqrt{a}, \qquad (1)$$

where $\omega_\gamma = 2.46 \times 10^{-5}$ is the photon density, $\omega_{er} = 1.68\omega_\gamma$ for 3 species of massless neutrinos and $\omega_m \equiv \omega_{hdm} + \omega_{cdm} + \omega_b$.

The angular diameter distance relation maps spatial structure at photon decoupling perpendicular to the line-of-sight with transverse wavenumber $k_\perp$ to angular structure, through $\ell = \mathcal{R}_{dec} k_\perp$. In terms of the comoving distance to photon decoupling (recombination), χ_{dec}, and the curvature scale d_k, $\mathcal{R}_{dec}$ is given by

$$\mathcal{R}_{dec} = \{d_k \sinh(\chi_{dec}/d_k), \chi_{dec}, d_k \sin(\chi_{dec}/d_k)\}, \text{ where } d_k = 3000|\omega_k|^{-1/2} \text{ Mpc},$$
$$\text{and } \chi_{dec} = 6000 \text{ Mpc} \int_{\sqrt{a_{dec}}}^{1} (\omega_m + \omega_Q a^{-6w_Q} + \omega_k a)^{-1/2} \, d\sqrt{a}. \tag{2}$$

The 3 cases are for negative, zero and positive mean curvature. Thus the mapping depends upon ω_k, ω_Q and w_Q as well as on ω_m. The location of the acoustic peaks $\ell_{pk,j}$ is proportional to the ratio of $\mathcal{R}_{dec}$ to r_s, hence depends upon ω_b through the sound speed as well. Thus $\ell_{pk,j}$ defines a functional relationship among these parameters, a *degeneracy* [27] that would be exact except for the integrated Sachs-Wolfe effect, associated with the change of Φ with time if Ω_Q or Ω_k is nonzero. (If $\dot{\Phi}$ vanishes, the energy of photons coming into potential wells is the same as that coming out, and there is no net impact of the rippled light cone upon the observed ΔT.)

Our 7th parameter is an astrophysical one, the Compton "optical depth" τ_C from a reionization redshift z_{reh} to the present. It lowers $\mathcal{C}_\ell$ by $\exp(-2\tau_C)$ at ℓ's in the Boomerang/CBI regime. For typical models of hierarchical structure formation, we expect $\tau_C \lesssim 0.2$. It is partly degenerate with σ_8 and cannot quite be determined at this precision by CMB data now.

The LSS also depends upon our parameter set: an important combination is the wavenumber of the horizon when the energy density in relativistic particles equals the energy density in nonrelativistic particles: $k_{Heq}^{-1} \approx 5\Gamma^{-1} h^{-1}$ Mpc, where $\Gamma \approx \Omega_m h \Omega_{er}^{-1/2}$. Instead of $\mathcal{P}_\Phi(k_n)$ for the amplitude parameter, we often use $\mathcal{C}_{10}$ at $\ell = 10$ for CMB only, and σ_8^2 when LSS is added. When LSS is considered in this paper, it refers to constraints on $\Gamma + (n_s - 1)/2$ and $\ln \sigma_8^2$ that are obtained by comparison with the data on galaxy clustering, cluster abundances and from weak lensing [23, 8, 4]. At the current time, the constraints from σ_8 from lensing and cluster abundances are stronger thoes from Γ, although, with the wealth of data emerging from the Sloan Digital Sky Survey and the 2dF redshift survey, shape should soon deliver more powerful information than overall amplitude. However, in the future, weak lensing will allow amplitude and shape to be simultaneously constrained without the uncertainties associated with the biasing of the galaxy distribution *wrt* the mass that the redshift surveys must deal with.

When we allow for freedom in ω_{er}, the abundance of primordial helium, tilts of tilts $(dn_{\{s,is,t\}}(k_n)/d\ln k, ...)$ for 3 types of perturbations, the parameter count would be 17, and many more if we open up full theoretical freedom in spectral shapes. However, as we shall see, as of now only 4 combinations can be determined with 10% accuracy with the CMB. Thus choosing 8 is adequate for the present; 7 of these are discretely sampled[28], with generous boundaries, though for drawing cosmological conclusions we adopt a weak prior probability on the Hubble parameter and age: we restrict h to lie in the 0.45 to 0.9 range, and the age to be above 10 Gyr.

Peaks, Dips and Ω_{tot}, Ω_Q and w_Q: For given ω_m and ω_b, we show the lines of constant $\ell_{pk,j} \propto \mathcal{R}_{dec}/r_s$ in the Ω_k–Ω_Q plane for $w_Q = -1$ in Fig. 5, and in the w_Q–Ω_Q plane for $\Omega_{tot} = 1$ in Fig. 7, using the formulas given above and in [27].

Our current best estimate [22] of the peak locations $\ell_{pk,j}$, using all current CMB data and the flat+wk-h+LSS prior, are 222 ± 3, 537 ± 6, 823 ± 45, 1138 ± 45, 1437 ± 59, obtained by forming $\exp < \ln \ell_{pk,j} >$, where the average and variance of $\ln \ell_{pk,j}$ are determined by integrating over the probability-weighted database described above. The interleaving dips are at 411 ± 5, 682 ± 48, 1023 ± 44, 1319 ± 51, 1653 ± 48. With just the data prior to April 2000, the first peak value was 224 ± 25, showing how it has localized. A quadratic fit sliding over the data can be used to estimate the peak and dip positions in a model-independent way. It gives numbers in good accord with those given here [12, 17], but of course with larger error bars, in some cases only one-sigma detections for the higher peaks and dips.

The critical spatial scale determining the positions of the peaks is r_s, found by averaging over the model space probabilities to be 145 ± 2 Mpc comoving, thus about 140 kpc as the physical sound horizon at decoupling. Converting peaks in k-space into peaks in ℓ-space is obscured by projection effects over the finite width of decoupling and the influence of sources other than sound oscillations such as the Doppler term. The conversion into peak locations in $\mathscr{C}_\ell$ gives $\ell_{pk,j} \sim f_j j \pi \mathscr{R}_{dec}/r_s$, where the numerically estimated f_j factor is ≈ 0.75 for the first peak, approaching unity for higher ones. Dip locations are determined by replacing j by $j + 1/2$. These numbers accord reasonably well with the ensemble-averaged $\ell_{pk/dip,j}$ given above.

The strength of the overall decline due to shear viscosity and the finite width of the region over which hydrogen recombination occurs can also be estimated, $R_D = 10 \pm 3$ Mpc, i.e., 10 kpc back then, corresponding to an angular damping scale $\ell_D = 1358 \pm 22$.

The constant $\ell_{pk,j}$ lines look rather similar to the contours in Figs. 6, 8, showing that the $\mathscr{R}_{dec}/r_s$ degeneracy plays a large role in determining the contours. The figures also show how adding other cosmological information such as H_0 estimation can break the degeneracies. The contours hug the $\Omega_k = 0$ line more closely than the allowed $\ell_{pk,j}$ band does for the maximum probability values of ω_m and ω_b, because of the shift in the allowed $\ell_{pk,j}$ band as ω_m and ω_b vary in this plane. The ω_b dependence in r_s would lead to a degeneracy with other parameters in terms of peak/dip positions. However, relative peak/dip heights are extremely significant for parameter estimation as well, and this breaks the degeneracy. For example, increasing ω_b beyond the nucleosynthesis (and CMB) estimate leads to a diminished height for the second peak that is not in accord with the data.

Marginalized Estimates of our Basic 8 Parameters: Table 1 shows there are strong detections with only the CMB data for Ω_{tot}, ω_b and n_s in the minimal inflation-based 8 parameter set. The ranges quoted are Bayesian 50% values and the errors are 1-sigma, obtained after projecting (marginalizing) over all other parameters. With "all-data", ω_{cdm} begins to localize, but more so when LSS information is added. Indeed, even with just the COBE-DMR+LSS data, ω_{cdm} is already localized. That Ω_Q is not well determined is a manifestation of the Ω_k–Ω_Q near-degeneracy discussed above, which is broken when LSS is added because the CMB-normalized σ_8 is quite different for open cf. pure Q-models. Supernova at high redshift give complementary information to the CMB, but with CMB+LSS (and the inflation-based paradigm) we do not need it: the CMB+SN1 and CMB+LSS numbers are quite compatible. In our space, the Hubble parameter,

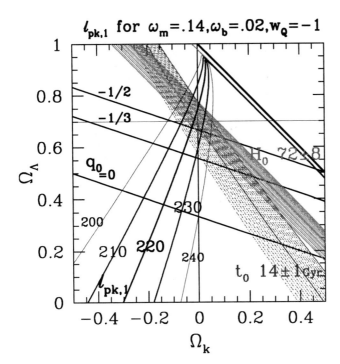

FIGURE 5. This shows lines of constant $\ell_{pk,1}$ in the Ω_k–Ω_Q plane (assuming $w_Q = -1$, i.e., a cosmological constant) for the $\{\omega_m, \omega_b\}$ shown, near their most probable values. The data give $\ell_{pk,1} = 222 \pm 3$. The higher peaks and dips have similar curves, scaled about the probable values listed above. Lowering ω_b increases the sound speed, decreasing $\ell_{pk,j}$, and varying ω_m also shifts it. The $0.64 < h < 0.82$ (heavier shading, H_0) and $13 <$ age < 15 (lighter shading, t_0) ranges and decelerations $q_0 = 0, -1/3, -1/2$ are also noted. The sweeping back of the $\ell_{pk,j}$ curves into the closed models as Ω_Λ is lowered shows that even if $\Omega_{tot} = 1$, the phase space results in a 1D projection onto the Ω_{tot} axis that would be skewed to $\Omega_{tot} > 1$. This plot explains much of the structure in the probability contour maps derived from the data, Fig. 6.

$h = (\sum_j (\Omega_j h^2))^{1/2}$, and the age of the Universe, t_0, are derived functions of the $\Omega_j h^2$: representative values are given in the Table caption. CMB+LSS does not currently give a useful constraint on w_Q, though $w_Q \lesssim -0.7$ with SN1. The values do not change very much if rather than the weak prior on h, we use 0.72 ± 0.08, the estimate from the Hubble key project [31]. Indeed just the CMB data plus this restricted range for h and the restriction to $\Omega_{tot} = 1$ results in a strong detection of Ω_Q. Allowing for a neutrino mass [32] changes the value of Ω_Q downward as the mass increases, but not so much as to make it unnecessary.

The Future, Forecasts for Parameter Eigenmodes: We can also forecast dramatically improved precision with future LDBs, ground-based single dishes and interferometers, MAP and Planck. Because there are correlations among the physical variables we wish to determine, including a number of near-degeneracies beyond that for

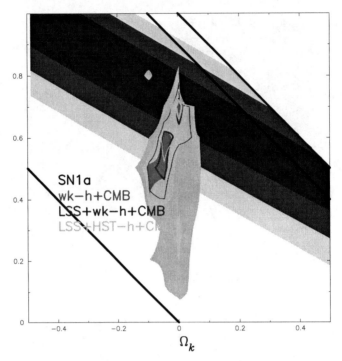

FIGURE 6. 1,2,3-sigma likelihood contour shadings for "all-data" and the weak-H+age prior probability in the $\Omega_k - \Omega_Q$ plane. The first interior lines are the 1,2-sigma ones when the LSS constraint is added, the most interior are the contours when the Hubble key project constraint is applied. The supernova contour shadings [29] are also plotted. Note that the contours are near the $\Omega_k = 0$ line, but also follow a weighted average of the $\ell_{pk,1} \sim 220$ lines. This approximate degeneracy implies Ω_Q is poorly constrained for CMB-only, but it is broken when LSS is added, giving a solid SN1-independent Ω_Q "detection". When the Hubble key project constraint on H_0 is added, partial breaking of this degeneracy occurs as well, as is evident from Fig. 5, and from this figure.

$\Omega_k - \Omega_Q$ [27], it is useful to disentangle them, by making combinations which diagonalize the error correlation matrix, "parameter eigenmodes" [26, 27]. For this exercise, we will add ω_{hdm} and n_t to our parameter mix, but set $w_Q = -1$, making 9. (The ratio $\mathcal{P}_{GW}(k_n)/\mathcal{P}_\Phi(k_n)$ is treated as fixed by n_t, a reasonably accurate inflation theory result.) The forecast for Boomerang based on the 800 sq. deg. patch with four 150 GHz bolometers used is 4 out of 9 linear combinations should be determined to ± 0.1 accuracy. This is indeed what was obtained in the full analysis of CMB only for Boomerang+DMR. The situation improves for the satellite experiments: for MAP, with 2 years of data, we forecast 6/9 combos to ± 0.1 accuracy, 3/9 to ± 0.01 accuracy; for Planck, 7/9 to ± 0.1 accuracy, 5/9 to ± 0.01 accuracy. While we can expect systematic errors to loom as the real arbiter of accuracy, the clear forecast is for a very rosy decade of high precision CMB cosmology that we are now fully into.

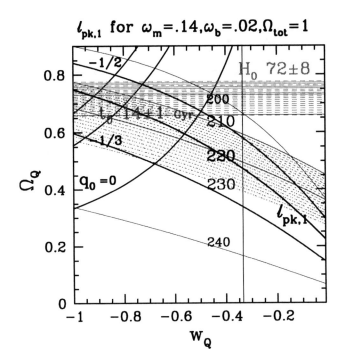

FIGURE 7. Lines of constant $\ell_{pk,1}$ in the w_Q–Ω_Q quintessence plane (with $\Omega_{tot}=1$) are shown for the most probable values of $\{\omega_m, \omega_b\}$. Lines of constant deceleration parameter $q_0 = (\Omega_m + (1+3w_Q)\Omega_Q)/2$, H_0 and age t_0 in the ranges indicated are also shown.

REFERENCES

1. Bond, J.R. *et al.*, in "Cosmology & Particle Physics", Proc. CAPP 2000, ed. J. Garcia-Bellida, R. Durrer, M. Shaposhnikov (Washington: American Inst. of Physics) (2000), astro-ph/0011379
2. Mather, J.C. et al., Ap. J. 512, 511 (1999)
3. Bond, J.R., Jaffe, A.H. & Knox, L. Ap. J. 533, 19 (2000)
4. Bond, J.R., & Jaffe, A., Phil. Trans. R. Soc. London, 357, 57 (1998), astro-ph/9809043
5. Miller, A.D. et al., Ap. J. 524, L1 (1999) TOCO
6. Mauskopf, P. et al., apjl 536, L59 (2000) BOOM-NA.
7. de Bernardis, P. et al., Nature 404, 995 (2000) BOOM00
8. Lange, A. et al., Phys. Rev. D 63, 042001 (2001)
9. Jaffe, A.H. et al., Phys. Rev. Lett. 86, 3475 (2001)
10. Hanany, S. et al., Ap. J. Lett. 545, 5 (2000) Maxima00
11. Netterfield, B. *et al.*, Ap. J. 571, 604 (2002) BOOM01
12. deBernardis, P. *et al.*, Ap. J. 564, 559 (2002)
13. Lee, A.T. *et al.*, Phys. Rev. D 561, L1 (2002) Maxima01
14. Ruhl, J. *et al.*, in preparation (2002) BOOM02
15. Halverson, N.W. *et al.*, Ap. J. 568, 38 (2002) DASI
16. Mason, B. *et al.*, submitted to Ap. J. (2002), astro-ph/0205384 CBIdeep
17. Pearson, T.J. *et al.*, submitted to Ap. J. (2002), astro-ph/0205388 CBImosaic

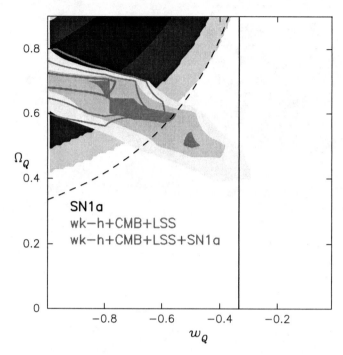

FIGURE 8. 1,2,3-sigma likelihood contour regions for "all-data" with the weak-H+age prior probability, $\Omega_{tot}=1$, and the LSS prior. 1,2,3-sigma SN1 contour regions [30] at upper right are rather similar to the constant deceleration lines. The $q_0 = 0$ line is the dashed one. The combined CMB+LSS+SN1 contours are shown as lines. The constraint from CMB on Ω_Q is reasonably good if a strong H_0 prior (Hubble key project result) is applied, but gets significantly more localized when LSS is added. In neither case is w_Q well determined, and w_Q localization with CMB+LSS+SN1 is mainly because of SN1. This implies the current limits are not that sensitive to the details of how quintessence impacts low ℓ in the DMR regime, which is an uncertainty in the theory.

18. Seljak, U. & Zaldarriaga, M. Ap. J. 469, 437 (1996) CMBFAST; Lewis, A. Challinor, A., & Lasenby, A. 2000, Ap. J. 538, 473 CAMB, an F90 alternative
19. Leitch, E.M. *et al.*, submitted to Ap. J. (2002), astro-ph/0209476; Kovac, J. *et al.*, submitted to Ap. J. (2002), astro-ph/0209478 DASIpol
20. Bond, J.R., Jaffe, A.H., & Knox, L., Phys. Rev. D 57, 2117 (1998)
21. Myers, S.T. *et al.*, submitted to Ap. J. (2002), astro-ph/0205385
22. Sievers, J. *et al.*, submitted to Ap. J. (2002), astro-ph/0205387
23. Bond, J.R. *et al.*, submitted to Ap. J. (2002), astro-ph/0205386
24. Dawson, K.S., Holzapfel, W.L., Carlstrom, J.E., LaRoque, S.J., Miller, A., Nagai, D. & Joy, M., submitted to Ap. J. (2002), astro-ph/0206012 BIMA
25. Bond, J.R. & Crittenden, R.G., in Proc. NATO ASI, Structure Formation in the Universe, eds. R.G. Crittenden & N.G. Turok (Kluwer) (2001), astro-ph/0108204
26. Bond, J.R., in *Cosmology and Large Scale Structure*, Les Houches Session LX, eds. R. Schaeffer J. Silk, M. Spiro & J. Zinn-Justin (Elsevier Science Press, Amsterdam), pp. 469-674 (1996), http://www.cita.utoronto.ca/~bond/papers/houches/LesHouches96.ps.gz
27. Efstathiou, G. & Bond, J.R., Mon. Not. R. Astron. Soc. 304, 75 (1999). Many other near-degeneracies between cosmological parameters are also discussed here.
28. The specific discrete parameter values used for the $\mathscr{C}_\ell$-database in this analysis were: ($\Omega_Q =$

TABLE 1. Cosmological parameter values and their 1-sigma errors are shown, determined after marginalizing over the other 6 cosmological and the various experimental parameters, for "all-data" and the weak prior (0.45 ≤ h ≤ 0.9, age > 10 Gyr). The LSS prior was also designed to be weak. The detections are clearly very stable if extra "prior" probabilities for LSS and SN1 are included. Similar tables are given in [22]. If Ω_{tot} is varied, but $w_Q = -1$, parameters derived from our basic 8 come out to be: age=15.2 ± 1.3 Gyr, h = 0.55 ± 0.08, Ω_m = 0.46 ± .11, Ω_b = 0.070 ± .02. Restriction to $\Omega_{tot} = 1$ and $w_Q = -1$ yields: age=14.1 ± 0.6 Gyr, h = 0.65 ± .05, $\Omega_m = 0.34 \pm .05$, $\Omega_b = 0.05 \pm .006$; allowing w_Q to vary yields quite similar results.

	cmb	+LSS	+SN1	+SN1+LSS
	Ω_{tot}	variable	$w_Q = -1$	CASE
Ω_{tot}	$1.03^{+.04}_{-.05}$	$1.03^{+.03}_{-.04}$	$1.01^{+.07}_{-.03}$	$1.00^{+.07}_{-.03}$
$\Omega_b h^2$	$.022^{+.004}_{-.002}$	$.022^{+.004}_{-.002}$	$.023^{+.005}_{-.003}$	$.023^{+.004}_{-.003}$
$\Omega_{cdm} h^2$	$.13^{+.03}_{-.03}$	$.12^{+.02}_{-.03}$	$.11^{+.03}_{-.03}$	$.11^{+.03}_{-.03}$
n_s	$0.94^{+.11}_{-.04}$	$0.93^{+.11}_{-.04}$	$0.97^{+.15}_{-.06}$	$0.99^{+.14}_{-.07}$
Ω_Q	$0.53^{+.14}_{-.13}$	$0.57^{+.13}_{-.08}$	$0.70^{+.07}_{-.08}$	$0.70^{+.06}_{-.07}$
	Ω_{tot}	=1	$w_Q = -1$	CASE
$\Omega_b h^2$	$.021^{+.003}_{-.002}$	$.022^{+.003}_{-.002}$	$.022^{+.003}_{-.002}$	$.022^{+.003}_{-.002}$
$\Omega_{cdm} h^2$	$.14^{+.03}_{-.02}$	$.12^{+.01}_{-.01}$	$.12^{+.01}_{-.01}$	$.12^{+.01}_{-.01}$
n_s	$0.93^{+.05}_{-.03}$	$0.95^{+.09}_{-.04}$	$0.95^{+.09}_{-.04}$	$0.97^{+.08}_{-.05}$
Ω_Q	$0.61^{+.09}_{-.38}$	$0.66^{+.05}_{-.06}$	$0.68^{+.03}_{-.05}$	$0.68^{+.03}_{-.05}$
	Ω_{tot}	=1	w_Q variable	CASE
$\Omega_b h^2$	$.021^{+.003}_{-.002}$	$.022^{+.003}_{-.002}$	$.022^{+.003}_{-.002}$	$.022^{+.003}_{-.002}$
$\Omega_{cdm} h^2$	$.14^{+.03}_{-.02}$	$.12^{+.01}_{-.01}$	$.12^{+.01}_{-.01}$	$.12^{+.01}_{-.01}$
n_s	$0.95^{+.05}_{-.04}$	$0.96^{+.07}_{-.04}$	$0.96^{+.07}_{-.04}$	$0.98^{+.07}_{-.05}$
Ω_Q	$0.54^{+.12}_{-.28}$	$0.61^{+.07}_{-.07}$	$0.69^{+.03}_{-.05}$	$0.69^{+.03}_{-.05}$
w_Q (95%)	< -0.43	< -0.46	< -0.71	< -0.71

0,.1,.2,.3,.4,.5,.6,.7,.8,.9,1.0,1.1), (Ω_k = .9,.7,.5,.3,.2,.15,.1,.05,0,-.05,-.1,-.15,-.2,-.3,-.5) & (τ_c =0, .025, .05, .075, .1, .15, .2, .3, .5) when $w_Q = -1$; (Ω_Q = 0,.1,.2,.3,.4,.5,.6,.7,.8,.9), (w_Q = -1,-.9,-.8,-.7,-.6,-.5,-.4,-.3,-.2,-.1,-.01) & (τ_c =0, .025, .05, .075, .1, .15, .2, .3, .5) when Ω_k=0. For both cases, (ω_c = .03, .06, .12, .17, .22, .27, .33, .40, .55, .8), (ω_b = .003125, .00625, .0125, .0175, .020, .025, .030, .035, .04, .05, .075, .10, .15, .2), (n_s =1.5, 1.45, 1.4, 1.35, 1.3, 1.25, 1.2, 1.175, 1.15, 1.125, 1.1, 1.075, 1.05, 1.025, 1.0, .975, .95, .925, .9, .875, .85, .825, .8, .775, .75, .725, .7, .65, .6, .55, .5), σ_8^2 was continuous, and there were a number of experimental parameters for calibration and beam uncertainties.

29. Perlmutter, S. *et al.*, Ap. J. 517, 565 (1999); Riess, A. *et al.*, AJ 116, 1009 (1998)
30. Perlmutter, S., Turner, M. & White, M. Phys. Rev. Lett. 83, 670 (1999)
31. Freedman, W. *et al.*, Phys. Rep. 568, 13 (2000).
32. The simplest interpretation of the superKamiokande data on atmospheric ν_μ is that $\Omega_{\nu_\tau} \sim 0.001$, about the energy density of stars in the universe, which implies a cosmologically negligible effect. Degeneracy between e.g., ν_μ and ν_τ could lead to the cosmologically very interesting $\Omega_{mv} \equiv \Omega_{hdm} \sim$ .1, although the coincidence of closely related energy densities for baryons, CDM, HDM and dark energy required would be amazing.

Black holes in high-energy collisions

Steven B. Giddings

Department of Physics, University of California, Santa Barbara, CA 93106 USA

Abstract. In TeV-scale gravity scenarios, the fundamental Planck scale could be as low as a TeV. If so, black hole production should be a prominent feature of scattering experiments at higher energies. Black hole events would have outstanding experimental signatures. The advent of black hole creation also appears to spell the end of our high-energy exploration of short distance physics, but the beginning of the exploration of extra dimensions of space.

INTRODUCTION

Black holes are perhaps the most profoundly mysterious objects in physics. We have long pondered what happens at the core of a black hole. The answer likely involves radically new physics, including breakdown of space and time, and is still beyond the reach of current approaches to quantum gravity such as string theory. Moreover, Hawking's discovery of black hole radiance[1] and proposal that black holes violate quantum mechanics[2] has led us to the sharp paradox of black hole information, which drives at the very heart of the problem of reconciling quantum mechanics and gravity. There is no clear way out:[1] information loss associated with breakdown of quantum mechanics apparently leads to disastrous violations of energy conservation; information cannot escape a black hole without violating locality; and the third alternative, black hole remnants, lead to catastrophic instabilities. String theorists have recently investigated the second alternative, via holography, but the jury is still out as no one has managed to understand how holographic theories can reproduce the approximately local physics that we see in our everyday world.

Experimental clues to the physics of black hole decay would be welcome. Unfortunately, manufacture of microscopic black holes apparently requires scattering energies above the four-dimensional Planck mass, $M_4 \sim 10^{19}$ GeV, placing this possibility far in our future.

However, recently there has been a revolution in thinking about the relationship between the Planck scale and the weak scale, $M_W \sim 1$ TeV. The longstanding "hierarchy problem" is to explain the large ratio of these; one would naturally expect $M_W \sim M_4$. The new idea is that the weak scale and the *fundamental* Planck scale, M_P, are indeed the same size, but four-dimensional gravity is weak (hence M_4 is large) due to dilution of gravity in large or warped extra dimensions. If this is the case, we may encounter the spectacular physics of black hole production in near-term experiments.

---

[1] For reviews, see [3, 4, 5, 6].

In outline, I'll start by describing some of the basic ideas of TeV-scale gravity, which make this remarkable scenario feasible. I'll then turn to a description of black holes on brane worlds and their production in high-energy collisions. Next is a discussion of black hole decay, largely through Hawking radiation. I'll close by describing some of the other consequences of this scenario, including what appears to be the end of investigation of short-distance physics, but may be the beginning of the exploration of the extra dimensions of space. For a more in-depth treatment of the subject of black hole production in TeV-scale gravity (and more complete references), the reader should consult the original references: [7] for the overall story (see also [8]); the more recent paper [9], which treats the classical problem of black-hole formation in high-energy collisions, and [10], which serves as another review, with further discussion of black hole creation in cosmic ray collisions with the upper atmosphere[11, 12, 13, 14, 15]

TeV-scale gravity

The idea that the fundamental Planck scale could be as low as the TeV scale is the essential new idea that inevitably leads to black hole production at energies above this threshhold. TeV-scale gravity is a novel approach to the long-standing *hierarchy problem*: we observe what appear to be two centrally important scales in physics, the weak scale $M_W \sim 1 TeV$, and the four-dimensional Planck scale, $M_4 \sim \frac{1}{\sqrt{G}} \sim 10^{19} GeV$, where G is Newton's gravitational constant; what explains the huge ratio between them? Traditional views invoke supersymmetry and its breaking, but the new idea is that the *fundamental* scale in physics is the TeV scale, and that the observed weakness of gravity, corresponding to the high value of M_4, results from dilution of gravitational effects in extra dimensions of space.

To explain further, suppose that there are D total dimensions of spacetime, with coordinates x^μ, $\mu = 0, 1, 2, 3$ parametrizing the ones we see, and y^m, $m = 4, \ldots, D-4$ parametrizing the small compact ones we don't. The most general spacetime metric consistent with the very nearly Poincaré invariant world we see is

$$ds^2 = e^{2A(y)} \eta_{\mu\nu} dx^\mu dx^\nu + g_{mn}(y) dy^m dy^n , \tag{1}$$

where A, conventionally called the "warp factor," is an arbitrary function of the unseen coordinates, $\eta_{\mu\nu}$ is the Minkowski metric, and g_{mn} is an arbitrary metric for the compact dimensions. Gravitational dynamics is governed by the D-dimensional Einstein-Hilbert action, and the action for four dimensional gravity is found by inserting (1), with a general four-dimensional metric $g_{\mu\nu}$, into this:

$$S_D = \frac{M_P^{D-2}}{4(2\pi)^{D-4}} \int d^D x \sqrt{-g} R \to \frac{M_4^2}{4} \int d^4 x \sqrt{-g_4(x)} R_4 + \cdots . \tag{2}$$

Here the extra terms on the right hand side are cancelled by whatever matter lagrangian is necessary to make the metric (1) a solution to the D-dimensional Einstein equations. Define the 'warped volume" of the extra dimensions,

$$V_W = \int d^{D-4} y \sqrt{g_{D-4}} e^{2A} . \tag{3}$$

The critical equation is

$$\frac{M_4^2}{M_P^2} = \frac{M_P^{D-4}}{(2\pi)^{D-4}} V_W : \quad (4)$$

the ratio of the observed and fundamental Planck scales is given by the warped volume in units of the fundamental Planck length.

We now have two options. The first one is the conventional one: assume that $M_P \sim M_4 \sim 10^{19} GeV$, which means

$$V_W \sim \frac{1}{M_P^{D-4}}. \quad (5)$$

The new alternative arises if $M_P \sim 1 TeV \ll M_4$, and this can be attained if the warped volume is for some reason very large:

$$V_W \gg \frac{1}{M_P^{D-4}} \sim \frac{1}{TeV^{D-4}}. \quad (6)$$

There are two approaches to achieving this. The first is the original idea of [16]: simply take the volume to be large. A second approach is to take the warp factor to be large; a toy model of this type was introduced in [17]. An obvious objection then arises, which is particularly clear in the large volume scenario: the size of the extra dimensions ranges from around a millimeter for $D = 6$ to $10 fm$ for $D = 10$, and gauge interactions have already been tested far past this, to around $10^{-3} fm$, for example in the context of precise electroweak measurements. Fortunately string theory has a made-to-order solution to this problem, which is the notion of a D-brane. For example, suppose that there are six extra dimensions, but that there are some three-branes present within them; ordinary matter and gauge fields may be composed of open strings, whose ends are restricted to move in the three-dimensional space defined by the brane, whereas gravity, which is always transmitted by closed string exchange, will propagate in all of the dimensions. String theory realizations of such "brane-world" scenarios with large volume were described in [18], and string solutions with large warping were derived in [19]; for discussion of their properties see [20]. We still lack completely realistic solutions with all the features to reproduce the known physics of the Standard Model at low energies, but ideas on this subject are still developing, or it may even be that such a scenario is realized outside of string theory.

Black holes on brane worlds

Suppose, therefore, that we live on such a brane world. If we collide two quarks with sufficiently high energy, they should form a black hole.[2] Being a gravitational object, this black hole will extend off the brane, as pictured in fig. 1. Study of such black holes

---

[2] For early discussion of this in large extra dimensions scenarios see [21], and for further discussion of some properties of black holes in such scenarios see [22]. Black hole evaporation was discussed in [23], and production of black holes in warped scenarios in [24].

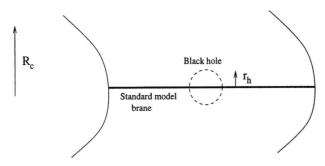

FIGURE 1. A schematic representation of a black hole on a brane world. Gauge fields and matter are confined to the brane, but the black hole extends into all of the dimensions. We consider the approximation where the black hole size is small as compared to characteristic geometric scales.

is greatly simplified by using two approximations. The first is to assume that the size of the black hole, r_h, is much less than the size of the extra dimensions or curvature scales of the extra dimensions, denoted R_c in the figure. This is typically true since the extra dimensions are "large." The second is the "probe-brane" approximation: in general the brane produces a gravitational field, but we neglect this field. This is justified if the black hole is massive as compared to the brane tension scale, which we expect to be approximately the Planck mass. So for large, but not too large, black holes, these two approximations reduce the problem to that of describing solutions in D flat dimensions.

Black holes created in particle collisions will typically have some angular momentum; spinning black hole solutions in D dimensions were first studied by Myers and Perry, in [25]. The horizon radius, Hawking temperature, and entropy of these black holes is given in terms of their mass M and spin J; in the $J \to 0$ limit, these take the form

$$r_h(M,J) \to \text{constant} \cdot M^{\frac{1}{D-3}}, \tag{7}$$

$$T_H(M,J) \to \frac{\text{constant}}{r_h} \tag{8}$$

and

$$S_{BH} \to \text{constant} \cdot M^{\frac{D-2}{D-3}}. \tag{9}$$

We'd like to estimate the rate at which such black holes would be produced in high-energy accelerators. The energy frontier is currently at proton machines. At the fundamental level, proton collisions are collisions among their constituents, quarks and gluons, generically called partons. In order to compute a rate, we need to know the density of partons of a given energy in the proton, or *parton distribution function*, and the cross-section for a pair of partons to make a black hole. Estimates of the parton distribution function are well known, but the black hole cross-section is not.

The cross section can, however, be estimated by making a key observation: for large center-of-mass energies of the partons, the formation process should be essentially classical. Indeed, consider a collision of two partons, each of energy $E/2$. If they pass closely enough, we expect them to form a big black hole, with horizon radius determined by E. For $E \gg M_P$, the curvature should be very weak at the black hole horizon, quantum gravity effects should be negligible, and a classical treatment should hold.

An estimate of the cross section follows from Thorne's hoop conjecture[26], which suggests that if energy E is concentrated in a region less than the corresponding Schwarzschild radius, $r_h(E)$, a black hole forms. If black holes form for impact parameters less than $r_h(E)$, this then indicates that that the cross section is given approximately by

$$\sigma \sim \pi r_h^2(E) \, . \tag{10}$$

Indeed, recently Eardley and I have revisited the classical problem of black hole production in high energy collisions[9], which was investigated some years ago for the case of zero impact parameter by Penrose[27] and D'Eath and Payne[28, 29, 30]. In particular, Penrose found a closed-trapped surface in the geometry describing the head-on collision of two Aichelberg-Sexl solutions, and this implies that a black hole of mass at least $E/\sqrt{2}$ forms. We extended this analysis to non-zero impact parameter. In the case of four dimensions, we explicitly found that a trapped surface forms for impact parameter $b \lesssim 1.6E$; this gives an estimated cross-section about 65% the naïve value (10). We also find the area of the trapped surface, providing a lower bound on the mass of the resulting black hole. While the $D > 4$ problem has not been explicitly solved, we did reduce it to boundary-value problem for Poisson's equation, which corresponds to a small displacement soap bubble problem. We expect this to yield similar results to those in $D = 4$. One can also explicitly see that the trapped surface forms before the collision – and consists of two disks in the collision surface connected by a catenoid between them[31, 32], as well as showing that the trapped surfaces can be deformed away from the curvature singularity at the center of the Aichelberg-Sexl solution[9]. This builds a fairly convincing case for black hole formation, with a cross section not too far from the estimate (10).

Current experimental bounds[33, 34, 35] on the Planck mass place it at $M_P \gtrsim 1.1 - .8 TeV$ for $D = 6 - 10$. The LHC has a design center-of-mass energy of 14 TeV. Suppose we make the optimistic assumption that indeed $M_P \sim 1$ TeV (and $D = 10$). One other important point is that one does not expect to create legitimate semiclassical black holes until one is a ways above the Planck scale, say at a minimum mass of $5 - 10$ TeV. The results are quite impressive[7, 8]: if the minimum mass for a black hole is 5 TeV, then LHC should produce black holes at the rate of about one *per second*. This would qualify LHC as a black hole factory, without bending standard nomenclature too far. If the minimum mass is 10 TeV, then we'd still produce black holes at the rate of about three per day.

Furthermore, from (10) and (7) we see that the cross-section for black hole production grows as

$$\sigma \sim E^{2/D-3} \, , \tag{11}$$

and so becomes even more dominant at higher energies.

Black hole decay and signatures

The resulting black holes will decay in several phases. When the horizon first forms, it will be very asymmetric and time dependent. The first thing that should happen is that the black hole sheds its hair, in a phase we term *balding*. It does so by classically emitting gauge and gravitational radiation. The amount of energy that can be emitted in this phase will be bounded given the bound on the minimum mass of the black hole resulting from the calculation of the area of the trapped surface. This may be improved by perturbative methods, as was done by D'Eath and Payne in [28, 29, 30]. A rough estimate is that somewhere between 15-40% if the initial collision energy of the partons is shed in this phase. The relevant decay time for this phase should be $\mathcal{O}(r_h)$. We expect the black hole to rapidly lose any charge as well. At the end of this phase we are left with a Kerr black hole with some mass and angular momentum.

Next quantum emission becomes relevant: the black hole will Hawking radiate. As shown by Page[36], it first does so by preferentially radiating particles in its equatorial plane, shedding its spin. We call the corresponding phase *spindown*. Page's four-dimensional calculations indicate that about 25% of the original mass of the black hole is lost during spin-down; we might expect the higher-dimensional situation to be similar. However, an important homework assignment is for someone to redo Page's analysis of decay of a spinning black hole in the higher-dimensional context, and fill in the details.[3]

At the end of spindown, a Schwarzschild black hole remains, and will continue to evaporate through the *Schwarzschild phase*. This phase ends when Hawking's caluculations fail, once the Schwarzschild radius becomes comparable to the D-dimensional Planck length, and quantum gravity effects become important.[4] Based on Page's analysis, roughly 75% of the original black hole mass might be emitted in this phase. As Hawking taught us, a prominent hallmark of this phase is that the emission is thermal, up to gray-body factors, at any given time. Based on this, one may estimate the resulting energy spectrum of particles in the final state, as well as estimating the ratios of different kinds of particles produced: hadrons, leptons, photons, *etc.*

The decay comes to an end with the *Planck phase*, the final decay of the Planck-size remnant of the black hole. This decay is sensitive to full-blown strongly coupled quantum gravity, and thus its details cannot yet be predicted. A reasonable expectation is that the Planck-size remnant would emit a few quanta with characteristic energy $\sim M_P$. This end of the spectrum is where much of the fascination lies: we can hope for experimental input on quantum gravity and/or string theory, and may see concrete evidence for the breakdown of spacetime structure, or as Hawking has advocated[2], even of quantum mechanics.

By putting this all together, we can infer some of the signatures that would evidence black hole production if it takes place at a future accelerator. Decay of a black hole should produce of order S_{BH} primary hard particles, with typical energies given by the Hawking temperature T_H, thus ranging over roughly 100 GeV - 1 TeV. Creation

---

[3] Some recent progress on this has been made in [37].

[4] In string theory there is another scale, the string length, at which the evaporation may be modified even earlier, due to effects stemming from the finite string size.

of primary particles is essentially democratic among species: we create an equal number of each color, spin, and flavor of quark, of each flavor and spin of lepton, of each helicity state of the gauge bosons, *etc*. These ratios are then changed through QCD jet formation, or decays of the primary particles. For example, this leads to a rough estimate that we would see five times as many hadrons as leptons. Simply the presence of the hard leptons would be one notable signature. Moreover, most of the Hawking radiation is isotropic in the black hole's rest frame, which can't be highly boosted with respect to the lab frame. These events would have a high sphericity. Finally, closer study may reveal the dipole pattern characteristic of the spindown phase.

So far no one has thought of events based on Standard model physics or any of its extensions that would mimic these signatures: if black holes are produced, their decays should stand out and be discovered.

Nature already provides us with particle collisions exceeding the reach of the LHC: cosmic rays hit the upper atmosphere with center-of-mass energies ranging up to roughly 400 TeV. We might ask if we could even see black holes produced by cosmic rays[10]. Unfortunately, the observed flux of ultra-high energy cosmic rays is believed to consist of either protons or nuclei, and even at the relevant energies, QCD cross sections dominate the cross section for black hole production by a factor of roughly a billion. So most of the hadronic cosmic rays will scatter via QCD processes before they can make black holes. A rough estimate is that these cosmic rays would produce 100 black holes per year over the entire surface of the Earth, which is clearly too small of a rate to measure[10].

However, it is also believed that there should be a neutrino component of the ultra-high energy cosmic ray background; this would for example arise from ultra-high energy protons scattering off the 3K photons in the microwave background to resonantly produce Δ's, which produce neutrinos in their decays[38, 39, 40]. Neutrinos interact only weakly, and at ultra-high energies it turns out that black hole production is roughly competitive with rates for neutrinos to interact via Standard Model processes. Taking existing neutrino flux estimates at face value, this suggests that neutrinos could produce black holes at rates around[11, 12, 13, 14][10][15]

$$\frac{\text{several black holes}}{\text{yr km}^3 (H_2O)}. \tag{12}$$

Detectors that are currently or soon to be operating, such as the HiRes Fly's Eye, Auger, Icecube, and OWL/Airwatch, are at a level of sensitivity where they might start to see black hole events, if the assumptions about the neutrino fluxes are correct.

The future of high energy physics

High energy physics is a logical extension of a longstanding human quest to understand nature at an ever more fundamental level. Once we reach the Planck scale, things may change; shorter distances than the Planck length may well not make sense. However, physics is an experimental subject, and ultimately we might expect to address the question of shorter-distance physics experimentally. However, once black hole production commences, exploration of shorter distances seems to come to an end.

Specifically, if we want to investigate physics at a distance scale Δx that is shorter than the Planck length, the uncertainty principle tells us we should scatter particles at energies $E \sim 1/\Delta x > M_P$. But if they indeed scatter at distances Δx, they will be inside a black hole. Once a black hole forms, the outside observer cannot witness the scattering process directly – all we see is the Hawking radiation that the black hole sheds. Short distance physics is cloaked by black hole formation, and thus investigation of short distances through high energy physics comes to an end.

Some of our experimental colleagues might consider this a dismal future. However, there is another prospect that we can offer them. As they increase the energy, they will be making bigger and bigger black holes. At some distance scale, these black holes will start to become sensitive to the shapes and sizes of the extra dimensions, or to other features such as parallel branes. When black holes start extending far enough off our brane to probe these features, their properties, such as their production rate, their decay spectrum, and other properties, change. So by doing black hole physics at increasing energies, experimentalists can reach further off the brane that we are otherwise confined to, and start to explore the geography of the extra dimensions of space. This could certainly continue to yield exciting experimental discoveries!

ACKNOWLEDGMENTS

I'd like to thank my collaborators D. Eardley, E. Katz, and S. Thomas for the opportunity to explore these fascinating ideas together, and the Aspen Center for Physics where this contribution was written. This work was supported in part by the Department of Energy under Contract DE-FG-03-91ER40618.

REFERENCES

1. Hawking, S.W., *Commun. Math. Phys.* **43**, 199 (1975).
2. Hawking, S.W., *Phys. Rev.* **D14**, 2460 (1976).
3. Giddings, S.B., "The Black hole information paradox," arXiv:hep-th/9508151.
4. Strominger, A., "Les Houches lectures on black holes," arXiv:hep-th/9501071.
5. Giddings, S.B., "Quantum mechanics of black holes," arXiv:hep-th/9412138.
6. 't Hooft, G., *Nucl. Phys. Proc. Suppl.* **43**, 1 (1995).
7. Giddings, S.B., and Thomas, S., *Phys. Rev.* **D65**, 056010 (2002).
8. Dimopoulos, S., and Landsberg, G., *Phys. Rev. Lett.* **87**, 161602 (2001).
9. Eardley, D.M., and Giddings, S.B., "Classical black hole production in high-energy collisions," arXiv:gr-qc/0201034.
10. Giddings, S.B., "Black hole production in TeV-scale gravity, and the future of high energy physics," in *Proc. of the APS/DPF/DPB Summer Study on the Future of Particle Physics (Snowmass 2001)* ed. Davidson, R., and Quigg, C., arXiv:hep-ph/0110127.
11. Feng, J.L., and Shapere, A.D., "Black hole production by cosmic rays," arXiv:hep-ph/0109106.
12. Dorfan, D., Giddings, S.B., Rizzo, T., and Thomas, S., unpublished.
13. Anchordoqui, L., and Goldberg, H., "Experimental signature for black hole production in neutrino air showers," arXiv:hep-ph/0109242.

14. Emparan, R., Masip, M., and Rattazzi, R., "Cosmic rays as probes of large extra dimensions and TeV gravity," arXiv:hep-ph/0109287.
15. Ringwald, A., and Tu, H., *Phys. Lett.* **B525**, 135 (2002).
16. Arkani-Hamed, N., Dimopoulos, S., and Dvali, G.R., *Phys. Lett.* **B429**, 263 (1998).
17. Randall, L., and Sundrum, R., *Phys. Rev. Lett.* **83**, 3370 (1999).
18. Antoniadis, I., Arkani-Hamed, N., Dimopoulos, S., and Dvali, G.R., *Phys. Lett.* **B436**, 257 (1998).
19. Giddings, S.B., Kachru, S., and Polchinski, J., "Hierarchies from fluxes in string compactifications," arXiv:hep-th/0105097.
20. DeWolfe, O., and Giddings, S.B., "Scales and hierarchies in warped compactifications and brane worlds," arXiv:hep-th/0208123.
21. Banks, T., and Fischler, W.,"A model for high energy scattering in quantum gravity," arXiv:hep-th/9906038.
22. Argyres, P.C., Dimopoulos, S., and March-Russell, J., *Phys. Lett.* **B441**, 96 (1998).
23. Emparan, R., Horowitz, G.T., and Myers, R.C., *Phys. Rev. Lett.* **85**, 499 (2000).
24. Giddings, S.B., and Katz, E., *J. Math. Phys.* **42**, 3082 (2001).
25. Myers, R.C., and Perry, M.J., *Annals Phys.* **172**, 304 (1986).
26. Thorne, K.S., "Nonspherical Gravitational Collapse: A Short Review," in Klauder, J.R., *Magic Without Magic, San Francisco 1972, 231-258.*
27. *Penrose, R., unpublished (1974).*
28. *D'Eath, P.D., and Payne, P.N., Phys. Rev.* **D46**, *658 (1992).*
29. *D'Eath, P.D., and Payne, P.N., Phys. Rev.* **D46**, *675 (1992).*
30. *D'Eath, P.D., and Payne, P.N., Phys. Rev.* **D46**, *694 (1992).*
31. *Giddings, S.B., Lindblom, L., Scheel, M., and Thorne, K.S., unpublished.*
32. *Yoshino, H., and Nambu, Y., "High-energy head-on collisions of particles and hoop conjecture," arXiv:gr-qc/0204060.*
33. *Peskin, M.E., "Theoretical summary," arXiv:hep-ph/0002041.*
34. *Giudice, G.F., Rattazzi, R., and Wells, J.D., Nucl. Phys.* **B544**, *3 (1999).*
35. *Mirabelli, E.A., Perelstein, M., and Peskin, M.E., Phys. Rev. Lett.* **82**, *2236 (1999).*
36. *Page, D.N., Phys. Rev.* **D14**, *3260 (1976).*
37. *Kanti, P., and March-Russell, J., "Calculable corrections to brane black hole decay. I: The scalar case," arXiv:hep-ph/0203223, and in progress.*
38. *Greisen, K., Phys. Rev. Lett.* **16**, *748 (1966).*
39. *Stecker, F.W., Astrophys. J.* **228**, *919 (1979).*
40. *Hill, C.T., and Schramm, D.N., Phys. Rev.* **D31**, *564 (1985).*

Direct Evidence for Neutrino Flavor Transformation from Neutral-Current Interactions in SNO

A.B. McDonald [§§§,*], Q.R. Ahmad [†], R.C. Allen [**], T.C. Andersen [‡],
J.D. Anglin [§], J.C. Barton [1,¶], E.W. Beier [||], M. Bercovitch [§], J. Bigu [††],
S.D. Biller [¶], R.A. Black [¶], I. Blevis [‡‡], R.J. Boardman [¶], J. Boger [§§],
E. Bonvin [*], M.G. Boulay [*,¶], M.G. Bowler [¶], T.J. Bowles [¶¶],
S.J. Brice [¶,¶¶], M.C. Browne [¶¶,†], T.V. Bullard [†], G. Bühler [**],
J. Cameron [¶], Y.D. Chan [***], H.H. Chen [2,**], M. Chen [*], X. Chen [¶,***],
B.T. Cleveland [¶], E.T.H. Clifford [*], J.H.M. Cowan [††], D.F. Cowen [||],
G.A. Cox [†], X. Dai [¶], F. Dalnoki-Veress [‡‡], W.F. Davidson [§],
P.J. Doe [¶¶,**,†], G. Doucas [¶], M.R. Dragowsky [***,¶¶], C.A. Duba [†],
F.A. Duncan [*], M. Dunford [||], J.A. Dunmore [¶], E.D. Earle [¶¶¶,*],
S.R. Elliott [¶¶,†], H.C. Evans [*], G.T. Ewan [*], J. Farine [‡‡,††], H. Fergani [¶],
A.P. Ferraris [¶], R.J. Ford [*], J.A. Formaggio [†], M.M. Fowler [¶¶],
K. Frame [¶], E.D. Frank [||], W. Frati [||], N. Gagnon [¶¶,***,†,¶],
J.V. Germani [†], S. Gil [†††], K. Graham [*], D.R. Grant [‡‡], R.L. Hahn [§§],
A.L. Hallin [*], E.D. Hallman [††], A.S. Hamer [*,¶¶], A.A. Hamian [†],
W.B. Handler [*], R.U. Haq [††], C.K. Hargrove [‡‡], P.J. Harvey [*],
R. Hazama [†], K.M. Heeger [†], W.J. Heintzelman [||], J. Heise [¶¶,†††],
R.L. Helmer [†††,‡‡], J.D. Hepburn [*], H. Heron [¶], J. Hewett [††], A. Hime [¶¶],
M. Howe [†], J.G. Hykawy [††], M.C.P. Isaac [***], P. Jagam [‡], N.A. Jelley [¶],
C. Jillings [*], G. Jonkmans [¶¶,††], K. Kazkaz [†], P.T. Keener [||], J.R. Klein [||],
A.B. Knox [¶], R.J. Komar [†††], R. Kouzes [§§§], T. Kutter [†††], C.C.M. Kyba [||],
J. Law [‡], I.T. Lawson [‡], M. Lay [¶], H.W. Lee [*], K.T. Lesko [***],
J.R. Leslie [*], I. Levine [‡‡], W. Locke [¶], S. Luoma [††], J. Lyon [¶], S. Majerus
[¶], H.B. Mak [*], J. Maneira [*], J. Manor [†], A.D. Marino [***], N. McCauley [¶,||],
D.S. McDonald [||], K. McFarlane [‡‡], G. McGregor [¶], R. Meijer Drees [†],
C. Mifflin [‡‡], G.G. Miller [¶¶], G. Milton [¶¶¶], B.A. Moffat [*], M. Moorhead
[¶], C.W. Nally [†††], M.S. Neubauer [||], F.M. Newcomer [||], H.S. Ng [†††],
A.J. Noble [‡‡,‡‡‡], E.B. Norman [***], V.M. Novikov [‡‡], M. O'Neill [‡‡],
C.E. Okada [***], R.W. Ollerhead [‡], M. Omori [¶], J.L. Orrell [†], S.M. Oser [||],
A.W.P. Poon [†,†††,¶,***], T.J. Radcliffe [*], A. Roberge [††], B.C. Robertson
[*], R.G.H. Robertson [¶¶,†], S.S.E. Rosendahl [***], J.K. Rowley [§§],
V.L. Rusu [||], E. Saettler [††], K.K. Schaffer [†], M.H. Schwendener [††],
A. Schülke [***], H. Seifert [†,¶¶,††], M. Shatkay [‡‡], J.J. Simpson [‡],
C.J. Sims [¶], D. Sinclair [‡‡‡,‡‡], P. Skensved [*], A.R. Smith [***],

CP646, *Theoretical Physics: MRST 2002*, edited by V. Elias et al.
© 2002 American Institute of Physics 0-7354-0101-2/02/$19.00

M.W.E. Smith [†], T. Spreitzer [‖], N. Starinsky [‡‡], T.D. Steiger [†],
R.G. Stokstad [***], L.C. Stonehill [†], R.S. Storey [§], B. Sur [*, ¶¶¶],
R. Tafirout [††], N. Tagg [¶, ‡], N.W. Tanner [¶], R.K. Taplin [¶], M. Thorman [¶],
P.M. Thornewell [¶], P.T. Trent [¶], Y.I. Tserkovnyak [†††], R. Van Berg [‖],
R.G. Van de Water [‖, ¶¶], C.J. Virtue [††], C.E. Waltham [†††], J.-X. Wang [‡],
D.L. Wark [♡, ¶¶, 1], N. West [¶], J.B. Wilhelmy [¶¶], J.F. Wilkerson [¶¶, †],
J.R. Wilson [¶], P. Wittich [‖], J.M. Wouters [¶¶] and M. Yeh [§§]

[*] *Department of Physics, Queen's University, Kingston, Ontario K7L 3N6*
[†] *Center for Experimental Nuclear Physics and Astrophysics, and Department of Physics, University of Washington, Seattle, WA 98195*
[**] *Department of Physics, University of California, Irvine, CA 92717*
[‡] *Physics Department, University of Guelph, Guelph, Ontario N1G 2W1*
[§] *National Research Council of Canada, Ottawa, ON K1A 0R6*
[¶] *Department of Physics, University of Oxford, Keble Road, Oxford, OX1 3RH, UK*
[‖] *Department of Physics and Astronomy, University of Pennsylvania, Philadelphia, PA 19104-6396*
[††] *Department of Physics and Astronomy, Laurentian University, Sudbury, Ontario P3E 2C6*
[‡‡] *Carleton University, Ottawa, Ontario K1S 5B6*
[§§] *Chemistry Department, Brookhaven National Laboratory, Upton, NY 11973-5000*
[¶¶] *Los Alamos National Laboratory, Los Alamos, NM 87545*
[***] *Institute for Nuclear and Particle Astrophysics and Nuclear Science Division, Lawrence Berkeley National Laboratory, Berkeley, CA 94720*
[†††] *Department of Physics and Astronomy, University of British Columbia, Vancouver, BC V6T 1Z1*
[‡‡‡] *TRIUMF, 4004 Wesbrook Mall, Vancouver, BC V6T 2A3*
[§§§] *Department of Physics, Princeton University, Princeton, NJ 08544*
[¶¶¶] *Atomic Energy of Canada, Limited, Chalk River Laboratories, Chalk River, ON K0J 1J0*
[♡] *Rutherford Appleton Laboratory, Chilton, Didcot, Oxon, OX11 0QX, UK and University of Sussex, Physics and Astronomy Department, Brighton BN1 9QH, UK*

Abstract. The Sudbury Neutrino Observatory (SNO) is a 1,000 tonne heavy water Cerenkov-based neutrino detector situated 2,000 meters underground in INCO's Creighton Mine near Sudbury, Ontario, Canada. For the neutrinos from ^{8}B decay in the Sun SNO observes the Charged Current neutrino reaction sensitive only to electron neutrinos and others (Neutral Current and Elastic Scattering) sensitive to all active neutrino types and thereby can search for direct evidence of neutrino flavor change. Using these reactions and assuming the standard ^{8}B shape, the ν_e component of the ^{8}B solar flux is $\phi_e = 1.76^{+0.05}_{-0.05}$(stat.)$^{+0.09}_{-0.09}$ (syst.) $\times 10^6$ cm^{-2}s^{-1} for a kinetic energy threshold of 5 MeV. The non-ν_e component is $\phi_{\mu\tau} = 3.41^{+0.45}_{-0.45}$(stat.)$^{+0.48}_{-0.45}$ (syst.) $\times 10^6$ cm^{-2}s^{-1}, 5.3σ greater than zero, providing strong evidence for solar ν_e flavor transformation. The total flux measured with the NC reaction is $\phi_{NC} = 5.09^{+0.44}_{-0.43}$(stat.)$^{+0.46}_{-0.43}$ (syst.) $\times 10^6$ cm^{-2}s^{-1}, consistent with solar models. For charged current events, assuming an undistorted ^{8}B spectrum, the night minus day rate is $14.0\% \pm 6.3\%^{+1.5}_{-1.4}\%$ of the average rate. If the total flux of active neutrinos is additionally constrained to have no asymmetry, the ν_e asymmetry is found to be $7.0\% \pm 4.9\%^{+1.3}_{-1.2}\%$. A global solar neutrino analysis in terms of matter-enhanced oscillations of two active flavors strongly favors the Large Mixing Angle (LMA) solution.

The Sudbury Neutrino Observatory (SNO) [1] is a water Cherenkov detector located

---

[1] Permanent Address:Birkbeck College, University of London, Malet Road, London WC1E 7HX, UK
[2] Deceased

at a depth of 6010 m of water equivalent in the INCO, Ltd. Creighton mine near Sudbury, Ontario, Canada. The detector uses ultra-pure heavy water contained in a transparent acrylic spherical shell 12 m in diameter to detect solar neutrinos. Cherenkov photons generated in the heavy water are detected by 9456 photomultiplier tubes (PMTs) mounted on a stainless steel geodesic sphere 17.8 m in diameter. The geodesic sphere is immersed in ultra-pure light water to provide shielding from radioactivity in both the PMT array and the cavity rock.

SNO detects ^{8}B solar neutrinos through the reactions:

$$\begin{aligned} \nu_e + d &\rightarrow p + p + e^- \quad \text{(CC)}, \\ \nu_x + d &\rightarrow p + n + \nu_x \quad \text{(NC)}, \\ \nu_x + e^- &\rightarrow \nu_x + e^- \quad \text{(ES)}. \end{aligned}$$

The charged current reaction (CC) is sensitive exclusively to electron-type neutrinos, while the neutral current reaction (NC) is equally sensitive to all active neutrino flavors ($x = e, \mu, \tau$). The elastic scattering reaction (ES) is sensitive to all flavors as well, but with reduced sensitivity to ν_μ and ν_τ. Sensitivity to these three reactions allows SNO to determine the electron and non-electron active neutrino components of the solar flux [2].

The SNO experimental plan calls for three phases wherein different techniques will be used for the detection of neutrons from the NC reaction. During the first phase, with pure heavy water, neutrons are observed through the Cerenkov light produced when neutrons are captured in deuterium, producing 6.25 MeV gammas. In this phase, the capture probability for such neutrons is about 25% and the cerenkov light is relatively close to the threshold of about 5 MeV electron energy, imposed by radioactivity in the detector. For the second phase, underway since June 2001, about 2.5 tonnes of NaCl has been added to the heavy water and neutron detection is enhanced through capture on Cl, with about 8.6 MeV gamma ray energy and about 83% capture efficiency. For the third phase, the salt will be removed and an array of ^{3}He-filled proportional counters will be installed to provide direct detection of neutrons with a capture efficiency of about 45%.

The CC and ES reaction results have been presented previously [3]. This summary discusses the first NC results and updated CC and ES results from the pure heavy water phase, recently submitted for publication [4]. The data reported here were recorded between Nov. 2, 1999 and May 28, 2001 and represent a total of 306.4 live days, spanning the entire first phase of the experiment, in which only D_2O was present in the sensitive volume. The analysis procedure was similar to that described in [3]. PMT times and hit patterns were used to reconstruct event vertices and directions and to assign to each event a most probable kinetic energy, T_{eff}. The total flux of active ^{8}B solar neutrinos with energies greater than 2.2 MeV (the NC reaction threshold) was measured with the NC signal (Cherenkov photons resulting from the 6.25 MeV γ ray from neutron capture on deuterium.) The analysis threshold was $T_{\text{eff}} \geq 5$ MeV, providing sensitivity to neutrons from the NC reaction. Above this energy threshold, there were contributions from CC events in the D_2O, ES events in the D_2O and H_2O, capture of neutrons (both from the NC reaction and backgrounds), and low energy Cherenkov background events.

A fiducial volume was defined to only accept events which had reconstructed vertices within 550 cm from the detector center to reduce external backgrounds and systematic uncertainties associated with optics and event reconstruction near the acrylic vessel. The neutron response and systematic uncertainty was calibrated with a ^{252}Cf source. The deduced efficiency for neutron captures on deuterium is $29.9 \pm 1.1\%$ for a uniform source of neutrons in the D$_2$O. The neutron detection efficiency within the fiducial volume and above the energy threshold is 14.4%. The energy calibration was updated from [3] with the ^{16}N calibration source [5] data and Monte Carlo calculations. The energy response for electrons, updated for the lower analysis threshold, was characterized as a Gaussian function with resolution $\sigma_T = -0.0684 + 0.331\sqrt{T_e} + 0.0425 T_e$, where T_e is the true electron kinetic energy in MeV. The energy scale uncertainty is 1.2%.

The primary backgrounds to the NC signal are due to low levels of uranium and thorium decay chain daughters (^{214}Bi and ^{208}Tl) in the detector materials. These activities generate free neutrons in the D$_2$O, from deuteron photodisintegration (pd), and low energy Cherenkov events. *Ex-situ* assays and *in-situ* analysis of the low energy (4 − 4.5 MeV) Cherenkov signal region provide independent uranium and thorium photodisintegration background measurements.

Two *ex situ* assay techniques were employed to determine average levels of uranium and thorium in water. Radium ions were directly extracted from the water onto either MnO$_x$ or hydrous Ti oxide (HTiO) ion exchange media. Radon daughters in the U and Th chains were subsequently released, identified by α spectroscopy, or the radium was concentrated and the number of decay daughter β-α coincidences determined. Typical assays circulated approximately 400 tonnes of water through the extraction media. These techniques provide isotopic identification of the decay daughters and contamination levels in the assayed water volumes, presented in Fig. 1 (a). Secular equilibrium in the U decay chain was broken by the ingress of long-lived (3.8 day half-life) ^{222}Rn in the experiment. Measurements of this background were made by periodically extracting and cryogenically concentrating ^{222}Rn from water degassers. Radon from several tonne assays was subsequently counted in ZnS(Ag) scintillation cells [6]. The Radon results are presented (as mass fractions in g(U)/g(D$_2$O)) in Fig. 1(b).

Independent measurements of U and Th decay chains were made by analyzing Cherenkov light produced by the radioactive decays. The β and β-γ decays from the U and Th chains dominate the low energy monitoring window. Events in this window monitor γ rays that produce photodisintegration in these chains (E$_\gamma$ > 2.2 MeV). Cherenkov events fitted within 450 cm from the detector center and extracted from the neutrino data set provide a time-integrated measure of these backgrounds over the same time period and within the fiducial volume of the neutrino analysis. Statistical separation of *in situ* Tl and Bi events was obtained by analyzing the Cherenkov signal isotropy. Tl decays always result in a β and a 2.614 MeV γ, while in this energy window Bi decays are dominated by decays with only a β, and produce, on average, more anisotropic hit patterns.

Results from the *ex situ* and *in situ* methods are consistent with each other as shown on the right hand side of Figs. 1(a) and 1(b). For the ^{232}Th chain, the weighted mean (including additional sampling systematic uncertainty) of the two determinations was used for the analysis. The ^{238}U chain activity is dominated by Rn ingress which is highly time dependent. Therefore the *in-situ* determination was used for this activity as it provides the appropriate time weighting. The average rate of background neutron production from activities in the D$_2$O region is 1.0 ± 0.2 neutrons per day, leading to 44^{+8}_{-9} detected background events. The production rate from external activities is

TABLE 1. Neutron and Cherenkov background events.

Source	Events
D_2O photodisintegration	44^{+8}_{-9}
H_2O + AV photodisintegration	27^{+8}_{-8}
Atmospheric ν's and sub-Cherenkov threshold μ's	4 ± 1
Fission	$\ll 1$
$^2H(\alpha,\alpha)pn$	2 ± 0.4
$^{17}O(\alpha,n)$	$\ll 1$
Terrestrial and reactor $\bar{\nu}$'s	1^{+3}_{-1}
External neutrons	$\ll 1$
Total neutron background	78 ± 12
D_2O Cherenkov	20^{+13}_{-6}
H_2O Cherenkov	3^{+4}_{-3}
AV Cherenkov	6^{+3}_{-6}
PMT Cherenkov	16^{+11}_{-8}
Total Cherenkov background	45^{+18}_{-12}

$1.3^{+0.4}_{-0.5}$ neutrons per day, which leads to 27 ± 8 background events since the neutron capture efficiency is reduced for neutrons born near the heavy water boundary. The total photodisintegration background corresponds to approximately 12% of the number of NC neutrons predicted by the standard solar model from ^8B neutrinos.

Low energy backgrounds from Cherenkov events in the signal region were evaluated by using acrylic encapsulated sources of U and Th deployed throughout the detector volume and by Monte Carlo calculations. Probability density functions (pdfs) in reconstructed vertex radius derived from U and Th calibration data were used to determine the number of background Cherenkov events from external regions which either entered or mis-reconstructed into the fiducial volume. Cherenkov event backgrounds from activities in the D_2O were evaluated with Monte Carlo calculations.

Table 1 shows the number of photodisintegration and Cherenkov background events (including systematic uncertainties) due to activity in the D_2O (internal region), acrylic vessel (AV), H_2O (external region), and PMT array. Other sources of free neutrons in the D_2O region are cosmic ray events and atmospheric neutrinos. To reduce these backgrounds, an additional neutron background cut imposed a 250-ms deadtime (in software) following every event in which the total number of PMTs which registered a hit was greater than 60. The number of remaining NC atmospheric neutrino events and background events generated by sub-Cherenkov threshold muons is estimated to be small, as shown in Table 1.

The data recorded during the pure D_2O detector phase are shown in Figure 2. These data have been analyzed using the same data reduction described in [3], with the addition of the new neutron background cut, yielding 2928 events in the energy region selected for analysis, 5 to 20 MeV. Fig. 2(a) shows the distribution of selected events in the cosine of the angle between the Cherenkov event direction and the direction from the sun ($\cos\theta_\odot$) for the analysis threshold of $T_{\text{eff}} \geq 5$ MeV and fiducial volume selection of

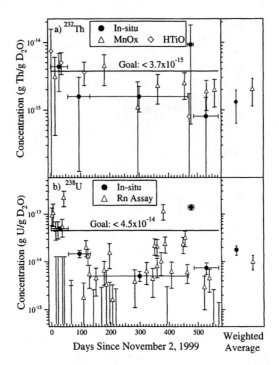

FIGURE 1. Thorium (a) and uranium (b) backgrounds (equivalent equilibrium concentrations) in the D_2O deduced by *in situ* and *ex situ* techniques. The MnO_x and HTiO radiochemical assay results, the Rn assay results, and the *in situ* Cherenkov signal determination of the backgrounds are presented for the period of this analysis on the left-hand side of frames (a) and (b). The right-hand side shows time-integrated averages including an additional sampling systematic uncertainty for the *ex situ* measurement.

$R \leq 550$ cm, where R is the reconstructed event radius. Fig. 2(b) shows the distribution of events in the volume-weighted radial variable $(R/R_{AV})^3$, where $R_{AV} = 600$ cm is the radius of the acrylic vessel. Figure 2(c) shows the kinetic energy spectrum of the selected events.

In order to test the null hypothesis, the assumption that there are only electron neutrinos in the solar neutrino flux, the data are resolved into contributions from CC, ES, and NC events above threshold using pdfs in T_{eff}, $\cos\theta_\odot$, and $(R/R_{AV})^3$, derived from Monte Carlo calculations generated assuming no flavor transformation and the standard ^{8}B spectral shape [7]. Background event pdfs are included in the analysis with fixed amplitudes determined by the background calibration. The extended maximum likelihood method used in the signal decomposition yields $1967.7^{+61.9}_{-60.9}$ CC events, $263.6^{+26.4}_{-25.6}$ ES events, and $576.5^{+49.5}_{-48.9}$ NC events [3], where only statistical uncertainties are given.

---

[3] We note that this rate of neutron events also leads to a lower bound on the proton lifetime for "invisible" modes (based on the free neutron that would be left in deuterium (V.I. Tretyak and Yu.G. Zdesenko, Phys.

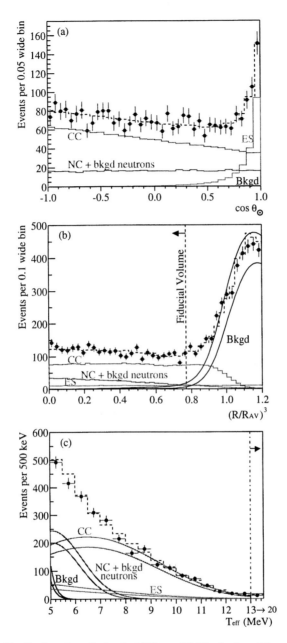

FIGURE 2. (a) Distribution of $\cos\theta_\odot$ for $R \leq 550$ cm. (b) Distribution of the volume weighted radial variable $(R/R_{AV})^3$. (c) Kinetic energy for $R \leq 550$ cm. Also shown are the Monte Carlo predictions for CC, ES and NC + bkgd neutron events scaled to the fit results, and the calculated spectrum of Cherenkov background (Bkgd) events. The dashed lines represent the summed components, and the bands show $\pm 1\sigma$ uncertainties. All distributions are for events with $T_{eff} \geq 5$ MeV.

TABLE 2. Systematic uncertainties on fluxes. The experimental uncertainty for ES (not shown) is -4.8,+5.0 percent. † denotes CC vs NC anti-correlation.

Source	CC Uncert. (percent)	NC Uncert. (percent)	$\phi_{\mu\tau}$ Uncert. (percent)
Energy scale †	-4.2,+4.3	-6.2,+6.1	-10.4,+10.3
Energy resolution †	-0.9,+0.0	-0.0,+4.4	-0.0,+6.8
Energy non-linearity †	±0.1	±0.4	±0.6
Vertex resolution †	±0.0	±0.1	±0.2
Vertex accuracy	-2.8,+2.9	±1.8	±1.4
Angular resolution	-0.2,+0.2	-0.3,+0.3	-0.3,+0.3
Internal source pd †	±0.0	-1.5,+1.6	-2.0,+2.2
External source pd	±0.1	-1.0,+1.0	±1.4
D_2O Cherenkov †	-0.1,+0.2	-2.6,+1.2	-3.7,+1.7
H_2O Cherenkov	±0.0	-0.2,+0.4	-0.2,+0.6
AV Cherenkov	±0.0	-0.2,+0.2	-0.3,+0.3
PMT Cherenkov †	±0.1	-2.1,+1.6	-3.0,+2.2
Neutron capture	±0.0	-4.0,+3.6	-5.8,+5.2
Cut acceptance	-0.2,+0.4	-0.2,+0.4	-0.2,+0.4
Experimental uncertainty	-5.2,+5.2	-8.5,+9.1	-13.2,+14.1
Cross section [8]	±1.8	±1.3	±1.4

Systematic uncertainties on fluxes derived by repeating the signal decomposition with perturbed pdfs (constrained by calibration data) are shown in Table 2.

Normalized to the integrated rates above the kinetic energy threshold of $T_{\text{eff}} \geq 5$ MeV, the flux of ^{8}B neutrinos measured with each reaction in SNO, assuming the standard spectrum shape [7] is (all fluxes are presented in units of 10^6 cm^{-2}s^{-1}):

$$\phi_{\text{CC}}^{\text{SNO}} = 1.76^{+0.06}_{-0.05}(\text{stat.})^{+0.09}_{-0.09}(\text{syst.})$$

$$\phi_{\text{ES}}^{\text{SNO}} = 2.39^{+0.24}_{-0.23}(\text{stat.})^{+0.12}_{-0.12}(\text{syst.})$$

$$\phi_{\text{NC}}^{\text{SNO}} = 5.09^{+0.44}_{-0.43}(\text{stat.})^{+0.46}_{-0.43}(\text{syst.}).$$

Electron neutrino cross sections are used to calculate all fluxes. The CC and ES results reported here are consistent with the earlier SNO results [3] for $T_{\text{eff}} \geq 6.75$ MeV. The excess of the NC flux over the CC and ES fluxes implies neutrino flavor transformations.

A simple change of variables resolves the data directly into electron (ϕ_e) and non-electron ($\phi_{\mu\tau}$) components. This change of variables allows a direct test of the null hypothesis of no flavor transformation ($\phi_{\mu\tau} = 0$) without requiring calculation of the CC, ES, and NC signal correlations.

$$\phi_e = 1.76^{+0.05}_{-0.05}(\text{stat.})^{+0.09}_{-0.09}(\text{syst.})$$

$$\phi_{\mu\tau} = 3.41^{+0.45}_{-0.45}(\text{stat.})^{+0.48}_{-0.45}(\text{syst.})$$

Lett. **B505**, 59 (2001)) in excess of 10^{28} years, approximately 3 orders of magnitude more restrictive than previous limits (J. Evans and R. Steinberg, Science, **197**, 989 (1977).) The possible contribution of this mechanism to the solar neutrino NC background is ignored.

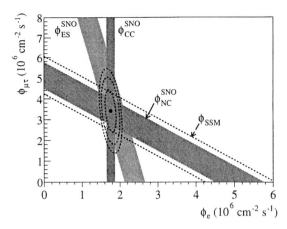

FIGURE 3. Flux of ^{8}B solar neutrinos which are μ or τ flavor vs flux of electron neutrinos deduced from the three neutrino reactions in SNO. The diagonal bands show the total ^{8}B flux as predicted by the SSM [12] (dashed lines) and that measured with the NC reaction in SNO (solid band). The intercepts of these bands with the axes represent the $\pm 1\sigma$ errors. The bands intersect at the fit values for ϕ_e and $\phi_{\mu\tau}$, indicating that the combined flux results are consistent with neutrino flavor transformation assuming no distortion in the ^{8}B neutrino energy spectrum.

assuming the standard ^{8}B shape. Combining the statistical and systematic uncertainties in quadrature, $\phi_{\mu\tau}$ is $3.41^{+0.66}_{-0.64}$, which is 5.3σ above zero, providing strong evidence for flavor transformation consistent with neutrino oscillations [9, 10]. Adding the Super-Kamiokande ES measurement of the ^{8}B flux [11] $\phi^{SK}_{ES} = 2.32 \pm 0.03 (\text{stat.})^{+0.08}_{-0.07}$ (syst.) as an additional constraint, we find $\phi_{\mu\tau} = 3.45^{+0.65}_{-0.62}$, which is 5.5σ above zero. Figure 3 shows the flux of non-electron flavor active neutrinos vs the flux of electron neutrinos deduced from the SNO data. The three bands represent the one standard deviation measurements of the CC, ES, and NC rates. The error ellipses represent the 68%, 95%, and 99% joint probability contours for ϕ_e and $\phi_{\mu\tau}$.

Removing the constraint that the solar neutrino energy spectrum is undistorted, the signal decomposition is repeated using only the $\cos\theta_\odot$ and $(R/R_{AV})^3$ information. The total flux of active ^{8}B neutrinos measured with the NC reaction is

$$\phi^{SNO}_{NC} = 6.42^{+1.57}_{-1.57}(\text{stat.})^{+0.55}_{-0.58}\ (\text{syst.}) \qquad (1)$$

which is in agreement with the shape constrained value above and with the standard solar model prediction [12] for ^{8}B, $\phi_{SSM} = 5.05^{+1.01}_{-0.81}$.

The neutrino flavor conversion can be explained by neutrino oscillation models based on flavor mixing [9, 10]. For some values of the mixing parameters, spectral distortions and a measurable dependence on solar zenith angle are expected [13, 14, 15]. The latter might be caused by interaction with the matter of the Earth (the MSW effect) and would depend not only on oscillation parameters and neutrino energy, but also on the path length and e^- density through the Earth.

To seek evidence for neutrino interactions with the matter of the Earth, the data was analyzed in two parts corresponding to day and night periods. The total livetimes for day and night are 128.5 and 177.9 days, respectively. The time-averaged inverse-square distance to the Sun $\langle(\frac{1AU}{R})^2\rangle$ was 1.0002 (day) and 1.0117 (night). During the development of this analysis, the data were partitioned into two sets of approximately equal livetime (split at July 1, 2000), each having substantial day and night components. Analysis procedures were refined during the analysis of Set 1 and fixed before Set 2 was analysed. The latter thus served as an unbiased test. Unless otherwise stated, the analysis presented in this paper is for the combined data set.

The data reduction described above was used for this analysis as well. For each event, the number, pattern, and timing of the hit photomultiplier tubes (PMTs) were used to reconstruct effective recoil electron kinetic energy T_{eff}, radial position R, and scattering angle $\theta_\odot$ with respect to the Sun-Earth direction. The charged current (CC), elastic scattering (ES) and neutral current (NC) reactions each have characteristic probability density functions (pdfs) of T_{eff}, R, and $\theta_\odot$. A maximum likelihood fit of the pdfs to the data determined the flux from each of these reactions.

The measured night and day fluxes ϕ_N and ϕ_D were used to form the asymmetry ratio for each reaction: $\mathscr{A} = 2(\phi_N - \phi_D)/(\phi_N + \phi_D)$. The CC interaction is sensitive only to ν_e. The NC interaction is equally sensitive to all active neutrino flavors, so active-only neutrino models predict $\mathscr{A}_{NC} = 0$ [16]. The same models allow $\mathscr{A}_{CC} \neq 0$. The ES reaction has additional contributions from $\nu_{\mu\tau}$ leading to a reduction in its sensitivity to ν_e asymmetries.

SNO used calibration sources [17] to constrain variations in detector response [18] that can lead to day-night asymmetries. A ^{16}N source [19], which produces 6.1-MeV gamma rays, revealed a 1.3% per year drift in the energy scale. Due to seasonal variation in day and night livetime, this drift can create an artificial asymmetry. The analysis corrected for this drift and a systematic uncertainty was assigned using worst-case drift models. Gamma rays from the ^{16}N source were also used to constrain directional dependences in SNO's response.

A set of signals that are continuously present in the detector was used to probe possible diurnal variations in detector response. The detector was triggered at 5 Hz with a pulser, verifying livetime accounting. Muons provide an almost constant signal and, through interactions with D_2O, produce secondary neutrons. After applying a cut to remove bursts with high neutron multiplicity, these muon-induced neutrons were used to limit temporal variations in detector response. A more sensitive study focused on a solitary point of high background radioactivity, or "hot spot", on the upper hemisphere of the SNO acrylic vessel, apparently introduced during construction. Its event rate was stable and sufficient to make an excellent test of diurnal variations. It also provides a sensitive test for changes in reconstruction. A limit of 3.5% on the hot spot rate asymmetry was determined, which because of its steeply falling energy spectrum constrained the day and night energy scales to be the same within 0.3%. An east/west division of the neutrino data based on the Sun's position should show no rate variations from matter effects. As expected, the CC rates for east and west data were consistent. The rate asymmetries for each test are shown in Fig. 4.

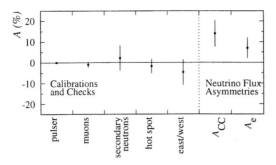

FIGURE 4. Various event classes used to determine systematic differences between day and night measurements. Also shown are measured asymmetries on the CC flux, and on the electron neutrino flux derived from the CC, ES, and NC rates when the total neutrino flux is constrained to have zero asymmetry.

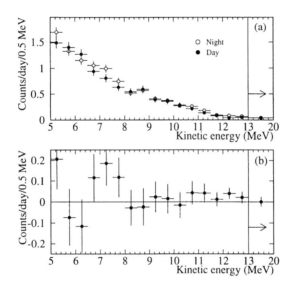

FIGURE 5. (a) Energy spectra for day and night. All signals and backgrounds contribute. The final bin extends from 13.0 to 20.0 MeV. (b) Difference, *night - day*, between the spectra. The day rate was 9.23 ± 0.27 events/day, and the night rate was 9.79 ± 0.24 events/day.

Backgrounds were subtracted separately for day and night as part of the signal extraction. The results were normalized for an Earth-Sun distance of 1 AU, yielding the results in Table 3. Day and night fluxes are given separately for data Sets 1 and 2, and for the combined data. A χ^2 consistency test of the six measured fluxes between Sets 1 and 2 yielded a chance probability of 8%. A similar test done directly on the three asymmetry parameters gave a chance probability of 2%. No systematic has been identified, in either signal or background regions, that would suggest that the differences between Set 1 and Set 2 are other than a statistical fluctuation. For the combined analysis, $\mathcal{A}_{CC}$ is $+2.2\sigma$

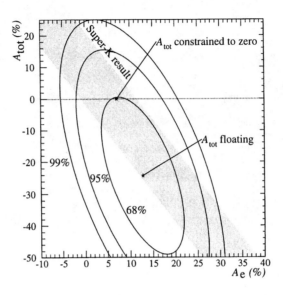

FIGURE 6. Joint probability contours for $\mathcal{A}_{tot}$ and $\mathcal{A}_e$. The points indicate the results when $\mathcal{A}_{tot}$ is allowed to float and when it is constrained to zero. The diagonal band indicates the 68% joint contour for the Super-K $\mathcal{A}_{ES}$ measurement.

TABLE 3. The results of signal extraction, assuming an undistorted ^{8}B spectrum. The systematic uncertainties (combined set) include a component that cancels in the formation of the $\mathcal{A}$. Except for the dimensionless $\mathcal{A}$, the units are 10^6 cm^{-2} s^{-1}. Flux values have been rounded, but the asymmetries were calculated with full precision.

signal	Set 1		Set 2	
	ϕ_D	ϕ_N	ϕ_D	ϕ_N
CC	1.53 ± 0.12	1.95 ± 0.10	1.69 ± 0.12	1.77 ± 0.11
ES	2.91 ± 0.52	1.59 ± 0.38	2.35 ± 0.51	2.88 ± 0.47
NC	7.09 ± 0.97	3.95 ± 0.75	4.56 ± 0.89	5.33 ± 0.84

signal	Combined		$\mathcal{A}$ (%)
	ϕ_D	ϕ_N	
CC	1.62 ± 0.08 ± 0.08	1.87 ± 0.07 ± 0.10	$+14.0 \pm 6.3^{+1.5}_{-1.4}$
ES	2.64 ± 0.37 ± 0.12	2.22 ± 0.30 ± 0.12	$-17.4 \pm 19.5^{+2.4}_{-2.2}$
NC	5.69 ± 0.66 ± 0.44	4.63 ± 0.57 ± 0.44	$-20.4 \pm 16.9^{+2.4}_{-2.5}$

from zero, while $\mathcal{A}_{ES}$ and $\mathcal{A}_{NC}$ are -0.9σ and -1.2σ from zero, respectively. Note that $\mathcal{A}_{CC}$ and $\mathcal{A}_{NC}$ are strongly statistically anti-correlated ($\rho = -0.518$), while $\mathcal{A}_{CC}$ and $\mathcal{A}_{ES}$ ($\rho = -0.161$) and $\mathcal{A}_{ES}$ and $\mathcal{A}_{NC}$ ($\rho = -0.106$) are moderatedly anti-correlated. Table 4 gives the systematic uncertainties on the asymmetry parameters. The day and night energy spectra for all accepted events are shown in Fig. 5.

Table 5 (a) shows the results for $\mathcal{A}_e$ derived from the CC day and night rate measurements, i.e., $\mathcal{A}_e = \mathcal{A}_{CC}$. The day and night flavor contents were then extracted by

TABLE 4. Effect of systematic uncertainties on $\mathscr{A}$ (%). For presentation, uncertainties have been symmetrized and rounded.

Systematic	$\delta\mathscr{A}_{CC}$	$\delta\mathscr{A}_{ES}$	$\delta\mathscr{A}_{NC}$
Long-term energy scale drift	0.4	0.5	0.2
Diurnal energy scale variation	1.2	0.7	1.6
Directional energy scale var.	0.2	1.4	0.3
Diurnal energy resolution var.	0.1	0.1	0.3
Directional energy resolution var.	0.0	0.1	0.0
Diurnal vertex shift var.	0.5	0.6	0.7
Directional vertex shift var.	0.0	1.1	0.1
Diurnal vertex resolution var.	0.2	0.7	0.5
Directional angular recon. var.	0.0	0.1	0.1
PMT β-γ background	0.0	0.2	0.5
AV+H$_2$O β-γ bkgd.	0.0	0.6	0.2
D$_2$O β-γ, neutrons bkgd.	0.1	0.4	1.2
External neutrons bkgd.	0.0	0.2	0.4
Cut acceptance	0.5	0.5	0.5
Total	1.5	2.4	2.4

TABLE 5. Measurement of the ϕ_e and ϕ_{tot} asymmetry for various constraints. All analyses assume an undistorted ^{8}B spectrum.

Constraints	Asymmetry (%)
a) no additional constraint	$\mathscr{A}_{CC} = 14.0 \pm 6.3^{+1.5}_{-1.4}$ $\mathscr{A}_{NC} = -20.4 \pm 16.9^{+2.4}_{-2.5}$ (see text for correlations)
b) $\phi_{ES} = (1-\varepsilon)\phi_e + \varepsilon\phi_{tot}$	$\mathscr{A}_e = 12.8 \pm 6.2^{+1.5}_{-1.4}$ $\mathscr{A}_{tot} = -24.2 \pm 16.1^{+2.4}_{-2.5}$ correlation = -0.602
c) $\phi_{ES} = (1-\varepsilon)\phi_e + \varepsilon\phi_{tot}$ $\mathscr{A}_{tot} = 0$	$\mathscr{A}_e = 7.0 \pm 4.9^{+1.3}_{-1.2}$
d) $\phi_{ES} = (1-\varepsilon)\phi_e + \varepsilon\phi_{tot}$ $\mathscr{A}_{tot} = 0$ $\mathscr{A}_{ES}(SK) = 3.3\% \pm 2.2\%^{+1.3}_{-1.2}\%$	$\mathscr{A}_e(SK) = 5.3 \pm 3.7^{+2.0}_{-1.7}$ (derived from SK $\mathscr{A}_{ES}$ and SNO total ^{8}B flux)

changing variables to $\phi_{CC} = \phi_e$, $\phi_{NC} = \phi_{tot} = \phi_e + \phi_{\mu\tau}$ and $\phi_{ES} = \phi_e + \varepsilon\phi_{\mu\tau}$, where $\varepsilon \equiv 1/6.48$ is the ratio of the average ES cross sections above 5 MeV for $\nu_{\mu\tau}$ and ν_e. Table 5 (b) shows the asymmetries of ϕ_e and ϕ_{tot} with this additional constraint from the ES rate measurements. This analysis allowed for an asymmetry in the total flux of ^{8}B neutrinos (non-zero $\mathscr{A}_{tot}$), with the measurements of $\mathscr{A}_e$ and $\mathscr{A}_{tot}$ having a strong anti-correlation. Fig. 6 shows the $\mathscr{A}_e$ vs. $\mathscr{A}_{tot}$ joint probability contours. Forcing $\mathscr{A}_{tot} = 0$, as predicted by active-only models, yielded the result in Table 5 (c) of $\mathscr{A}_e = 7.0\% \pm 4.9\%$ (stat.)$^{+1.3}_{-1.2}\%$ (sys.).

The Super-Kamiokande (SK) collaboration measured $\mathscr{A}_{ES}(SK) = 3.3\% \pm 2.2\%$ (stat.)$^{+1.3}_{-1.2}\%$ (sys.) [20]. The ES measurement includes a neutral current component, which reduces the asymmetry for this reaction relative to $\mathscr{A}_e$ [21]. $\mathscr{A}_{ES}(SK)$ may be converted to an equivalent electron flavor asymmetry using the total neutrino flux

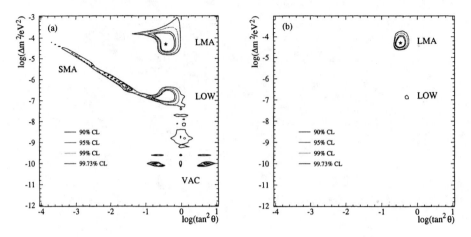

FIGURE 7. Allowed regions of the MSW plane determined by a χ^2 fit to (a) SNO day and night energy spectra and (b) with additional experimental and solar model data. The star indicates the best fit. See text for details.

measured by SNO, yielding $\mathscr{A}_e(SK)$ (Table 5 (d)). This value is in good agreement with SNO's direct measurement of $\mathscr{A}_e$, as seen in Fig. 6.

SNO's day and night energy spectra (Fig. 5) have also been used to produce MSW exclusion plots and limits on neutrino flavor mixing parameters. MSW oscillation models [22] between two active flavors were fit to the data. For simplicity, only the energy spectra were used in the fit, and the radial R and direction $\cos\theta_\odot$ information was omitted. This procedure preserves most of the ability to discriminate between oscillation solutions. A model was constructed for the expected number of counts in each energy bin by combining the neutrino spectrum [23], the survival probability, and the cross sections [24] with SNO's response functions [18]. For this analysis, the dominant systematics are those for the combined fluxes, as detailed in Table 2 and [18], and not the diurnal systematics of Table 4.

There are 3 free parameters in the fit: the total ^{8}B flux ϕ_B, the difference Δm^2 between the squared masses of the two neutrino mass eigenstates, and the mixing angle θ. The flux of higher energy neutrinos from the solar *hep* reaction was fixed at 9.3×10^3 cm^{-2} s^{-1} [25]. Contours were generated in Δm^2 and $\tan^2\theta$ for $\Delta\chi^2(c.l.) = 4.61$ (90%), 5.99 (95%), 9.21 (99%), and 11.83 (99.73%). Fig. 7(a) shows allowed mixing parameter regions using only SNO data with no additional experimental constraints or inputs from solar models. By including flux information from the Cl [26] and Ga experiments [27, 28, 29, 30, 31], the day and night spectra from the SK experiment [20], along with solar model predictions for the more robust pp, pep and ^{7}Be neutrino fluxes [25], the contours shown in Fig. 7(b) were produced. This global analysis strongly favors the Large Mixing Angle (LMA) region (see Table 6), and $\tan^2\theta$ values < 1. While the absolute chi-squared per degree of freedom is not particularly large for the LOW solution, the difference between chi-squared values still reflects the extent to which one region of MSW parameter space is favored compared to another. Repeating

TABLE 6. Best fit points in the MSW plane for global MSW analysis using all solar neutrino data. ϕ_B is the best-fit ^{8}B flux for each point, and has units of 10^6 cm^{-2} s^{-1}. Δm^2 has units of eV^2. $\mathscr{A}_e$ is the predicted asymmetry for each point.

Region	χ^2_{min}/dof	ϕ_B	$\mathscr{A}_e$(%)	Δm^2	$\tan^2\theta$	c.l.(%)
LMA	57.0/72	5.86	6.4	5.0×10^{-5}	0.34	—
LOW	67.7/72	4.95	5.9	1.3×10^{-7}	0.55	99.5

the global analysis using the total SNO energy spectrum instead of separate day and night spectra gives nearly identical results.

In summary, the results presented here are the first direct measurement of the total flux of active ^{8}B neutrinos arriving from the sun and provide strong evidence for neutrino flavor transformation. The CC and ES reaction rates are consistent with the earlier results [3] and with the NC reaction rate under the hypothesis of flavor transformation. The total flux of ^{8}B neutrinos measured with the NC reaction is in agreement with the SSM prediction. The day-night asymmetries of the CC, NC, and ES reaction rates have also been measured. From these results the first direct measurements of the day-night asymmetries in the v_e flux and the total v flux from the Sun have been deduced. A global fit to SNO's day and night energy spectra and data from other solar neutrino experiments strongly favors the LMA solution in a 2-flavor MSW neutrino oscillation analysis.

ACKNOWLEDGMENTS

This research was supported by: Canada: NSERC, Industry Canada, NRC, Northern Ontario Heritage Fund Corporation, Inco, AECL, Ontario Power Generation; US: Dept. of Energy; UK: PPARC. We thank the SNO technical staff for their strong contributions.

REFERENCES

1. The SNO collaboration, *Nucl. Instr. and Meth.* **A449**, 172 (2000).
2. Chen, H.H., *Phys. Rev. Lett.* **55**, 1534 (1985).
3. Ahmad, Q.R., *et al.*, *Phys. Rev. Lett.* **87**, 071301 (2001).
4. Ahmad, Q.R., *et al.*, *Phys. Rev. Lett.* (2002), submitted for publication.
5. Dragowsky, M.R., *et al.*, *Nucl. Instr. and Meth.* **A481**, 284 (2002).
6. Liu, M.-Q., Lee, H.W., and McDonald, A.B., *Nucl. Inst. Meth.* **A329**, 291 (1993).
7. Ortiz, C.E., *et al.*, *Phys. Rev. Lett.* **85**, 2909 (2000).
8. Cross section uncertainty includes g_A uncertainty (0.6%), difference between NSGK (Nakamura, S., Sato, T., Gudkov, V., and Kubodera, K., *Phys. Rev.* **C63**, 034617 (2001)) and BCK (Butler, M., Chen, J.-W., and Kong, X., *Phys. Rev.* **C63**, 035501 (2001)) in SNO's calculations (0.6%), radiative correction uncertainties (0.3% for CC, 0.1% for NC, Kurylov, A., Ramsey-Musolf, M.J., and Vogel, P., *Phys. Rev.* **C65**, 055501 (2002)), uncertainty associated with neglect of real photons in SNO (0.7% for CC), and theoretical cross section uncertainty (1%, Nakamura, S., *et al.*, arXiv:nucl-th/0201062, (to be published)).
9. Maki, Z., Nakagawa, N., and Sakata, S., *Prog. Theor. Phys.* **28**, 870 (1962).
10. Gribov, V., and Pontecorvo, B., *Phys. Lett.* **B28**, 493 (1969).
11. Fukuda, S., *et al.*, *Phys. Rev. Lett.* **86**, 5651 (2001).

12. Bahcall, J.N., Pinsonneault, M.H., and Basu, S., *Astrophys. J* **555**, 990 (2001).
13. Mikheyev, S.P., and Smirnov, A.Y., in *'86 Massive Neutrinos in Astrophysics and in Particle Physics, Proceedings of the Moriond Workshop*, edited by Fackler, O., and Tran Thanh Van, J., (Editions Frontières, Gif-sur-Yvette, 1986), 335.
14. Baltz, A.J., and Weneser, J., *Phys. Rev.* **D37**, 3364 (1988).
15. Gonzalez-Garcia, M.C., Peña-Garay, C., and Smirnov, A.Y., *Phys. Rev.* **D63** 113004 (2001).
16. A non-zero value for $\mathscr{A}_{NC}$ might be evidence for sterile neutrinos.
17. The SNO Collaboration, *Nucl. Instr. and Meth.* **A449**, 172 (2000).
18. Details of SNO response functions are available from the SNO web site: http://sno.phy.queensu.ca
19. Dragowsky, M., *et al.*, *Nucl. Instr. and Meth.* **A481**, 284 (2002).
20. Fukuda, S., *et al.*, *Phys. Rev. Lett.* **86**, 5651 (2001).
21. Bahcall, J.N., Krastev, P., and Smirnov, A.Y., *Phys. Rev. D* **62**, 093004 (2000); Maris, M., and Petcov, S.T., *Phys. Rev. D* **62**, 093006 (2000); Bahcall, J.N., Gonzalez-Garcia, M.C., and Peña-Garay, C., e-print hep-ph/0111150.
22. Parke, S.J., *Phys. Rev. Lett.* **57**, 1275 (1986); Petcov, S.T., *Phys. Rev.* **B200**, 373 (1988); Fogli, G.L., and Lisi, E., *Astropart. Phys.* **3**, 185 (1995); Lisi, E., Marrone, A., Montanino, D., Palazzo, A., and Petcov, S.T., *Phys. Rev.* **D63**, 093002 (2001).
23. Ortiz, C. E., *et al.*, *Phys. Rev. Lett.* **85**, 2909 (2000).
24. Nakamura, S., *et al.*, e-print nucl-th/0201062.
25. Bahcall, J.N, Pinsonneault, H. M., and Basu, S., *Astrophys. J* **555**, 990 (2001).
26. Cleveland, B. T., *et al.*, *Astrophys. J* **496**, 505 (1998).
27. Abdurashitov, J.N., *et al.*, *Phys. Rev.* **C60**, 055801 (1999).
28. Abdurashitov, J. N., *et al.*, e-print astro-ph/0204245.
29. Altmann, M., *et al.*, *Phys. Lett.* **B490**, 16 (2000).
30. Hampel, W., *et al.*, *Phys. Lett.* **B447**, 127 (1999).
31. C.M. Cattadori *et al.*, in *Proceedings of the TAUP 2001 Workshop*, (September 2001), Assergi, Italy.

The ground state of quantum gravity with positive cosmological constant

Lee Smolin[1]

Perimeter Institute for Theoretical Physics, Waterloo, Canada N2J 2W9
and
Department of Physics, University of Waterloo

Abstract. The main results of loop quantum gravity, in the connection representation, are summarized, for the case $\Lambda > 0$. These include the existence of a ground state, discoverd by Kodama, which both is an exact solution to the constraints of quantum gravity and has a semiclassical limit which is deSitter spacetime. The long wavelength excitations of this state are known to reproduce both gravitons and, when matter is included, quantum field theory on deSitter spacetime. Quantum gravity for $\Lambda > 0$ also has an intrinsic thermality, due to the existence of a periodicity on the full configuration space, which generalizes and explains the thermal behavior of quantum fields on deSitter spacetime.

INTRODUCTION

The cosmological constant, Λ, is the source of some of the most recalcitrant puzzles facing any quantum theory of gravity. In recent years the mystery has been deepened by the accumulation of observational evidence that there is at present a tiny-but non-zero value for the cosmological constant. The problem is first to explain why the value of this apparently fundamental parameter is on the order of 10^{-120} in natural units. Even more disturbing, the measured value is positive, which creates big problems for theories like string theory whose consistency is based on supersymmetry. It is not even clear if perturbative string theory can be made consistent with a positive value of the renormalized cosmological constant.

There is, however, one approach to quantum gravity, which is improved by the existence of a cosmological constant. This is loop quantum gravity[1]-[6], which is an approach to quantum gravity based on the quantization, using the Dirac procedure, of Einstein's theory of general relativity, in the form discovered by Sen[7] and formalized by Ashtekar[8]. As I will review briefly here, the case of $\Lambda > 0$ is the best understood case for loop quantum gravity. For this case there is a state, called the Kodama state[9] which is not only an exact solution to the full equations of the theory, it has a good low energy limit, which reproduces the physics of quantum fields around deSitter spacetime. As a result, I believe it is possible to claim that, when $\Lambda \neq 0$, loop quantum gravity provides an apparently complete and consistent quantum theory of gravity.

---

[1] lsmolin@perimeterinstitute.ca, www.perimeterinstitute.ca , www.qgravity.org

The purpose of this short contribution is to review, without details, the main results concerning loop quantum gravity with cosmological constant[2].

Among these results are the following.

- Small, long wavelength perturbations of the Kodama state do, for small $\lambda = \Lambda G\hbar$ reproduce the spectrum of gravitons on a background of deSitter spacetime[10].
- The theory may be coupled to arbitrary matter fields. For the same conditions, long wavelength and small λ, perturbations of the Kodama state in the matter sector reproduce QFT on the deSitter background[11, 12].
- When one extends the approximation to higher order terms in $l_{Pl}E$ one finds corrections to the energy-momentum relations for matter fields, of a form[10, 35],

$$E^2 = p^2 + m^2 + \alpha E^3 + \ldots \quad (1)$$

Such corrections are in fact amenable to experimental test[13, 14, 15, 16] in present and near future experiments.

- There is a natural boundary term[17] which may be added to study the case of horizons[18, 19] or timelike boundaries[20, 21]. This leads to an explicit construction of the boundary Hilbert space. One consequence is that the Bekenstein bound is satisfied automatically[17]. Another is a new understanding[10], in purely background independent terms, of the N bound conjectured by Banks[22, 23].
- The quantum theory of gravity with Λ is intrinsically thermal, due to the fact that the Euclidean continuation of the configuration space of the full theory is periodic when expressed in Ashtekar-Sen variables[11]. This explains and generalizes the thermality of quantum field theory on a fixed deSitter background[38].

Due to space limitations, we do not describe here results gotten from the loop representation[5, 6, 17, 30, 10], nor do we describe the path integral approach in loop quantum gravity[25]-[31].

GRAVITY AS A GAUGE THEORY

The Sen-Ashtekar approach to general relativity is based on the embedding of the variables of Einstein's theory into those of an $SU(2)$ Yang-Mills theory, in which the connection is extended to complex values. The variables are then an $SU(2)$ connection, A_{ai} and its conjugate electric field E^{ai}. The latter is kept real, but the former is extended to complex values. The Poisson brackets have the form,

$$\{A_a^i(x), E_j^b(y)\} = \imath G \delta_a^b \delta_j^i \delta^3(y,x) \quad (2)$$

where G is Newton's constant[3].

---

[2] This contribution can be considered an outline, and an invitation to, a longer and more pedagogical paper[10].
[3] The reader should also note the factor of $\imath$ present.

E_i^a is related to the three metric q_{ab} by

$$\sqrt{q}q^{ab} = E^{ai}E_i^b \tag{3}$$

The determinant is there because the expression is a density of weight two.

The $SU(2)$ connection A_a turns out to be, for solutions, the self-dual part of the spacetime connection. For Lorentzian solutions this is complex, and its real and imaginary parts each have a geometrical interpretation.

$$A_{ai} = \text{3d spin connection}_{ai} + \frac{\iota}{\sqrt{q}} K_{ab} E_i^b \tag{4}$$

where K_{ab} is the extrinsic curvature of the 3 manifold Σ embedded in the spacetime, which in turn is essentially the time derivative of the three metric.

We consider spacetimes of the form $\mathcal{M} = \Sigma \times R$ where Σ is a compact three manifold, with or without boundary. The action of the theory can be put in Hamiltonian form,

$$I^{GR} = \int dt \int_\Sigma \iota E^{ai} \dot{A}_{ai} - N\mathcal{H} - N^a H_a - w_i \mathcal{G}^i - h \tag{5}$$

where $\mathcal{H}_a$ generates the diffeomorphisms of Σ, $\mathcal{H}$ must be the so-called Hamiltonian constraint that generates the rest of the diffeomorphism group of the spacetime (and hence changes of the slicings of the spacetime into spatial slices) while $\mathcal{G}^i$ generates the local $SU(2)$ gauge transformations. h represents the terms in the hamiltonian that are not proportional to constraints. However, there is a special feature of gravitational theories, which is there is no way *locally* to distinguish the changes in the local fields under evolution from their changes under a diffeomorphism that changes the time coordinate. Hence h is always just a boundary term, in a theory of gravity.

As in Yang-Mills theory we know that the constraint that generates local gauge transformations under (2) is just Gauss's law

$$\mathcal{G}^i = \mathcal{D}_a E^{ai} = 0 \tag{6}$$

Note that E_i^a is a vector *density*, so there is no metric used in either the Poisson brackets or Gauss's law[4].

Three more constraints generate the spatial diffeomorphisms

$$\mathcal{H}_a = E_i^b F_{ab}^i = 0 \tag{7}$$

where F_{ab}^i is the Yang-Mills field strength. The Hamiltonian constraint, that generates the dynamics, is

$$\mathcal{H} = \varepsilon_{ijk} E^{ai} E^{bj} (F_{ab}^k + \frac{\Lambda}{3} \varepsilon_{abc} E^{ck}) = 0 \tag{8}$$

---

[4] One thing to get used to in this field is that as there is no background metric, while in the quantum theory the metric is a composite operator, one must be completely explicit about all places the metric appears and all density weights.

Given the forms of the constraints, the Einstein equations are the hamiltons equations of motion. The simplest way to evolve is with the shift, $N^a = 0$ which corresponds to the time-space components of the metric vanishing. The equations of motion are then,

$$\dot{A}_{ai} = \{A_{ai}, \int N\mathcal{H}\} = N\iota G \varepsilon_{ijk} E^{bj}(2F^k_{ab} + \Lambda \varepsilon_{abc} E^{ck}) \tag{9}$$

$$\dot{E}^{ai} = \{E^{ai}, \int N\mathcal{H}\} = \iota G \varepsilon^{ijk} \mathcal{D}_b(N E^a_j E^b_k) \tag{10}$$

Note that as $\mathcal{H}$ has density weight two, N is an inverse density.

THE DESITTER SOLUTION AS A GAUGE FIELD

To understand the results that follow on the Kodama state, we need to know how to represent deSitter spacetime in this langauge. The key point is that deSitter spacetime, being self-dual, satisfies,

$$F^i_{ab} = -\frac{\Lambda}{3} \varepsilon_{abc} E^{ci} \tag{11}$$

In fact, as is easy to check (11) is sufficient to solve all 7 constraints of the theory. Thus, all self-dual spactimes are solutions. But, here is a key fact, while there are many Euclidean self-dual solutions, deSitter spacetime is the only real Lorentzian self-dual spacetime. Thus, any complexified $SU(2)$ gauge field that satisfies (11) for real E^{ai} represents a piece of the lorentzian DeSitter spacetime. For example, the following gauge field corresponds to the flat slicing of deSitter spacetime,

$$A_{ai} = \iota\sqrt{\frac{\Lambda}{3}} f(t) \delta_{ai} \mapsto F_{abi} = -f^2(t) \frac{\Lambda}{3} \varepsilon_{abi} \tag{12}$$

where $f(t)$ is a function of the time coordinates.

Taking A^i_a to be purely imaginary makes sense in light of (4), it means that we are making an ansatz that the three geometry is flat, so the three dimensional spin connection vanishes. The metric can then be taken to be homogeneous, as must also be its time derivative, which is the imaginary part of A^i_a.

By the self-dual initial conditions we see that

$$E^{ai} = f^2 \delta^{ai} \mapsto q_{ab} = f^2 \delta_{ab} \tag{13}$$

As we have satisfied the self-dual condition all the constraints are satisfied. We merely have to plug into the equations of motion (9,10) to find the evolution equations for f. Both equations of motion agree that

$$\dot{f} = N\sqrt{\frac{\Lambda}{3}} f^4 \tag{14}$$

Remembering that N is an inverse density, we should take $\mapsto N \approx det(q)^{-1/2} = f^{-3}$. This gives us

$$\dot{f} = N\sqrt{\frac{\Lambda}{3}} f^4 = \sqrt{\frac{\Lambda}{3}} f \tag{15}$$

so that $f = e^{\sqrt{\frac{\Lambda}{3}}t}$.

With the identifications we have made this gives the deSitter spacetime in spatially flat coordinates[5]:

$$ds^2 = -dt^2 + e^{2\sqrt{\frac{\Lambda}{3}}t}(dx^a)^2 \tag{16}$$

HAMILTON-JACOBI THEORY, DESITTER SPACETIME AND CHERN-SIMONS THEORY

One way of thinking about the solution just described is that it comes from a Hamilton Jacobi function, S. To see this note that Hamilton-Jacobi theory requires,

$$E^{ai} = -\frac{\delta S(A)}{\delta A_{ai}} \tag{17}$$

We found that all seven constraints are solved with the self-dual ansatz (11). This means that the Hamilton-Jacobi function must satisfy a first order differential equation,

$$F^i_{ab} = -\frac{\Lambda}{3}\varepsilon_{abc}E^{ci} = \frac{\Lambda}{3}\varepsilon_{abc}\frac{\delta S(A)}{\delta A_{ci}} \tag{18}$$

This integrates immediately to

$$S_{CS} = \frac{2}{3\Lambda}\int Y_{CS} \tag{19}$$

Here Y_{CS} is the Chern-Simons invariant, given by

$$Y_{CS} = \frac{1}{2}Tr(A \wedge dA + \frac{2}{3}A^3). \tag{20}$$

It satisfies $\frac{\delta \int Y_{CS}}{\delta A_{ai}} = 2\varepsilon^{abc}F^i_{bc}$

Thus, *the self-dual solutions follow trajectories in configuration space which are gradients of the Chern-Simons invariant.* Not only is deSitter spacetime one of these, there is the remarkable fact that, while there are an infinite number of self-dual solutions for Euclidean signature, there is only one for Lorentzian signature and it is deSitter spacetime.

This suggests that the semiclassical state that describes deSitter is

$$\Psi_K(A) = \mathcal{N}e^{\frac{3}{2\Lambda}\int Y_{CS}} \tag{21}$$

$\mathcal{N}$ is a normalization depending only on topology[12].

In fact, *this is an exact quantum state* as was shown in 1990 by Hideo Kodama[9]. We now go on to study this state.

---

[5] A good review of the different coordinatizations of deSitter spacetime is in [24].

THE KODAMA STATE

We begin with a very brief review of how diffeomorphism invariant theories are to be quantized in the Hamiltonian approach. For more details on the basic approach, see [4, 5, 1, 3, 2].

The approach taken here is Dirac quantization. This means that the whole unconstrained configuration space is quantized. This defines a *kinematical state space* $H^{kinematical}$. The constraints are imposed as operator relations on the states, as in

$$\mathscr{C}|\Psi> = 0 \tag{22}$$

where $\mathscr{C}$ stands for operators representing all the first class constraints of the theory. The solutions to the constraints define subspaces of the Hilbert space. A physical state must be a simultaneous solution to all the constraints.

From a naive point of view, we could take the canonical commutation relations (2) as the basis of the quantization. Thus, we heuristically define the connection representation by the relations

$$<A|\Psi> = \Psi(A) \quad E^{ai} = -\hbar G \frac{\delta}{\delta A_{ai}} \tag{23}$$

These satisfy the commutation relations,

$$[A_a^i(x), E_j^b(y)] = \hbar G \delta_a^b \delta_j^i \delta^3(y,x) \tag{24}$$

Note that because there is an ι in the classical commutation relation (2) no ι appears here[6]. Unless explicitly mentioned, from now on we are working with the Lorentzian theory.

For a more careful quantization, that takes into account regularization, one proceeds a bit differently. But I will use the naive approach here, and refer the reader to details for the more careful approaches[5, 6, 3, 32, 2].

The quantum constraints have the form,

- Gauss's law:

$$\mathscr{G}^i \Psi(A) = \mathscr{D}_a \frac{\delta}{\delta A_{ai}} \Psi(A) = 0 \tag{25}$$

- Diffeomorphism constraint

$$\mathscr{H}_a \Psi(A) = F_{ab}^i \frac{\delta}{\delta A_{bi}} \Psi(A) = 0 \tag{26}$$

- Hamiltonian constraint:

$$\mathscr{H}\Psi(A) = \varepsilon_{ijk} \frac{\delta}{\delta A_{ai}} \frac{\delta}{\delta A_{bj}} (F_{ab}^k + \frac{\lambda}{3} \varepsilon_{abc} \frac{\delta}{\delta A_{ck}}) \Psi(A) = 0 \tag{27}$$

---

[6] Were we working instead with the Euclidean theory there would be an ι here.

Note that with the ordering given here, the quantum algebra of constraints can be shown, after a suitable regularization procedure, to be consistent[8, 4, 32]. This means that the commutators give terms proportional to operators, which are of the form of (new operator)× operator constraints. Thus, there are a non-trivial space of states in the simultaneous kernel of all the constraints. In fact, infinite dimensional spaces of simultaneous solutions to all the regularized constraints have been found and studied[5, 33]. Among them is the Kodama state (21).

To show that (21) is indeed a solution to all the constraints, one makes use of the identity,

$$\frac{\delta \Psi_K(A)}{\delta A_{ck}} = \frac{3}{2\lambda} \varepsilon^{abc} F^i_{ab} \Psi_K(A) \qquad (28)$$

Thus, the Kodama state is in the kernel of the operator

$$\mathscr{J}^i_{ab} = F^k_{ab} - \frac{\lambda}{3} \varepsilon_{abc} \frac{\delta}{\delta A_{ck}} \qquad (29)$$

and satisfies the Hamiltonian constraint because we have chosen an ordering such that

$$\mathscr{H} = \varepsilon_{ijk} \frac{\delta}{\delta A_{ai}} \frac{\delta}{\delta A_{bj}} \cdot \mathscr{J}^k_{ab} \qquad (30)$$

$\mathscr{J}^i_{ab}$ is of course an operator version of the self-dual condition.

The Kodama state solves the other constraints because it is manifestly invariant under diffeomorphisms of Σ and *small* gauge transformations. (Note that only small gauge transformations are generated by constraints.)

Invariance under large $SU(2)$ (real) gauge transformations is gotten by choosing $\mathscr{N}$ to be a topological invariant also sensitive to them. For details of this, see the paper by Soo [12].

It will also be important to note that as A_a is complex, so is its Chern-Simons invariant. Hence the Kodama state is complex.

We now study some of the properties of the Kodama state.

THE RECOVERY OF QFT ON DESITTER SPACETIME

The first thing we can do to probe the Kodama state is to add matter, and then see what happens if we excite the matter in the presence of the state[7].

Adding matter fields is straightforward. In the language of loop quantum gravity it is simple to add all kinds of matter: gauge fields, fermions, scalars, antisymmetric tensor gauge fields. For what we are doing here we do not need any details, so we will refer to all matter fields as ϕ, their canonical momenta as π and the matter hamiltonian as $H^{matter}(\phi, \pi)$.

---

[7] The material in this section comes from ref. [11] to which the reader is referred for more details.

All the constraints get new terms in the matter fields. For the Hamiltonian constraint we have

$$\mathcal{H}^{total} = \mathcal{H}^{grav}(A,E) + H^{matter}(A,E,\phi,\pi) \tag{31}$$

We will work in an extended connection representation in which the states are functionals $\Psi[A,\phi]$, π is represented by $-\imath\hbar\delta/\delta\phi$ and so forth.

As in the pure gravity case, the gauge and diffeomorphism constraints, applied to the states, require that the states are gauge invariant and invariant under diffeomorphisms of Σ.

This is straightforward, so we focus here on the hamiltonian constraint.

To study perturbations of the Kodama state, we follow the proposal of Banks[34], which is to study the semiclassical approximation in quantum cosmology by a version of the Born-Oppenheim approximation, in which the gravitational degrees of freedom play the role of the heavy, nuclear degrees of freedom, while the matter degrees of freedom play the role of the light, electron degrees of freedom.

Thus, we consider a product state of the form

$$\Psi(A,\phi) = \Psi_K(A)\chi(A,\phi) \tag{32}$$

The exact Hamiltonian constraint is then of the form

$$\left(H^{grav} + H^{matter}\right)\Psi_K(A)\chi(A,\phi) = 0 \tag{33}$$

The idea is to make an approximation to the exact equations, which is described in terms of quantum matter fields propagating on a classical background spacetime (A^0, E^0). This approximation is gotten by expanding the Wheeler-DeWitt equation (33) in a nieghborhood of a classical solution. We use the fact that the Kodama state can be understood as a *WKB* state as well as an exact solution. This tells us that the classical background (A^0, E^0) must be deSitter spacetime, as it is the unique solution gotten by taking S_{CS} to be the Hamiltonian-Jacobi function, consistent with the requirement that the lorentzian metric be real.

As shown in [11], we find that an approximation to (33) takes the form of a Tomonaga-Schwinger equation:

$$\imath\frac{\delta\chi}{\delta\tau_{CS}} = \frac{1}{\Lambda}H^{matter}_{E^{ai}=(3/\Lambda)\varepsilon^{abc}F^i_{bc}}\chi + O(l_{Pl}E) \tag{34}$$

In this equation, the matter Hamiltonian is evaluated with classical gravitational fields satisfying the self-dual condition $E^{ai} = (3/\Lambda)\varepsilon^{abc}F^i_{bc}$. As we just said, the reality conditions then tell us that the background is deSitter. We have neglected higher order terms in $l_{Pl}E$, where E is the energy of the matter fields measured with respect to the background metric.

The approximation procedure picks out a time coordinate called τ_{CS}, related to the Chern-Simons invariant. It is first of all a coordinate on the phase space of the theory, defined by

$$\delta\tau_{CS}(x) = \frac{1}{2}\mathcal{I}m\varepsilon^{abc}F^i_{bc}\delta A_{ai}(x) \tag{35}$$

Thus, integrated over the spatial manifold Σ, we have

$$\int_\Sigma \delta\tau_{CS}(x) = \delta \mathscr{I}m \int Y_{CS}(A) \tag{36}$$

If we take the integral, we can define

$$T_{CS} = \int_\Sigma \tau_{CS} = Im \int_\Sigma Y_{CS} \tag{37}$$

This can be argued to provide a provides a good global time parameter on the configuration space[11, 39]. This is because its derivative is always orthogonal, in the tangent space of the configuration space, to both the gauge directions and the directions that parameterize the physical degrees of freedom.

When evaluated on a background solution, this gives rise to a time coordinate on the spacetime. One can then show that, to leading order in λ, $Y_{CS} = \imath\sqrt{det(q)}\mathscr{K} + O(\sqrt{\lambda})$, where $\mathscr{K}$ is the trace of the extrinsic curvature K_{ab}. Thus. this choice of time coordinate agrees, to leading order in λ, with that proposed by York[11]. This time coordinate has been shown to have many good properties that an intrinsic time coordinate should have.

Thus, QFT on deSitter *is* a good approximation to the physics of $\Psi(A,\phi) = \Psi_K(A)\chi(A,\phi)$ when $\lambda = \hbar G\Lambda$ and $l_{Planck}E$ are small. This stands as a first piece of evidence that $\Psi_K(A)$ may be indeed a good ground state.

There are additional terms in $l_{Planck}E$, where E is the matter energy on the deSitter background. These terms are studied in [10, 35]. They result in corrections to the energy-momentum relations of the matter fields of the form,

$$E^2 = p^2 + m^2 + \alpha l_{Pl} E^3 + \ldots \tag{38}$$

where α is a computable constant[8].

These represent new predictions of the quantum theory of gravity[36, 37, 10, 35].

GRAVITONS FROM PERTURBATIONS AROUND THE KODAMA STATE

To further probe the properties of the Kodama state we should also investigate its gravitational excitations. To do this we return to the case of pure gravity and consider states of the form

$$\Psi[A] = \mathscr{N} e^{\frac{3}{2\lambda}\int Y_{CS} + \lambda S'(A)} \tag{39}$$

We will show that there are solutions of this form, and that they do describe long wavelength gravitons moving on the classical background of deSitter spacetime. We begin by expanding

$$A_{ai} = \imath\sqrt{\Lambda}f(t)\delta_{ai} + \lambda a_{ai}; \quad E^{ai} = f^2\delta^{ai} + \lambda e^{ai} \tag{40}$$

---

[8] These kinds of corrections have also been derived in loop quantum gravity by[36, 37].

It is not hard to see that the gauge and diffeomorphism constraints tell us that the state just depends on the tracefree-symmetric components of a_{ai}, which are the graviton degrees of freedom. The leading order approximation to the Hamiltonian constraint takes the form,

$$i\frac{\partial S'}{\partial t} = \left[\hat{H}^2\right] S' + O(l_{Planck} E) + O(\sqrt{\lambda}) \qquad (41)$$

where $\hat{H}^2$ is the free Hamiltonian for gravitons.

Thus for long wavelength perturbations, but only so long as $l_{Pl} E \ll 1$, the linearized theory is recovered. However it must be emphasized that we have only obtained a correspondence with the standard linearized theory for low energy and small λ.

To go beyond this, which is necessary for example to compute graviton graviton scattering, we need to have exact expressions for the physical states which are obtained by local perturbations of the Kodama state. This requires solving the kinematical constraints exactly, rather than order by order in l_{Pl} and λ. To do this we must go to the loop representation, suitably modified for $\Lambda > 0$[17, 30, 10].

THE THERMAL NATURE OF QUANTUM GRAVITY WITH $\Lambda > 0$

It is well established that quantum field theory on deSitter spacetime must be interpreted as irreducibly thermal[38], with a temperature given by

$$\mathcal{T} = \frac{1}{2\pi}\sqrt{\frac{\Lambda}{3}} \qquad (42)$$

This can be understood as due to the presence of the horizon. Alternatively, one can show that any quantum field on deSitter spacetime satisfies the *KMS* condition for a thermal state. This is that the continuation of any correlation function to imaginary time coordinate $t_E = it$ be periodic, with period $\beta = 1/\mathcal{T}$.

What is not so well known, however, is that in loop quantum gravity the full background independent quantum theory of gravity plus arbitrary matter fields has also an irreducibly thermal nature. This is because it satisfies the *KMS* condition on the whole configuration space[9].

To apply the *KMS* condition to quantum gravity we need two things: 1) a definition of a time coordinate on the configuration space and 2) a definition of the continuation to Euclidean time. In a background independent theory we cannot use a time coordinate on a given classical spacetime, as that is just a given classical solution. We need instead a time coordinate on the configuration space of the theory.

We saw above that there is in fact a preferred time coordinate on the configuration space, which is picked out by the semiclassical expansion around the Kodama state. It is equal to the imaginary part of the Chern-Simons invariant, eq. (37). There are other arguments that confirm that (37) is a good time coordinate on the configuration

---

[9] The argument of this section is taken from [11].

space, for example it is always normal to the gauge directions in the tangent space of the configuration space. For details see [11, 39]. The Chern-Simons time coordinate is dimensionless, as we saw in the last sections when we evaluate it on a given solution we have to scale it appropriately.

It is interesting to note that (Lorentzian) Kodama state can be written

$$\Psi_K(A) = e^{\imath M T_{CS}} e^{\frac{k}{4\pi} \mathcal{R}e \int Y_{CS}(A)} \tag{43}$$

where the dimensionless "energy" is

$$M = \frac{k}{4\pi} = \frac{3}{2\lambda} \tag{44}$$

Now we need a definition of how to continue to Euclidean time. As we are dealing with a theory of spacetime we should continue the whole theory to Euclidean signature. This requires the following changes: The connection A_{ai} becomes a real, $SU(2)$ connection and there is now an $\imath$ in $E^{ai} = \imath \delta/\delta A_{ai}$. As a consequence of which the Chern-Simons state is now,

$$\Psi_K^{Euc}(A) = e^{\frac{\imath k}{4\pi} \int Y_{CS}(A)} \tag{45}$$

The Euclidean time coordinate is then just

$$T_{ECS} = \int Y_{CS}(A) \tag{46}$$

as can be seen directly, or by repeating the derivation from the semiclassical theory. Thus, the Euclidean wavefunction is,

$$\Psi_K^{Euc}(A) = e^{\imath \frac{k}{4\pi} T_{Euc}} \tag{47}$$

This is periodic in T_{Euc}. However, this is not enough to show that the *KMS* condition is satisfied, for that requires that *every correlation function* be periodic.

Interestingly enough, this is in fact the case whenever Σ is chosen so that $\pi^3(\Sigma)$ is nontrivial. In this case there are large gauge transformations that have the property that

$$\int Y_{CS}(A) \to \int Y_{CS}(A) + 8\pi^2 n \tag{48}$$

where n is the winding number of the large gauge transformation. This means that $T_{ECS} = \int Y_{CS}(A)$ is actually a periodic function on the configuration space. As a result, *every* correlation function will satisfy the *KMS* condition in T_{ECS}, no matter what the state. That is, by equating configurations of A_{ai} that differ by a large gauge transformations we reduce the topology of the configuration space to a circle, which is parameterized by T_{ECS}.

As a result of this universal periodicity there is a temperature, given in dimensionless units by $\mathcal{T}_{dimless} = \frac{1}{8\pi^2}$. This dimensionless temperature corresponds to the fact that the time coordinate on the configuration space, T_{CS} is dimensionless.

It is interesting to ask if this dimensionless temperature corresponds to the temperature on deSitter spacetime. To investigate this we may consider a trajectory in configuration space that corresponds to a slicing of deSitter spacetime with topology $S^3 \times R$. Such coordinates are given by

$$ds^2 = -(1 - \frac{\Lambda r^2}{3})dt^2 + \frac{1}{(1 - \frac{\Lambda r^2}{3})}dr^2 + d\Omega^2 \tag{49}$$

To work out the scaling of the coordinate t on the solution with the coordinate T_{CS} on the configuration space, we compute

$$\frac{\partial T_{CS}}{\partial t} = \int_{S^3} N\{T_{CS}(A), \mathcal{H}\} \tag{50}$$

where the (densitized) lapse N is read off from the solution (49). A simple calculation gives

$$\frac{\partial T_{CS}}{\partial t} = 4\pi\sqrt{\frac{\Lambda}{3}} \tag{51}$$

Thus, if the Euclidean continuation T_{ECS} is periodic with period $8\pi^2$, the Euclidean continuation of the time coordinate on the solution must be periodic with period $2\pi\sqrt{\frac{3}{\Lambda}}$. In fact, this is the periodicity of the Euclidean deSitter solution, in these coordinates! This leads to the temperature of deSitter spacetime, (42).

Thus, we learn that *the periodicity of the Euclidean deSitter spacetime is a consequence of that spacetime having an interpretation as a trajectory on the configuration space of $SU(2)$ connections.* The periodicity of the Euclidean Schwarzschild solution is a consequence of the fact that the whole configuration space is periodic due to the action of the large gauge transformations. This is yet another connection between the properties of the gauge theory and the physics of gravitation. Thus, the thermal nature of quantum field theory on deSitter spacetime is a consequence of a deeper and more general result, which is that the whole quantum theory with $\Lambda > 0$ is thermal.

Finally, we can deduce one more fact from these considerations. For the analysis we have just given to be relevant to the Kodama state, it must be that the Euclidean Kodama state is itself well defined under large gauge transformations. This will only be the case if k is an integer.

FURTHER RESULTS FOR $\Lambda \neq 0$

In this paper we have presented a number of results, some old and some new, which together support a claim that for $\lambda > 0$ and in $3+1$ dimensions loop quantum gravity gives a satisfactory quantum theory of gravity. These main results were stated in the introduction.

There are a number of related developments which I did not have space to mention here. These include

- The idea that gravity is closely related to topological field theory turns out to be very powerful when applied to the derivation of measures for the path integral formulations of quantum gravity, which are called spin foam models[25, 27, 28].

- A related model, called a causal spin foam model, has been developed for $\Lambda > 0$[30].
- The presence of a bare cosmological constant imposes a deformation on the space of states in the loop[5, 6] or spin network[40] representation[17]. The states are "quantum deformed" in the sense that the representation theory of the quantum group $su_q(2)$ governs the states. There is a relation between the level, k and the cosmological constant, given by

$$k = \frac{6\pi}{G\Lambda} \qquad (52)$$

The deformed state space appears to solve a key problem in loop quantum gravity[41, 42], as it eliminates a certain obstruction to the existence of a good low energy limit[30].
- One can study quantum gravity in the presence of boundaries, which may represent either timelike boundaries or horizons. A natural boundary condition to add involves the Chern-Simons invariant of the connection A_{ai} pulled back to the spacetime boundary[17]. One finds that the Bekenstein bound is automatically satisfied for all cases[17]. Further, in the case of $\Lambda \neq 0$ one finds a natural explanation[10] for the N-bound[22, 23] of Banks and Bousso[10].
- The Chern-Simons state has been studied in the context of reduced quantum cosmological models, in which only a few degrees of freedom are retained, such as in the quantum Bianchi models[43].
- The Chern-Simons state can be transformed to the triad, or frame field representation[44].
- The Kodama state and many of the related results, can be extended also to quantum supergravity, for the case $\Lambda < 0$, at least up to $N = 2$[47, 48, 21].
- Soo has described an expansion around the Kodama state in powers of λ [12].
- A strong coupling expansion, in powers of $\frac{1}{\lambda}$ was proposed in [45] and has been developed in some detail in [46].
- The Kodama state can be used to generate an infinite number of physical states at $\lambda = 0$[49].

ACKNOWLEDGMENTS

I wish first to thank my collaborators on different aspects of the problem of quantum gravity with a cosmological constant, Roumen Borissov, Yi Ling, Joao Magueijo, Seth Major, Fotini Markopoulou, Carlo Rovelli and Chopin Soo. I also have learned a lot about this and related subjects from discussions and correspondence with Giovanni Amelino Camelia, John Baez, Tom Banks, Raphael Bousso, Louis Crane, Laurent Freidel, Lou Kauffman, Michael Reisenberger and Artem Starodubstev. This work was supported generously by The National Science Foundation and the Jesse Phillips Foundation.

REFERENCES

1. C. Rovelli, Living Rev. Rel. 1 (1998) 1, gr-qc/9710008.
2. R. Gambini and J. Pullin, Loops, knots, gauge theories and quantum gravity Cambridge University Press, 1996.
3. L. Smolin: in Quantum Gravity and Cosmology, eds J Perez-Mercader et al, World Scientific, Singapore 1992; The future of spin networks gr-qc/9702030 in the Penrose Festschrift.
4. T. Jacobson and L. Smolin, Nucl. Phys. B 299 (1988) 295.
5. C. Rovelli and L. Smolin, Knot theory and quantum theory, Phys. Rev. Lett 61(1988)1155; Loop representation of quantum general relativity, Nucl. Phys. B331(1990)80-152.
6. R. Gambini and A. Trias, Phys. Rev. D23 (1981) 553, Lett. al Nuovo Cimento 38 (1983) 497; Phys. Rev. Lett. 53 (1984) 2359; Nucl. Phys. B278 (1986) 436; R. Gambini, L. Leal and A. Trias, Phys. Rev. D39 (1989) 3127.
7. A. Sen, On the existence of neutrino zero modes in vacuum spacetime J. Math. Phys. 22 (1981) 1781, Gravity as a spin system Phys. Lett. B11 (1982) 89.
8. Abhay Ashtekar, New variables for classical and quantum gravity," Phys. Rev. Lett. 57(18), 2244-2247 (1986).
9. H. Kodama, Prog. Theor. Phys. 80, 1024(1988); Phys. Rev. D42(1990)2548.
10. L. Smolin "Quantum gravity with a positive cosmological constant." hep-th/0209??.
11. L. Smolin and C. Soo, The Chern-Simons Invariant as the Natural Time Variable for Classical and Quantum Cosmology, Nucl. Phys. B449 (1995) 289, gr-qc/9405015.
12. C. Soo, Wave function of the Universe and Chern-Simons Perturbation Theory, gr-qc/0109046.
13. G. Amelino-Camelia et al, Int.J.Mod.Phys.A12:607-624,1997; G. Amelino-Camelia et al Nature 393:763-765,1998; J. Ellis et al, Astrophys.J.535:139-151,2000; J. Ellis, N.E. Mavromatos and D. Nanopoulos, Phys.Rev.D63:124025,2001; ibidem astro-ph/0108295.
14. G. Amelino-Camelia and T. Piran, Phys.Rev. D64 (2001) 036005.
15. Tomasz J. Konopka, Seth A. Major, "Observational Limits on Quantum Geometry Effects", New J.Phys. 4 (2002) 57. hep-ph/0201184; Ted Jacobson, Stefano Liberati, David Mattingly, " TeV Astrophysics Constraints on Planck Scale Lorentz Violation", hep-ph/0112207.
16. Subir Sarkar, " Possible astrophysical probes of quantum gravity", Mod.Phys.Lett. A17 (2002) 1025-1036, gr-qc/0204092.
17. L. Smolin, Linking topological quantum field theory and nonperturbative quantum gravity, J. Math. Phys. 36(1995)6417, gr-qc/9505028.
18. K. Krasnov, On Quantum Statistical Mechanics of a Schwarzschild Black Hole , grqc/9605047, Gen. Rel. Grav. 30 (1998) 53-68; C. Rovelli, "Black hole entropy from loop quantum gravity," grqc/9603063.
19. A. Ashtekar, J. Baez, K. Krasnov, Quantum Geometry of Isolated Horizons and Black Hole Entropy gr-qc/0005126; A. Ashtekar, J. Baez, A. Corichi, K. Krasnov, "Quantum geometry and black hole entropy," gr-qc/9710007, Phys.Rev.Lett. 80 (1998) 904-907.
20. L. Smolin, A holograpic formulation of quantum general relativity, Phys. Rev. D61 (2000) 084007, hep-th/9808191.
21. Y. Ling and L. Smolin, Supersymmetric spin networks and quantum supergravity, Phys. Rev. D61, 044008(2000), hep-th/9904016; Holographic Formulation of Quantum Supergravity, hep-th/0009018, Phys.Rev. D63 (2001) 064010.
22. T.Banks, "Cosmological Breaking of Supersymmetry?", hep-th/0007146.
23. Raphael Bousso, "Positive vacuum energy and the N-bound", hep-th/0010252, JHEP 0011 (2000) 038.
24. M. Spradlin, A. Strominger, A. Volovich1y, Les Houches Lectures on de Sitter Space, hep-th/0110007
25. M. P. Reisenberger. "Worldsheet formulations of gauge theories and gravity," in Proceedings of the 7th Marcel Grossman Meeting, ed. by R. Jantzen and G. MacKeiser, World Scientific, 1996; gr-qc/9412035; A lattice worldsheet sum for 4-d Euclidean general relativity. gr-qc/9711052.
26. 8] M. P. Reisenberger and C. Rovelli. Sum-over-surface form of loop quantum gravity," gr-qc/9612035, Phys. Rev. D 56 (1997) 3490; Spacetime as a Feynman diagram: the connection formulation. Class.Quant.Grav., 18:121140, 2001; Spin foams as Feynman diagrams, gr-qc/0002083.

27. J. Barrett and L. Crane, "Relativistic spin networks and quantum gravity", J. Math. Phys. 39 (1998) 3296-3302, gr-qc/9709028.
28. J. Baez, Spin foam models, Class. Quant. Grav. 15 (1998) 1827-1858, gr-qc/9709052; An introduction to spin foam models of quantum gravity and BF theory. Lect.Notes Phys., 543:2594, 2000.
29. Fotini Markopoulou, "Dual formulation of spin network evolution", gr-qc/9704013.
30. Fotini Markopoulou, Lee Smolin, "Quantum geometry with intrinsic local causality", Phys.Rev. D58 (1998) 084032, gr-qc/9712067.
31. J. Iwasaki, A reformulation of the Ponzano-Regge quantum gravity model in terms of surfaces," gr-qc/9410010; A definition of the Ponzano-Regge quantum gravity model in terms of surfaces," gr-qc/9505043, J. Math. Phys. 36 (1995) 6288; L. Freidel and K. Krasnov. Spin foam models and the classical action principle. Adv.Theor.Math.Phys., 2:11831247, 1999; R. De Pietri, L. Freidel, K. Krasnov, and C. Rovelli. Barrett-Crane model from a boulatov-ooguri field theory over a homogeneous space. Nucl.Phys. B, 574:785806, 2000.
32. B. Bruegmann, R. Gambini and J. Pullin, Phys. Rev. Lett. 68 (1992) 431- 434.
33. Thomas Thiemann, "Introduction to Modern Canonical Quantum General Relativity" gr-qc/0110034
34. T. Banks, "T C P, Quantum Gravity, The Cosmological Constant And All That", Nucl.Phys. B249 (1985) 332.
35. C. Rovelli and L. Smolin, in preparation.
36. Rodolfo Gambini, Jorge Pullin, "Nonstandard optics from quantum spacetime", Phys.Rev. D59 (1999) 124021, gr-qc/9809038;
37. Jorge Alfaro, Hugo A. Morales-Tõcotl, Luis F. Urrutia, "Loop quantum gravity and light propagation", Phys.Rev. D65 (2002) 103509, hep-th/0108061
38. G. W. Gibbons and S. W. Hawking, Cosmological Event Horizons, Thermodynamics, and Particle Creation," Phys. Rev. D 15, 2738 (1977).
39. L. N. Chang and C. Soo, Ashtekar's variables and the topological phase of quantum gravity, Proceedings of the XXth. Conference on Differential Geometric Methods in Physics, (Baruch College, New York, 1991), edited by S. Catto and A. Rocha (World Scientific, 1992); Phys. Rev. D46 (1992) 4257; C. Soo and L. N. Chang, Int. J. Mod. Phys. D3 (1994) 529.
40. C. Rovelli and L. Smolin, Spin networks and quantum gravity, gr-qc/9505006, Physical Review D 52 (1995) 5743-5759.
41. L. Smolin, The classical limit and the form of the hamiltonian constraint in nonperturbative quantum gravity, gr-qc/9607034.
42. J. Lewandowski, D. Marolf, Loop constraints: a habitat and their algebra," gr-qc/9710016; R. Gambini, J. Lewandowski, D. Marolf, J. Pullin, On the consistency of the constraint algebra in spin-network quantum gravity," gr-qc/9710018.
43. R. Graham and R. Paternoga, Phys. Rev. D 54, 2589 (1996); D 54, 4805 (1996); D 58 083501 (1998).
44. R. Paternoga and R. Graham, Triad representation of the Chern-Simons state in quantum gravity, gr-qc/0003111
45. C. Rovelli and L. Smolin, The physical hamiltonian in nonperturbative quantum gravity, Phys. Rev. Lett.72(1994)446; Spin Networks and Quantum Gravity, Phys. Rev. D52(1995)5743-5759.
46. R. Gambini and J. Pullin, Phys. Rev. Lett. 85 (2000) 5272, Class. Quant. Grav. 17 (2000) 4515.
47. T. Jacobson, New Variables for canonical supergravity, Class. Quant. Grav.5(1988)923- 935; D. Armand-Ugon, R. Gambini, O. Obregon, J. Pullin, Towards a loop representation for quantum canonical supergravity, hep-th/9508036, Nucl. Phys. B460 (1996) 615; L. F. Urrutia Towards a loop representation of connection theories defined over a super-lie algebra, hep-th/9609010; H. Kunitomo and T. Sano The Ashtekar formulation for canonical N=2 supergravity, Prog. Theor. Phys. suppl. (1993) 31; Takashi Sano and J. Shiraishi, The Nonperturbative Canonical Quantization of the N=1 Supergravity, Nucl. Phys. B410 (1993) 423, hep-th/9211104; The Ashtekar Formalism and WKB Wave Functions of N=1,2 Supergravities, hep-th/9211103; T. Kadoyoshi and S. Nojiri, N=3 and N=4 two form supergravities, Mod. Phys. Lett. A12:1165-1174,1997, hep-th/9703149; K. Ezawa, Ashtekar's formulation for N=1, N=2 supergravities as constrained BF theories, Prog. Theor. Phys.95:863-882, 1996, hep-th/9511047.
48. Yi Ling, "Introduction to supersymmetric spin networks", hep-th/0009020, J.Math.Phys. 43 (2002) 154-169
49. Rodolfo Gambini, Jorge Griego, Jorge Pullin, "Chern–Simons states in spin-network quantum gravity", gr-qc/9703042, Phys.Lett. B413 (1997) 260-266; C. Di Bartolo, R. Gambini, J. Griego, J. Pullin, "Consistent canonical quantization of general relativity in the space of Vassiliev knot invariants", gr-qc/9909063, Phys.Rev.Lett. 84 (2000) 2314-2317; "Canonical quantum gravity in the Vassiliev invariants arena: I. Kinematical structure", gr-qc/9911009, Class.Quant.Grav. 17 (2000) 3211-3238.

SUPERSTRINGS

String bits and the Myers effect

Pedro J. Silva

Physics Department, Syracuse University, Syracuse, New York 13244

Abstract. Based on the non-abelian effective action for D1-branes, a new action for matrix string theory in non-trivial backgrounds is proposed. Once the background fields are included, new interactions bring the possibility of non-commutative solutions i.e. The Myers effect for "string bits"

MATRIX STRINGS

Matrix string theory [1, 2, 3] is one of the most interesting outcomes of the different dualities in M-theory. Perhaps a simple way to define it is by looking at Matrix theory [4] with an extra compactified dimension. For example, following Dijkgraaf et. al. [1], if we consider M-theory on a torus of radii A and B, by first reducing on A and then making an infinite boost on B we get type IIA string theory on the discrete light cone (DLC) with D0-brane particles. If, on the other hand, we reduce on B first, boost and then consider t-duality on A, we get (1+1) super Yang-Mills (SYM) theory with fundamental string charge on the world-volume i.e. the low energy theory of D1-branes. One finds therefore, that Matrix string theory is a non-perturbative definition of string theory built in terms of a two dimensional SYM theory and a collection of scalar fields in the adjoint representation of the gauge group (see the original papers for an extended discussion of this derivation). Although we have discussed only type IIA string theory, there are other constructions similar to the one sketched before, where the other four superstring theories are written in terms of two dimensional SYM theory[1].

The Matrix string theory conjecture was originally formulated on flat backgrounds. Lately, using some techniques developed by Taylor and Raamsdonk [5] a generalization for closed strings on non-trivial weak backgrounds has appeared [6]. This talk is based on the paper [7], a further generalization of the original Matrix string theory to non-trivial weak backgrounds based on the non-abelian D1-brane action proposed by Myers.

In [6, 7], the possibility of non-abelian configurations of fundamental strings was pointed out. In particular, the appearance of a Myers-like effect was computed explicitly (By now the Myers effect [8] is a well known phenomenon where N D-branes adopt a non-abelian configuration that can be understood as a higher dimensional abelian D-brane). These configurations come about as the result of new interaction terms that appear in the non-abelian effective actions.

---

[1] Actually for the cases of type IIB and type I string theory there is an additional construction in terms of a three dimensional theory [3]

The appearance of strings describing D-branes is not new. There are computations of Dp-branes collapsing into fundamental strings [9] and fundamental strings blowing up into Dp-branes [10], always in terms of the abelian Born-Infeld actions of the corresponding D-branes. What is new in the matrix string formulation is that we have a formalism in which a two-dimensional action naturally includes matrix degrees of freedom representing the "string bits"[2], which also incorporate the description of higher dimensional objects of M-theory using non-commutative configurations.

One of the important properties of this new theoretical framework lies in the similarity of the mathematical language used to describe the fundamental objects of M-theory, bringing for first time the possibility of describing strings and D-branes in a unified framework, a "democracy of p-branes" [12].

MATRIX STRING AND NON-ABELIAN D1-BRANES

In the previous section we talked about a theoretical framework that describes fundamental strings in terms of matrix degrees of freedom. For example, in type IIA this action is a two dimensional supersymmetric gauge theory that contains DLC string theory and has extra degrees of freedom representing non-perturbative objects of string theory. Also, it is a second quantized theory as it is built from many strings.

We know that by means of different dualities the five superstring theories are described in the neighborhood of a 1+1 dimensional orbifold conformal field theory. In this language the strings are free in the conformal field theory limit, representing DLC string theory. The interactions between the strings are turned on by operators describing the splitting and joining of fundamental strings. These operators deform the theory away from the conformal fixed point.

To further clarify these ideas, let us follow a sketch of the derivation for the case of type IIA string theory. Consider type IIA strings in the DLC frame with string mass m_A, string coupling g_A and a null compact direction of radius R_A (where we identify the null coordinate as $x^- \approx x^- + R_A$).

Using the relation between the null compactification and a space-like compactification a la Seiberg-Sen [13], we get type IIA string theory on a space-like circle of radius R in the sector with momentum N, string mass m and string coupling g. The relation between these two heterotic string theories is given by

$$m^2 R = m_A^2 R_A \ , \ g = g_A \ , \ R \to 0. \tag{1}$$

Next, we perform a t-duality transformation on R, so that the new constants of the string theory (m', g', R') are given by

$$m' = m \ , \ g' = \frac{g}{mR} \ , \ R' = \frac{1}{m^2 R}. \tag{2}$$

---

[2] The idea is that the string can be seen as a chain of partonic degrees of freedom [11]

Finally, we perform a s-duality transformation to obtain type IIB string theory with N D1-strings and constants (m_b, g_b, R_b) given by the following expressions,

$$m_b = \frac{m'}{g'^{1/2}} , \quad g_b = \frac{1}{g'} , \quad R_b = R'. \tag{3}$$

In terms of the initial type IIA theory and R we get

$$m_b = \left[\frac{m_A^6 R_A^3}{g_A^2 R}\right]^{1/4} \to \infty,$$

$$g_b = \left[\frac{m_A^2 R_A R}{g_A^2}\right]^{1/2} \to 0,$$

$$R_b = \frac{1}{m_A^2 R}. \tag{4}$$

Therefore, we get the low energy theory of N D1-branes at weak coupling, where the gauge coupling constant g_{YM} is given by,

$$g_{YM} \propto m_A^2 R_A / g_A. \tag{5}$$

This is the 1+1 dimensional SYM theory with eight scalars in the adjoint representation of the gauge group. This effective action is obtained by the dimensional reduction of $N=1$ supersymmetry Yang-Mills theory in ten dimensions down to two dimensions.

To define the type IIB case, a possible route to take is to start with type IIB strings in the DLC frame, then perform a t-duality transformation on the null circle taking us to type IIA in the DLC. This is similar to the previous situation with the difference that winding modes are exchanged for momentum modes. The relation between the corresponding meaningful constants is,

$$m_b = \left[\frac{m_B^2}{g_B^2 R_B R}\right]^{1/4} \to \infty,$$

$$g_b = \left[\frac{m_B^2 R_B R}{g_B^2}\right]^{1/2} \to 0,$$

$$R_b = R_B, \tag{6}$$

where (m_B, g_B, R_B) are the initial type IIB string parameters and (m_b, g_b, R_b) are the final (also type IIB) string theory parameters. Again, we get a low energy weakly coupled string theory with N D1-branes. The gauge coupling constant g_{YM} is given by

$$g_{YM} = m_B / g_B. \tag{7}$$

The heterotic case is similar but some care has to be taken with the inclusion of Wilson lines [14]. On the other hand, type I theory is more subtle and is related to the low energy limit of type IA theory in the presence of D8-branes and D0-branes plus winding modes on the orbifold. Therefore it is a quantum mechanics system but with an infinite tower of winding modes.

In order to obtain the relevant action for one of the five matrix string theories, we start with the world-volume gauge theory of N D1-branes, and then go back along the chain of dualities until we reach the desired DLC string theory. For example, consider first an s-duality transformation on the D1-brane effective action, then a t-duality transformation and finally the boost relations of Seiberg-Sen. As a result we get type IIA matrix string theory. This can be written as

$$L_F^{IIA} \equiv B \circ T \circ S [L_{D1}]. \tag{8}$$

Other matrix string theories Lagrangians can be obtained by similar procedures. For example, $L_F^{IIB} \equiv T \circ B \circ T \circ S [L_{D1}]$.

As we mentioned in the introduction, there are generalizations of the matrix string action which include weak backgrounds. This time the calculations are based on the relation between matrix string and the matrix theory proposal. In particular, previous works of Taylor and Van Raamsdonk [5] are used to support these results. One of the positive outcomes of the above work is a proposal for the transformation of the D1-brane world-volume fields under s-duality. Thus, based on these different proposals we are able to actually construct the matrix actions using maps like the one in equation (8).

It is important to note that recently Myers wrote a non-abelian action of N Dp-branes in general backgrounds [8] which is fully covariant under t-duality. This action incorporates (in the limit of weak backgrounds), all the couplings derived previously by Taylor et. al. and also introduces some new ones. If we believe this effective action for the D1-branes, we are forced to conjecture that:

Matrix string theory is defined by Myers D1-brane world-volume action plus the web of dualities needed.

Note that since the non-abelian D1-brane action proposed by Myers does not capture the full physics of the infrared limit, we can only trust its expansion up to sixth-order in the field strength [15], and this problems is inherited by the above conjecture for the matrix theory action. Another technical problem comes from the chain of dualities, since it makes it difficult to give an explicit closed form for the final Lagrangian. In particular, the t-duality map mixes RR fields and NS fields. Nevertheless, we only have to use the Buscher rules [16] on the supergravity background fields as t-duality (once we have s-dualized), leaves the world-volume fields invariant. At last the action of Myers only tells us about the bosonic degrees of freedom, therefore the fermionic counterpart has to be calculated using supersymmetry.

For example, let us consider the type IIA case. Following equation (8), the action for the Matrix string is given in two parts, the first corresponding to the original Born-Infeld term of the D1-brane action,

$$S_{F1}^1 = \frac{1}{\lambda} \int d\xi^2 Str \left\{ \sqrt{-det(P[\widetilde{E} + \widetilde{E}(\widetilde{Q}^{-1} - \delta)\widetilde{E}] + \lambda e^{\widetilde{\phi}} \widetilde{g} F)det(\widetilde{Q})} \right\} \tag{9}$$

where

$$\widetilde{E}_{AB} = \widetilde{G}_{AB} - e^{\widetilde{\phi}} \widetilde{C}_{AB}^{(2)},$$

$$\tilde{Q}^i_j = \delta^i_j + i\lambda[\Phi^i, \Phi^k]\tilde{E}_{kj}(\tilde{g}e^{\tilde{\phi}})^{-1}, \qquad (10)$$

and the tilde represents the t-dual transformation of the background fields. For example the form of $\tilde{C}_{AB}$ is

$$\tilde{C}_{AB} = \begin{pmatrix} C_{\alpha\beta y} + 2C_{[\alpha}B_{\beta]y} - 2C_y B_{y[\alpha}G_{\beta]y} & C_\alpha - C_y G_{\alpha y}/G_{yy} \\ -C_\beta + C_y G_{\beta y}/G_{yy} & 0 \end{pmatrix}, \qquad (11)$$

where the space-time index A has been divided into the t-dualized direction y and the other directions α.

The second part, corresponding to the original Chern-Simons term of the D1-brane action is

$$S^2_{F1} = \frac{1}{\lambda}\int d\xi^2 STr\left\{P\left[e^{i\tilde{g}^{-1}\lambda i_\Phi i_\Phi}[(-\tilde{B}+\tilde{C}^{(4)})e^{-\tilde{C}^{(2)}}]\right]e^{\lambda\tilde{g}F}\right\}. \qquad (12)$$

This action contains the action of the matrix string theory of Dijkgraaf et. al. [1], since by construction in trivial backgrounds the D1-brane action of Myers gives the 1+1 SYM theory corresponding the dimensional reduction on N=1 SYM in ten dimensions down to two dimensions. Hence, by taking all of the background fields to be trivial, we recover the standard form of type IIA matrix string theory,

$$S^1_{F1} = \lambda \int d\xi^2 Tr\left\{\frac{\partial\Phi^2}{2} + \frac{1}{4g^2}[\Phi,\Phi]^2 + \frac{g^2}{4}F^2\right\}. \qquad (13)$$

Also, all of the linear couplings obtained by Schiappa [6] for the weak field case, are derivable from the action of equation (9) and (12). It has been checked that the D1-branes linear couplings found by Taylor et. al. are included in the non-abelian action of Myers and the t-duality and s-duality relations are the same as the ones used by Schiappa. Nevertheless, we have to keep in mind that there are new couplings not considered before.

Once the relevant action is obtained, we can search for non-commutative classical solutions. Given the similarity of the mathematical structure with D-brane physics, we expect to find relevant physical situations where these types of solutions appear. Nevertheless, in this framework the building blocks that make the higher dimensional objects are the "string bits" of the DLC. Remember that, the matix-value scalar in the action represent a large number of "long strings" and these are the basic objects that form the higher dimensional branes. Some examples of non-commutative configurations of strings can be found in [6, 7, 17, 18, 19].

ACKNOWLEDGMENTS

The author would like to thank the Perimeter Institute for making possible the conference *MRST 2002*, also Alfonso Ramallo, Cesar Gomes, Yoland Lozano, Joel Rozowsky, Simeon Hellerman and Don Marolf for useful discussions. This work was supported in part by NSF grant PHY-0098747 to Syracuse University and by funds from Syracuse University.

REFERENCES

1. R. Dijkgraaf, E. Verlinde, H. Verlinde, "Matrix String Theory", Nucl.Phys. B500 (1997) 43-61, hep-th/9703030.
2. Tom Banks, Nathan Seiberg, "Strings from Matrices", Nucl.Phys. B497 (1997) 41-55, hep-th/9702187.
  R. Dijkgraaf, E. Verlinde, H. Verlinde, "Notes on Matrix and Micro Strings", Nucl.Phys.Proc.Suppl. 62 (1998) 348-362, hep-th/9709107.
  Lubos Motl, "Proposals on nonperturbative superstring interactions", hep-th/9701025.
3. Clifford V. Johnson, "On Second-Quantized Open Superstring Theory", Nucl.Phys. B537 (1999) 144-160, hep-th/980615.
4. T. Banks, W. Fischler, S.H. Shenker and L. Susskind, "M Theory As A Matrix Model: A Conjecture , Phys.Rev. D55 (1997) 5112-5128, hep-th/9610043.
  L. susskind,"Another Conjecture about M(atrix) Theory", hep-th/9704080.
5. W. Taylor, M. Van Raamsdonk, "Multiple D0-branes in Weakly Curved Backgrounds ", Nucl.Phys. B558 (1999) 63-95, hep-th/9904095.
  W. Taylor, M. Van Raamsdonk, "Multiple Dp-branes in Weak Background Fields ", Nucl.Phys. B573 (2000) 703-734, hep-th/9910052.
6. Ricardo Schiappa, "Matrix Strings in Weakly Curved Background Fields", Nucl.Phys. B608 (2001) 3-50, hep-th0005145.
7. Pedro J. Silva, "Matrix string theory and the Myers effect, JHEP 0202 (2002) 004, *hep-th/0111121*.
8. R.C. Myers, "Dielectric-Branes", JHEP 9912 (1999) 022, hep-th/9910053.
9. C. G. Callan Jr. and J. M. Maldacena, "Brane Dynamics From the Born-Infeld Action", Nucl.Phys. B513 (1998) 198-212, hep-th/9708147.
  G. W. Gibbons, "Born-Infeld particles and Dirichlet p-branes", Nucl.Phys. B514 (1998) 603-639, hep-th/9709027.
10. R. Emparan, "Born-Infeld Strings Tunneling to D-branes", Phys.Lett. B423 (1998) 71-78, hep-th/9711106.
  R. Emparan, "Tubular Branes in Fluxbranes", Nucl.Phys. B610 (2001) 169-189, hep-th/0105062.
11. C. B. Thorn, Phys.Rev.D56:6619-6628,1997, hep-th/9707048. O. Bergman and C. B. Thorn, Phys.Rev.D52:5980-5996,1995, hep-th/9506125
12. P. Townsend, PASCOS/Hopkins 1995:0271-286, hep-th/9507048.
13. A. Sen, "D0 Branes on T^n and Matrix Theory", Adv.Theor.Math.Phys. 2 (1998) 51-59, hep-th/9709220.
  N. Seiberg, "Why is the Matrix Model Correct?", Phys.Rev.Lett. 79 (1997) 3577-3580,hep-th/9710009.
14. Tom Banks and Lubos Motl, "Heterotic Strings from Matrices", JHEP 9712 (1997) 004, *hep-th/9703218*. David A. Lowe, "Heterotic Matrix String Theory", Phys.Lett. B403 (1997) 243-249, *hep-th/9704041* Morten Krogh, "Heterotic Matrix theory with Wilson lines on the lightlike circle", Nucl.Phys. B541 (1999) 98-108, hep-th/9803088.
15. Akikazu Hashimoto, Washington Taylor IV, "Fluctuation Spectra of Tilted and Intersecting D-branes from the Born-Infeld Action", Nucl.Phys. B503 (1997) 193-219, hep-th/9703217.
  P. Bain, "On the non-abelian Born-Infeld action", hep-th/9909154.
16. T.H. Buscher, "Path integral derivation of quantum duality in onlinear sigma models", Phys.Lett.B201:466,1988.
17. Iosif Bena, "The polarization of F1 strings into D2 branes: "Aut Caesar aut nihil"",*hep-th/0111156*.
18. Dominic Brecher, Bert Janssen, Yolanda Lozano,"Dielectric Fundamental Strings in Matrix String Theory", *hep-th/0112180*.
19. J. M. Camino, A. Paredes, A.V. Ramallo, "Stable Wrapped Branes", JHEP 0105 (2001) 011, hep-th/0104082.

Fixed-topology solutions in the Myers effect

Garnik G. Alexanian*, A.P. Balachandran* and Pedro J. Silva*

*Physics Department, Syracuse University, Syracuse, New York 13244

Abstract. A study of the relation between topology change, energy and Lie algebra representations for fuzzy geometry in connection to M-theory is presented. We encounter two different types of topology change, related to the different features of the Lie algebra representation appearing in the matrix models of M-theory. From these studies, we propose a new method of obtaining non-commutative solutions for the non-Abelian D-brane action found by Myers. This mechanism excludes one of the two topology changing processes previously found in other non-commutative solutions of many matrix-based models in M-theory i.e. in M(atrix) theory, Matrix string theory and non-Abelian D-brane physics.

INTRODUCTION

During the last few years we have seen how non-commutative geometry has come to play an important role in string theory. It appears not only at the fundamental Planck distances where a smooth geometry can not be trusted, but also at the level of effective theories in D-brane physics, where the Chan-Paton factors result in matrix degrees of freedom. In these cases the effective Lagrangian of N-Dp-branes comes with built-in non-commutative features in the form of matrix models. Myers [1] proposed an effective action for these non-Abelian D-branes by demanding consistency with t-duality among the different Dp-branes. In particular, he started from the well known D9-brane action and proceeded by t-dualizing. The agreement with the weak background actions of Taylor and Van Raamsdonk [2] was used as a consistency check. Note that these linearized actions came from a very different theoretical framework associated with the BFSS M(atrix) theory proposal [3].

This remarkable characteristic of the built-in non-commutativity is not new in the framework of M-theory. We already have at least two other examples where this type of construction is found, namely M(atrix) theory [3] and Matrix string theory [4, 5, 6, 7, 8].

An essential characteristic of the matrix actions is their capability to describe non-commutative geometries that correspond to different extended objects of the theory. For example, higher dimensional Dq-branes may be formed by smaller Dp-branes ($q > p$). To be more precise, consider the dielectric effect [1] where N Dp-branes form a single $D(p+2)$-brane generating a configuration corresponding to a non-commutative two-sphere (fuzzy sphere [9, 10]). Another kind of construction corresponds to N D1-branes forming n D5-branes ($n < N$), this time using a fuzzy four-sphere [1] [11]. Also it is worth mentioning that there are other non-commutative manifolds (apart from the fuzzy

---

[1] This particular type of quantum geometry has been used extensively, see for example [13, 14, 7]

n-spheres) which are relevant to matrix models, e.g. tori, CP(n), RP(n), etc.

In this talk we will report on the main results of [12], where we have studied the relations between the discrete partonic picture (the non-commutative picture) and the smooth geometry that is obtained from it in the limit of large number of partons, i.e. the "reconstruction" of the geometry.

QUANTUM GEOMETRY AND NEW SOLUTIONS WITH THE OBSTRACTIONS TO TOPOLOGY CHANGE

In the original calculation [1], Myers considered N D0-branes in a constant four-form Ramond-Ramond (RR) field strength background of the form $F^{(4)}_{t123} = -2f\varepsilon_{123}$. The relevant D0-brane action is given by

$$S_{Do} = -\mu_0\lambda^2 \int dt \left(\frac{1}{2}\partial_t\Phi^i\partial_t\Phi_i + \frac{1}{4}[\Phi,\Phi]^2 + \frac{i}{3}F^{(4)}_{tijk}\Phi^i\Phi^j\Phi^k\right), \tag{1}$$

where Φ^i are matrix-valued scalars in the adjoint representation of the $U(N)$; they represent the nine directions transverse to the D0-brane, $i = (1..9)$, μ_0 is the charge of the D0-brane and $\lambda = 2\pi\alpha'$ (α' is the string length squared). For static configurations the kinetic term vanishes and center of mass degrees of freedom decouple. Hence, variation of the action gives the following polynomial equation in Φ,

$$[\Phi^j,[\Phi_j,\Phi_i]] + if\varepsilon_{ijk}[\Phi^j,\Phi^k] = 0. \tag{2}$$

This is solved by setting to zero all but the first three scalars Φ^i ($i = (1,2,3)$) which are replaced by Lie algebra generators T^i (in this case $su(2)$) times a scalar r. Then the above equation becomes an equation for r,

$$r - \frac{f}{2} = 0. \tag{3}$$

Therefore, three of the scalars Φ^i are of the form $\Phi^i = f/2 T^i$. The non-commutative geometry appears since these Φ also correspond to the first three cartesian coordinates transverse to the brane.

However, as it is explained in appendix B of [12], the above equations do not fully define the quantum geometry. Still, it is necessary to fix the representation of T^i as different representations will give different solutions and topologies. Each of these solutions has a characteristic energy that is a function of the quadratic Casimir given by,

$$V_N = -\frac{\mu_0\lambda^2 f^2}{6} \sum_{i=1}^{3} \text{Tr}[\Phi^i\Phi_i]. \tag{4}$$

Given a fixed size of the representation N, the irreducible representation corresponds to the lower bound (strictly speaking this is the fuzzy sphere), while the reducible representations have higher energies corresponding to more complicated topologies of the products of fuzzy spheres. An important characteristic of the above construction is that in the large N limit the algebra of functions defined on these solutions becomes the algebra of functions on the classical manifold (see appendix B of [12]).

This example contains all of the ingredients that define the process of finding the non-commutative solutions. Generalizations of this program have appeared, but the underlying structure is the same. These solutions are usually called "fuzzy spaces", although not all of them are properly well defined, the best known example being the so called fuzzy four-sphere[2].

We would like to emphasize two important aspects of these solutions. First, the representation in which the different scalars Φ^i act is not fixed by the equations of motion. Second, the existence of the possibility of the topology change process suggested by the natural decay of higher energy solutions into lower energy ones. This cascade process has already received some attention in [17], the main result being the discovery of unstable modes that trigger the topology change. These modes are related to the relative position of the different fuzzy manifolds that appear in the product of higher energy solutions. Hence, given a solution A we can always construct another solution (of higher energy and more complicated topology) by considering larger matrices of two or more copies of A. The different topology of the above type of solutions comes from the fact that they correspond to reducible representations.

Nevertheless, it is important to note that this is not the only way of obtaining topology change. There is also the possibility of having different quantum geometries defined in the same group structure, which are not related by reducible-irreducible relations. For example, $SU(4)$ contains many different fuzzy geometries among which we have $(CP(3), CP(2), S^2 \times S^2)$ (see appendix B of [12]).

Regarding the first aspect, it is obvious that if the representation is fixed by the equations of motion there is no room for topology change. The fact that we can choose the representation signals the existence of a degeneracy in the set of quantum geometries, a symmetry that allows the topology change. What we have seen in this section is that from the point of view of the D-brane physics, the equations of motion only define the group structure, while energy is (in all of the solutions found) the only quantity that differentiates between quantum geometries and therefore "chooses" the classical geometry in the large N limit via the decay processes.

Therefore, we have a clear relation between topology change and the fact that the equations of motion do not fix the representation of the group. Although all of the solutions that currently exist share this behavior, there is no reason to believe that there are no circumstances in which the equations of motion can fix the representation and hence preclude topology changing processes from happening. To investigate this matter we need to understand in greater depth the definition of fuzzy geometry and the relations between fuzzy coordinates, representations and invariants of the algebraic structure.

An intuitive way to understand a fuzzy geometry [18] is to define it by a modification of the algebra of functions on classical geometry. For example, take the functions on the

---

[2] In this case it is known that the algebra of functions defined on the "fuzzy S^4" does not close and some extra structure will be needed to properly define the quantum geometry [16].

sphere. It will be enough to consider the spherical harmonics. In a classical case there is an infinite number of these. Then truncate the basis at a given angular momentum j by projecting out all the higher angular momentum modes. The resulting algebra will represent basically the fuzzy sphere.

An important mathematical point is that this algebra of functions can be obtained from the symmetric irreducible representations of size $2j+1 = N$ of $su(2)$ using coherent-state techniques (see appendix B of [12]). Actually, there exist a precise relation between cartesian coordinates in R^3 (which defines the embedding of the sphere in R^3) and the $su(2)$ Lie algebra generators in the corresponding representation. These are the three matrices T^i appearing in the dielectric effect of Myers (note that $\text{Tr}(T^i T_i) = NR^2$, with R equal to the radius of the fuzzy sphere).

An interesting generalization of the fuzzy sphere with rich enough algebraic structure is the 2n-dimensional fuzzy complex projective planes (fuzzy $CP(n)$) [15]. These quantum geometries are strongly related to $su(n)$. It turns out that to define fuzzy $CP(n)$[3] one needs to consider only the *symmetric* representations of $SU(3)$ for which the following equation is sufficient,

$$d^{ijk} T_j T_k = w(N) T^i, \qquad (5)$$

where d_{ijk} is the only invariant rank three symmetric tensor in $su(n+1)$, and $w(N)$ is related to the representation used. We can now show how to find a concrete example where the D-brane equations of motion include an invariant tensorial equation (like equation 5) that determines the representation and hence enforces an obstruction to the topology change. In doing so we will use equation (5) as a hint and will search for a simple configuration where fuzzy $CP(2)$[4] could appear.

In order to do this, consider N D1-branes with a constant world-volume electric field $F_{[2]}$ in the presence of a five-form RR field strength F_{ijklm}, flat metric constant dilaton and zero B-field. The energy density for this system is obtained by expanding the non-Abelian action proposed by Myers[5] followed by the Hamiltonian transformation. Here we show the final form, once we have restricted our study to static and constant configurations.

$$\begin{aligned} E(\Phi, F_{[2]}, F_{[5]}) &= \mu_1 \lambda^2 \left[\frac{1}{4} \Phi^{ij} \Phi_{ji} - \frac{1}{4} F_{[2]}^2 \right] + \mu_1 \lambda^4 \left[\frac{1}{16} (\Phi^{ij} \Phi_{ji})^2 - \frac{1}{8} \Phi^{ij} \Phi_{jk} \Phi^{kl} \Phi_{li} \right. \\ &\left. - \frac{1}{16} F_{[2]}^2 \Phi^{ij} \Phi_{ji} + \frac{1}{10} \Phi^i \Phi^j \Phi^k \Phi^l \Phi^m F_{mlkji} F_{\tau\sigma} \right], \qquad (6) \end{aligned}$$

where we have used the convention $\Phi^{ij} = [\Phi^i, \Phi^j]$ and $i = (1,..,8)$. We also set the the five-form to be

$$F_{ijklm} = h \, d^{pq}_{[i} f^p_{jk} f^p_{lm]}, \qquad (7)$$

---

[3] Construction of fuzzy $CP(n)$ is presented in appendix B of [12]
[4] This is smallest non-trivial $CP(n)$ geometry, $CP(1)$ is the sphere.
[5] see appendix A of [12] for a detail derivation and conventions

where "h" represents the strength of the RR field. Note that this is the only invariant tensor with five indices in $su(3)$ ($F_{[5]}$ is related to $(U^{-1}dU)^5$, the only closed 5-form in $SU(3)$).

To find the extremal points of this potential we will use configurations such that each Φ^i is proportional to the generator of $su(3)$, T^i i.e.

$$\Phi^i = \rho T^i, \tag{8}$$

The detailed form of the equations of motion can be found in appendix A of [12]. These equations are complicated and do not shed extra light on the discussion. For our purposes it is enough to show the general structure. After some algebra, one can show (see appendix A of [12]) that they are similar to (5)

$$T^i = w d^i{}_{jk} T^j T^k, \tag{9}$$

where w is a function of $(N, h, F_{[2]}, \rho)$. These equations can only be solved if the $su(3)$ generators are in the specific representation of appendix B of [12]. Then, once this representation is chosen, the above expression becomes an algebraic equation defining ρ as a function of $(F_{[2]}, h, N)$. Also, all the matrix products in the field equations simplify. To study the stability of these solutions in terms of the variable ρ, it is sufficient to substitute the ansatz back into the expression for the energy density

$$E = \mu_1 \lambda^2 N \left[-\tfrac{1}{4} F^2 + \tfrac{3}{16}(4 - \lambda^2 F^2) c_2 \rho^4 + \tfrac{9\lambda^2}{32}(c_2 - 1) c_2 \rho^8 + \tfrac{\lambda^2}{40} F_{\tau\sigma} h (2n+3) c_2 \rho^5 \right],$$

where $N = (n+1)(n+2)/2$, $c_2 = (n^2 + 3n)/3$ is the quadratic Casimir of the symmetric representation of T^i and n is an arbitrary positive integer.

Clearly, this potential has a global minimum at some value of ρ that we will call ρ_0. It depends on the value of the electric field $F_{[2]}$, the strength of the background RR field h and N. In particular, for $F_{\tau\sigma} = \sqrt{2}$ and $\lambda = 1$ the expressions for ρ_0 and radius of S^7 in which it is embedded simplify and can are given by

$$\rho_0 = \left(\frac{h(2n-3)}{9\sqrt{2}(c_2-1)} \right)^{1/3}, \quad R = \lambda \sqrt{Tr(\Phi^i \Phi_i)/N} = \left(\frac{h(2n-3) c_2^{3/2}}{9\sqrt{2}(c_2-1)} \right)^{1/3} \tag{10}$$

that in the large N limit gives $R \sim \frac{\sqrt{2}h}{27} n^{2/3}$. In the above expression we can see how the radius increases with the number of D1-branes and the strength of the RR field. Also, note that in order to achieve this effect we needed a nontrivial electric field on the D1-brane, so that we have fundamental strings diluted into the D1-brane as a requirement.

Therefore, we have found new type of solutions corresponding to N D1-branes forming a fuzzy $CP(2)$, with an obstruction for the topology change. In the large N limit this configuration goes over a D5-brane with topology $R^{(1,1)} \times CP(2)$.

In fact, we can check this correspondence by looking at the different couplings of the non-abelian D1-branes to various RR fields. For example, using solution (8) one can show that in the large N limit it takes the form of the coupling of a single D5-brane [12].

There exist other types of examples of non-Abelian D-branes forming higher dimensional D-branes, where the resulting geometry is a fuzzy CP(2) manifold. These cases however, are different in nature from the one presented here since the resulting fuzzy geometry is determined by the lowest energy condition and not by the equations of motion. The dual picture corresponding to D-branes with CP(2) topology has already been studied, we refer the reader to [14] for further information.

ACKNOWLEDGMENTS

We would like to thanks V.P. Nair for useful discussions. The work of PJS was supported in part by NSF grant PHY-0098747 to Syracuse University and by funds from Syracuse University. GA and APB were supported in part by the DOE and NSF under contract numbers DE-FG02-85ER40231 and INT-9908763 respectively.

REFERENCES

1. R.C. Myers, JHEP 9912 (1999) 022, hep-th/9910053.
2. W. Taylor, M. Van Raamsdonk, Nucl.Phys. B558 (1999) 63-95, hep-th/9904095. W. Taylor, M. Van Raamsdonk, Nucl.Phys. B573 (2000) 703-734, hep-th/9910052.
3. T. Banks, W. Fischler, S.H. Shenker and L. Susskind, Phys.Rev. D55 (1997) 5112-5128, hep-th/9610043. L. susskind, hep-th/9704080.
4. R. Dijkgraaf, E. Verlinde, H. Verlinde, Nucl.Phys. B500 (1997) 43-61, hep-th/9703030.
5. Tom Banks, Nathan Seiberg, Nucl.Phys. B497 (1997) 41-55, hep-th/9702187. R. Dijkgraaf, E. Verlinde, H. Verlinde, Nucl.Phys.Proc.Suppl. 62 (1998) 348-362, hep-th/9709107. Lubos Motl, hep-th/9701025.
6. Ricardo Schiappa, Nucl.Phys. B608 (2001) 3-50, hep-th0005145.
7. Pedro J. Silva, JHEP 0202 (2002) 004,hep-th/0111121.
8. D Brecher, B Janssen and Y Lozano, hep-th/0112180.
9. J. Madore, *An Introduction to Noncommutative Differential Geometry and its Applications*, Cambridge University Press,
10. H. Grosse and P. Prešnajder, Lett.Math.Phys. **33**, 171 (1995) and references therein; H. Grosse, C. Klimčík and P. Prešnajder, Commun.Math.Phys. **178**,507 (1996); **185**, 155 (1997); H. Grosse and P. Prešnajder, Lett.Math.Phys. **46**, 61 (1998) and ESI preprint, 1999; H. Grosse, C. Klimčík, and P. Prešnajder, hep-th/9602115 and Commun.Math.Phys. **180**, 429 (1996); H. Grosse, C. Klimčík, and P. Prešnajder, in *Les Houches Summer School on Theoretical Physics*, 1995, hep-th/9603071; P. Prešnajder, hep-th/9912050 and J.Math.Phys. **41** (2000) 2789-2804; Cambridge, 1995; gr-qc/9906059.
11. Neil R. Constable, Robert C. Myers, Oyvind Tafjord, JHEP 0106 (2001) 023, hep-th/0102080.
12. G.Alexanian, A.P.Balachandran and P.Silva, hep-th/0207052.
13. J. Castelino, S. Lee, W. Taylor, Nucl.Phys. B526 (1998) 334-350, hep-th/9712105.
14. S. P. Trivedi, S. Vaidya, JHEP 0009 (2000) 041, hep-th/0007011.
15. G.Alexanian, A.P.Balachandran, G.Immirzi, B.Ydri, J.Geom.Phys. 42 (2002) 28-53, hep-th/0103023.
16. P. Ming Ho and S Ramgoolam, Nucl.Phys. B627 (2002) 266-288, hep-th/0111278.
17. D. P. Jatkar, G. Mandal, S. R. Wadia, K.P. Yogendran, JHEP 0201 (2002) 039, hep-th/0110172. K. Hashimoto, hep-th/0204203. J. Madore, L.A. Saeger, Class.Quant.Grav. 15 (1998) 811-826, gr-qc/9708053.
18. M. R. Douglas and N. A. Nekrasov, "Noncommutative field theory," Rev. Mod. Phys. **73**, 977 (2002) hep-th/0106048; A. Connes, "Noncommutative geometry: Year 2000," math.qa/0011193; A. P. Balachandran, "Quantum spacetimes in the year 1," hep-th/0203259.

A New Non-Commutative Field Theory

Konstantin Savvidy

Perimeter Institute, 35 King St N, Waterloo Ontario, N2J 2W9, Canada

Abstract. We investigate a new type of non-commutative field theory based on a constant skew-symmetric three-form parameter. In $3+1$ dimensions such a three-form parameter can be viewed as a short-distance regulator which nevertheless preserves spatial-rotation and at long range preserves Lorentz invariance approximately. For a scalar field theory with quartic self-interaction we obtain drastically improved ultra-violet behavior of the diagrams, due to the oscillatory dependence of the interaction vertex on the momenta. The radiative corrections to the coupling are rendered finite already at the one-loop level. The key finding of this paper is that what appears as the reemergence of UV divergences as IR singularity in $p \to 0$ limit, must be interpreted simply as the logarithmic running of the coupling. Thus at low energies the theory is virtually indistinguishable from the standard theory. Conversely at high energies the diagram converges exponentially fast, the running of the coupling stops and the theory avoids developing the Landau pole. Bare coupling defined at high energy can be kept small, and in this sense the theory is similar to asymptotically free theories.

INTRODUCTION

Non-commutative theories have been extensively studied recently in connection with string theory [1, 2]. Non-commutative geometry and field theory naturally arises as a limit of string theory in the presence of the anti-symmetric two-form NS-NS field $B_{\mu\nu}$. Such field theories have been also studied in their own right with an emphasis on the possibility of viewing the non-commutativity as a short-distance regulator [3, 4, 5]. Studies of perturbative dynamics have revealed the so-called UV-IR connection [6, 7], *i.e.* the strong IR effects appearing due to integrating out the high momentum modes.

The author was originally motivated by the intriguing problem of finding a generalization of non-commutative field theory in the case that the B-field is linear, *i.e.* its field strength $H = dB$ is constant. Unfortunately the string sigma model is no longer soluble exactly in this case and so far little progress has been made attacking the problem directly.

We propose in Section an ad-hoc construct for a non-local field theory with a three-form skew-symmetric parameter. Formally the development follows closely that of the conventional non-commutative theory, but in several important physical aspects the results deviate significantly.

Renormalizability of the theory is checked up to two loop order in Sections , . The four-point (coupling renormalization) amplitude is finite already at the one-loop order. We conjecture that the four-point function diagrams are finite at arbitrary loop-order after taking into account the renormalization of divergent two-point subdiagrams.

In conventional non-commutative theory the planar diagrams get overall phases [3] that do not depend on the internal momenta, the loop integrals being exactly the same as in the corresponding diagrams of the commutative theory. This results in non-local counterterms which nonetheless are of the same form as the original lagrangian. By contrast, in the present theory the divergent parts of the diagram do not depend on the quantum parameter θ, thus the necessary counterterms are local. These arise only for the case when the terms in the original lagrangian were local as well.

An important property of the theory is that the diagrams which are rendered finite by the non-commutative cutoff are nonetheless singular as the spatial part of external momentum tends to zero. We interpret this in terms of the conventional renormalization group approach as the logarithmic running of the coupling constant. The relation between the bare coupling at high momenta and the apparent physical coupling at low momenta is the same as in conventional ϕ^4 field theory. At high energies convergence of the diagram is exponential, so corrections to the bare coupling are small. This means that unlike conventional field theory the running of the coupling eventually stops, allowing to keep the bare coupling small and avoiding the Landau pole.

THE ⊙-PRODUCT

We would like to construct a quantum space equipped with an anti-symmetric three-form fuzziness parameter. Define a non-local ⊙-product[1] of three functions, while leaving the binary product intact,

$$(f \odot g \odot h)|_{x_0} = e^{\theta_{\mu\nu\rho} \frac{\partial}{\partial x_\mu} \frac{\partial}{\partial y_\nu} \frac{\partial}{\partial z_\rho}} f(x) g(y) h(z) |_{x=y=z=x_0} . \qquad (1)$$

Here $\theta_{\mu\nu\rho}$ is a completely anti-symmetric three-form in d dimensions, however we restrict ourselves to the case $d = 4$ throughout the paper. In four spacetime dimensions such a form has exactly four independent components: $\theta_{123}, \theta_{012}, \theta_{013}, \theta_{023}$. Provided that the vector dual to our three-form is time-like, i.e. $\theta_{123}^2 > \theta_{012}^2 + \theta_{013}^2 + \theta_{023}^2$, one can always choose a coordinate system in which only θ_{123} is nonzero. Thus without loss of generality we set $\theta_{ijk} = \theta \, \varepsilon_{ijk}$ – implying the absence of time-like components and consequently time derivatives in the definition of the sun-product (1). This will nevertheless allow for a construction of a Galilean theory: spatial rotation and translation invariance are preserved.

The θ parameter has mass dimension $[-3]$, therefore we introduce a convenient mass scale for the fuzziness of space $\mu = \theta^{-1/3}$. Naively, experiments in the infrared $p \ll \mu$ should not be able to discern the quantum structure of the underlying space, because the sun-product (1) goes over to ordinary product in this limit. In reality non-commutative field theory exhibits strong IR effects [6, 7, 8], and we find that the present theory is no exception.

We note also that the underlying quantum structure of the space is not clear as yet, in the sense of the equivalence between the Weyl operator ordering on the quantum space and the Moyal star-product on the commutative space. It is tempting to speculate that the algebra of the coordinates may be obtained by considering the sun-product of the

---

[1] Pronounced as "sun-product".

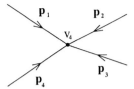

FIGURE 1. The basic four-point vertex of the theory.

coordinates themselves. In this way we arrive at the following anomalous Jacobi-like identity,

$$x_1[x_2,x_3] + x_2[x_3,x_1] + x_3[x_1,x_2] = \theta \quad (2)$$

We do not make any use of this and leave the investigation of this quantum space to the future. In this paper we concentrate instead on the perturbative properties of the quantum field theory on this space.

THE INTERACTION VERTEX

Consider the quantum field theory of one real scalar with quartic self-interaction:

$$\mathscr{L} = \int d^4x \left[\frac{1}{2} \partial_\mu \phi \, \partial^\mu \phi - \frac{1}{2} m^2 \phi^2 - \frac{\lambda}{4!} (\phi \odot \phi \odot \phi) \phi \right] . \quad (3)$$

It is clear that the kinetic and mass terms are not modified, for the same reasons as in the standard non-commutative theory. However, the interaction term is understood as a non-local momentum-dependent four-point vertex (Figure 1). Had we attempted a different prescription for the interaction from the one above, it would be reduced to the commutative case. For example, $\phi \odot \phi \odot (\phi^2)$ is equivalent to ϕ^4 due to conservation of momentum.

In momentum representation the interaction vertex is obtained by symmetrization with respect to permutations of the external legs

$$V_{(4)} = \frac{\lambda}{4!} \sum_{a \neq b \neq c} \exp(i\theta \, \varepsilon_{ijk} \, p_i^a p_j^b p_k^c) = \lambda \cos(\theta \, p_{[i}^1 p_j^2 p_{k]}^3) . \quad (4)$$

Because of momentum conservation all twenty-four terms above arising from reordering have the same magnitude, twelve occurring with plus sign and twelve with a minus, for example $p_{[i}^1 p_j^2 p_{k]}^3 = -p_{[i}^2 p_j^3 p_{k]}^4$. Moreover, it is possible to give the following geometrical interpretation of this phase (Figure 2). Construct a tetrahedron with the four (spatial) momenta being the vectors normal to the faces of the tetrahedron and equal to the area of the corresponding face. The phase in (4) is equal exactly to the square of the volume of the tetrahedron (see caption to Figure 2).

The non-locality and non-renormalizability introduced through the higher-derivative interactions both in standard non-commutative theory and in our sun-product theory

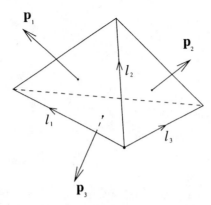

FIGURE 2. Given four arbitrary vectors that add up to nil, one can always construct a tetrahedron with the four vectors being the area-normals of the faces, $\vec{p}_1 = \frac{1}{2}\vec{l}_1 \times \vec{l}_2$ etc... Compute: $8\vec{p}_1 \cdot (\vec{p}_2 \times \vec{p}_3) = [\vec{l}_1 \times \vec{l}_2] \cdot \left[[\vec{l}_2 \times \vec{l}_3] \times [\vec{l}_3 \times \vec{l}_1]\right] = \left(\vec{l}_2 \cdot [\vec{l}_3 \times \vec{l}_1]\right)^2$, and this quantity is equal to the volume of the tetrahedron squared.

is potentially troublesome, in that the θ-expanded interactions are non-renormalizable but the very special exponential structure prevents the appearance of most problems generically associated with such non-renormalizable interactions.

Note that $|V^{(4)}| \leq \lambda$: the vertex is bounded from above by its standard value. Therefore, the convergence of all diagrams is improved, or at worst unchanged. This may enable us to use the standard theorems about renormalizability at higher loop order and the disentanglement of overlapping divergences [9]. One must worry that the necessary counterterms may not be of the same non-local form as in the original lagrangian, however it turns out that only local counterterms for the two-point function appear. We illustrate these issues in Sections , by computing the divergent and finite parts of the one- and two-loop diagrams.

ONE-LOOP RENORMALIZATION

Let us see what happens in this theory with the one-loop renormalization of mass. Since any three legs of the vertex in the diagram (Figure 3) have linearly dependent momenta, being $p, k, -p, -k$, the phase factor is trivial. This means that the diagram is still quadratically divergent. Perhaps this is a good thing, as the non-locality of the theory does not give rise to new infrared singularities in the propagator (at least in this order)

$$i\Gamma^{(2)} = \lambda \int \frac{d^4k}{(2\pi)^4} \frac{1}{k^2 + m^2} = \frac{\lambda}{32\pi^2}\left(\Lambda^2 - m^2 \ln\frac{\Lambda^2}{m^2}\right) \qquad (5)$$

Thus the counterterms necessary to cancel the divergence in this one-loop mass renormalization diagram are indeed of the same form as the lagrangian we started with. It

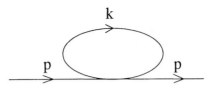

FIGURE 3. The one-loop mass renormalization diagram: the sun-phase vanishes due to the collinearity of any three of the legs at the vertex. Conventional regularization is necessary to cancel the quadratic divergence.

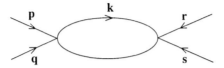

FIGURE 4. The one-loop coupling renormalization diagram: the sun-phase does not vanish at either of the two vertices for generic values of momenta, being $\mathbf{k}\cdot(\mathbf{q}\times\mathbf{p})$ and $\mathbf{k}\cdot(\mathbf{r}\times\mathbf{s})$ respectively. The contribution of this diagram is finite.

turns out that at two-loop order (see Section) the mass renormalization diagram contains both convergent and divergent contributions. The divergent part is of the same form as the original lagrangian and is local, while only the finite part contains nontrivial dependence on the quantum parameter θ.

Presently we consider the one-loop radiative correction to the coupling (Figure 4). Momenta in bold represent spatial vectors

$$i\Gamma^{(4)} = \frac{\lambda^2}{2}\int \frac{dk_0\, d^3\mathbf{k}}{(2\pi)^4}\frac{\cos\theta\,\mathbf{k}\cdot(\mathbf{p}\times\mathbf{q})}{k^2+m^2}\frac{\cos\theta\,\mathbf{r}\cdot(\mathbf{s}\times\mathbf{k})}{(p+q+k)^2+m^2}$$

$$= \frac{\lambda^2}{32\pi^2}\int_0^\infty d\alpha\, d\beta\,(\alpha+\beta)^{-2} e^{-(\alpha+\beta)m^2} e^{-\frac{\alpha\beta}{\alpha+\beta}(p+q)^2} e^{-\frac{\theta^2}{\alpha+\beta}(\mathbf{p}\times\mathbf{q}+\mathbf{r}\times\mathbf{s})^2} + [\mathbf{r}\leftrightarrows\mathbf{s}] \quad (6)$$

Upon introducing the convenient variables $z = \alpha + \beta$ and $u = \alpha/z$, obtain

$$i\Gamma^{(4)} = \frac{\lambda^2}{32\pi^2}\int_0^\infty z^{-1} dz \int_0^{+1} du\, e^{-zm^2} e^{-zu(1-u)(p+q)^2} e^{-\frac{\theta^2}{z}(\mathbf{p}\times\mathbf{q}+\mathbf{r}\times\mathbf{s})^2} + [\mathbf{r}\leftrightarrows\mathbf{s}] \quad . \quad (7)$$

This is now identical to the standard result, with the expression $\Lambda_{eff}^{-2} = \theta^2(\mathbf{p}\times\mathbf{q}\pm\mathbf{r}\times\mathbf{s})^2$ playing the role of the cutoff length scale. The integral in (7) would have been logarithmically divergent near $z \to 0$ without this cutoff. Thus, for sufficiently large Λ_{eff}^2, the z integral can be approximated by its leading logarithmic divergence,

$$i\Gamma^{(4)} = \frac{\lambda^2}{32\pi^2}\left[\ln\frac{(\mathbf{p}\times\mathbf{q}+\mathbf{r}\times\mathbf{s})^2}{\mu^4} + [\mathbf{r}\leftrightarrows\mathbf{s}] + \int_0^1 du\,\ln\frac{m^2+(p+q)^2 u(1-u)}{\mu^2}\right] \quad (8)$$

For small values of Λ_{eff}^2 the integral converges much faster, in fact exponentially.

At this point we recognize the reemergence of the UV-IR mixing of [6, 7]. The amplitude (8) is divergent for some accidental values of non-zero momenta, but more importantly it is singular as $\mathbf{p} \to 0$. In Section we show that this does not lead to genuine IR singularity because the effective action remains finite.

In conventional field theory the logarithmic divergence in the four-point function is interpreted in terms of renormalization of the coupling [10, 11]. We would like to make the connection between our theory and the standard one. For that, we should consider the first term in (8) as the correction to the effective coupling at low energy. Taking into account the additional contributions in the $\mathbf{t}, \mathbf{u}$ channels,

$$\lambda_{eff} = \lambda - \frac{3\lambda^2}{8\pi^2} \ln \frac{\mu}{|\mathbf{p}|} \quad \text{for small } \mathbf{p} \tag{9}$$

For large $\mathbf{p}$ the behavior of the integral is different, and λ_{eff} exponentially approaches the bare coupling λ from below. In this way, the running of the coupling stops once momenta reach the fuzziness scale $\mathbf{p} \gg \mu$. Thus in this theory there is no Landau pole singularity. However, there is instead the possibility of destabilization at nonperturbatively low momenta $|\mathbf{p}| \sim \mu \exp(-1/\lambda)$. We fully expect that higher loop diagrams will remove this apparent pathology, changing the behavior at such low momenta to

$$\lambda_{eff} = \frac{\lambda}{1 + \frac{3\lambda}{8\pi^2} \ln \frac{\mu}{|\mathbf{p}|}} \tag{10}$$

as is the modification of the one-loop result inferred from standard renormalization group approach [10, 11]. This correctly reproduces the fact that the theory is known to be free in the infrared.

TWO-LOOP MASS RENORMALIZATION

Two-loop renormalization in non-commutative field theory was considered in detail in [12, 13, 14]. We consider the two-loop order primarily because at one-loop order the two-point function does not receive non-trivial θ-dependent corrections.

In order to obtain radiative corrections to the mass at two loop order (Figure 5) we should compute the double integral

$$\lambda^2 \int \int \frac{d^4 k_1}{(2\pi)^4} \frac{d^4 k_2}{(2\pi)^4} \frac{\cos \theta \, \mathbf{p} \cdot (\mathbf{k}_1 \times \mathbf{k}_2)}{(k_1^2 + m^2)(k_2^2 + m^2)} \frac{\cos \theta \, \mathbf{p} \cdot (\mathbf{k}_1 \times \mathbf{k}_2)}{(p + k_1 + k_2)^2 + m^2}. \tag{11}$$

This integral has overall quadratic divergence by power counting, which is not completely cured by the appearance of the phases. The product of the two cosines has a zero-frequency component as well as a $2\theta \mathbf{p} \cdot (\mathbf{k}_1 \times \mathbf{k}_2)$ component. Such zero-frequency components do not seem to persist at higher loop order. We therefore conjecture that the two-point function at three and higher loop order are finite after the renormalization of divergent one- and two- loop subdiagrams.

The finite part of the integral can be computed as usual with the introduction of Schwinger parameters α, β, γ and performing the gaussian integrations over momenta

$$\int_0^\infty \frac{e^{-(\alpha+\beta+\gamma)m^2} e^{-\frac{\alpha\beta\gamma}{\alpha\beta+\beta\gamma+\alpha\gamma} p^2}}{(\alpha\beta+\beta\gamma+\alpha\gamma)(\alpha\beta+\beta\gamma+\alpha\gamma+\theta^2 \mathbf{p}^2)} d\alpha\, d\beta\, d\gamma = \frac{\mathscr{F}(m^2\theta|\mathbf{p}|, p^2\theta|\mathbf{p}|)}{\theta|\mathbf{p}|} \quad (12)$$

where $\mathscr{F}$ is a function weakly dependent on its dimensionless arguments.

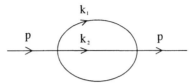

FIGURE 5. Two-loop mass renormalization diagram. The vertex phase is nontrivial, but is the same for both vertices. This leads to both a finite and an infinite contribution. Since the infinite contribution is θ independent, the infinity can be cancelled by a local counterterm.

Again, the result is analogous to the standard one, if we identify $\theta|\mathbf{p}|$ with the effective cutoff. Aside from the leading $1/\theta|\mathbf{p}|$ behaviour there is also subleading term $\sim p^2 \ln p^2 \theta|\mathbf{p}|$ corresponding to the necessary wavefunction renormalization at two loops in ordinary field theory. Again, the divergence of this diagram, which is normally manifest in the ultra-violet region, here reappears in the infrared: $\mathbf{p} \to 0$.

EFFECTIVE ACTION

The interaction of UV and IR in non-commutative field theories leads to troublesome IR singularities. It has been suggested that these IR effects can be explained through the appearance of closed string modes in string theory. Another possibility might have been to introduce extra IR cutoffs. In [8] a suggestion was made to solve this problem through a non-local field redefinition, however the field transformation is potentially singular, and in any case the problem reappears at the second loop order. Therefore we are compelled to investigate this issue in the present theory.

The appropriate tool is to check whether the effective action is finite, after the removal of the UV cutoff Λ. The quadratic one-loop effective action in our theory is

$$\Gamma^{(2)}_{eff} = \int d^4 p\, \phi(p)\phi(-p) \left[p^2 + m^2 + \frac{\lambda}{32\pi^2}\left(\Lambda^2 - m^2 \ln \frac{\Lambda^2}{m^2} \right) \right] \quad (13)$$

The divergent part as $\Lambda \to \infty$ is removed as usual by reabsorbing into the definition of the mass.

The one-loop effective four-point function is read off from (8)

$$i\Gamma^{(4)}_{eff} = \frac{\lambda^2}{32\pi^2} \int d^4 p_1 d^4 p_2 d^4 p_3 d^4 p_4\; \phi(p_1)\phi(p_2)\phi(p_3)\phi(p_4)\, \delta^4(p_1+p_2+p_3+p_4)$$
$$\left[\ln \frac{(\mathbf{p}_1 \times \mathbf{p}_2 + \mathbf{p}_3 \times \mathbf{p}_4)^2}{\mu^4} + \int_0^1 d\alpha\, \ln \frac{m^2 + (p_1+p_2)^2 \alpha(1-\alpha)}{\mu^2} \right]$$

Despite the fact that there is a logarithmic divergence in the integrand near $\mathbf{p} \to 0$, this is more than compensated by the measure in the integral. The IR singularity in the two-loop two-point function is potentially more troublesome, since it is not a logarithmic one as above, but is a pole as $\mathbf{p} \to 0$.

$$\Gamma^{(2)}_{two\ loop} = \int dp_o\, d^3\mathbf{p}\, \phi(p)\, \phi(-p) \left[p^2 + M^2 + \lambda^2 \frac{\mathscr{F}(m^2 \theta |\mathbf{p}|, p^2 \theta |\mathbf{p}|)}{\theta |\mathbf{p}|} \right] \quad (14)$$

The crucial difference with standard non-commutative theory is that the effective cutoff appearing above involves momenta along all three spatial directions, while in star-product theory only momenta in the non-commutative plane appear. Thus, the integral above is actually convergent near $\mathbf{p} \to 0$. The theory is free of the IR singularity.

ACKNOWLEDGMENTS

I would like to thank J. Ambjørn, L. Alvarez-Gaume, J. Brodie, L. Freidel, R. Myers, L. Smolin, N. Seiberg and R.J. Szabo for extensive discussions and useful comments. Also I am grateful to the organizers of the MRST-2002 conference for the invitation to present this work.

REFERENCES

1. Connes, A., Douglas, M. R., and Schwarz, A., *JHEP*, **02**, 003 (1998).
2. Seiberg, N., and Witten, E., *JHEP*, **09**, 032 (1999).
3. Filk, T., *Phys. Lett.*, **B376**, 53–58 (1996).
4. Krajewski, T., and Wulkenhaar, R., *Int. J. Mod. Phys.*, **A15**, 1011–1030 (2000).
5. Varilly, J. C., and Gracia-Bondia, J. M., *Int. J. Mod. Phys.*, **A14**, 1305 (1999).
6. Minwalla, S., Van Raamsdonk, M., and Seiberg, N., *JHEP*, **02**, 020 (2000).
7. Van Raamsdonk, M., and Seiberg, N., *JHEP*, **03**, 035 (2000).
8. Grimstrup, J. M., Grosse, H., Popp, L., Putz, V., Schweda, M., Wickenhauser, M., and Wulkenhaar, R. (2002).
9. Chepelev, I., and Roiban, R., *JHEP*, **05**, 037 (2000).
10. Callan, C. G., in *Les Houches 1975, Proceedings, Methods In Field Theory*, Amsterdam 1976, 41-77.
11. Gross, D. J., in *Les Houches 1975, Proceedings, Methods In Field Theory*, Amsterdam 1976, 141-250.
12. Aref'eva, I. Y., Belov, D. M., and Koshelev, A. S., *Phys. Lett.*, **B476**, 431–436 (2000).
13. Micu, A., and Sheikh Jabbari, M. M., *JHEP*, **01**, 025 (2001).
14. Huang, W.-H., *Phys. Lett.*, **B496**, 206–211 (2000).

TOPICS IN QUANTUM FIELD THEORY

Trace Anomaly and Quantization of Maxwell's Theory on Non-Commutative Spaces

S. I. Kruglov

International Educational Centre, Toronto, Canada M3J 3G9

Abstract. The canonical and symmetrical energy-momentum tensors and their non-zero traces in Maxwell's theory on non-commutative spaces have been found. Dirac's quantization of the theory under consideration has been performed. I have found the extended Hamiltonian and equations of motion in the general gauge covariant form.

INTRODUCTION

Non-commutative (NC) gauge theories are of interest now because they appear in the superstring theory [1]. In the presence of the external background magnetic field, NC coordinates emerge naturally [1]. The gauge theories on NC spaces are equivalent (the Seiberg-Witten map) to effective commutative theories with the additional deformation parameters $\theta_{\mu\nu}$. This allows us to formulate a Lagrange field theory on NC spaces with the same degrees of freedom in terms of ordinary fields. The effective Lagrangian is formulated in terms of ordinary fields and the parameters $\theta_{\mu\nu}$. The coordinates of NC spaces obey the commutation relation [2], [3]: $[\widehat{x}_\mu, \widehat{x}_\nu] = i\theta_{\mu\nu}$. We use here the Lorentz-Heaviside units, and set $\hbar = c = 1$.

The effective Lagrangian density in terms of ordinary fields, which is equivalent to Maxwell's Lagrangian density on NC spaces, in the leading order of θ is [4]

$$\mathscr{L} = -\frac{1}{4}F_{\mu\nu}^2 + \frac{1}{8}\theta_{\alpha\beta}F_{\alpha\beta}F_{\mu\nu}^2 - \frac{1}{2}\theta_{\alpha\beta}F_{\mu\alpha}F_{\nu\beta}F_{\mu\nu} + \mathscr{O}(\theta^2). \tag{1}$$

The field strength tensor is given by

$$F_{\mu\nu} = \partial_\mu A_\nu - \partial_\nu A_\mu, \tag{2}$$

where $A_\mu = (\mathbf{A}, iA_0)$ is the electromagnetic field vector-potential; $E_i = iF_{i4}$ is the electric field; $B_i = \varepsilon_{ijk}F_{jk}$ ($\varepsilon_{123} = 1$) is the magnetic induction field.

The BRST-shift symmetry and the vacuum polarization of photons at the one-loop level of the theory under consideration have been investigated in [5]. Authors of [6], [7] showed that the velocity of photon propagation in the direction that is perpendicular to a background magnetic induction field in such a θ-deformed Maxwell theory differs from c. The Lagrangian (1) can also be rewritten as follows:

$$\mathscr{L} = \frac{1}{2}\left(\mathbf{E}^2 - \mathbf{B}^2\right)[1 + (\boldsymbol{\theta} \cdot \mathbf{B})] - (\boldsymbol{\theta} \cdot \mathbf{E})(\mathbf{E} \cdot \mathbf{B}) + \mathscr{O}(\theta^2), \tag{3}$$

where $\theta_i = (1/2)\varepsilon_{ijk}\theta_{jk}$, and we put $\theta_{i4} = 0$ which guarantees the unitarity [8], [9]. Parameters $\theta_{\mu\nu}$ are the CP violating variables (at CP transformations $\theta_{\mu\nu} \to -\theta_{\mu\nu}$), and, as a result, particles possess the dipole moments [10], [11]. Quantum electrodynamics on NC spaces is a renormalizable theory [12], [4] with the conservation of CPT symmetry.

From Eq. (1) equations of motion [6], [7] read

$$\frac{\partial}{\partial t}\mathbf{D} - \text{rot}\mathbf{H} = 0, \qquad \text{div}\mathbf{D} = 0, \tag{4}$$

where $(\text{rot}\mathbf{H})_i = \varepsilon_{ijk}\partial_j H_k$ and $\text{div}\mathbf{D} = \partial_i D_i$. The displacement (**D**) and magnetic (**H**) fields are given by

$$\mathbf{D} = \mathbf{E} + \mathbf{d}, \quad \mathbf{d} = (\boldsymbol{\theta}\cdot\mathbf{B})\mathbf{E} - (\boldsymbol{\theta}\cdot\mathbf{E})\mathbf{B} - (\mathbf{E}\cdot\mathbf{B})\boldsymbol{\theta},$$
$$\mathbf{H} = \mathbf{B} + \mathbf{h}, \quad \mathbf{h} = (\boldsymbol{\theta}\cdot\mathbf{B})\mathbf{B} + (\boldsymbol{\theta}\cdot\mathbf{E})\mathbf{E} - \frac{1}{2}\left(\mathbf{E}^2 - \mathbf{B}^2\right)\boldsymbol{\theta}. \tag{5}$$

Eq. (2) leads to $\partial_\mu \widetilde{F}_{\mu\nu} = 0$ (where $\widetilde{F}_{\mu\nu} = (1/2)\varepsilon_{\mu\nu\alpha\beta}F_{\alpha\beta}$, $\varepsilon_{\mu\nu\alpha\beta}$ is an antisymmetric tensor, $\varepsilon_{1234} = -i$) which is rewritten as

$$\frac{\partial}{\partial t}\mathbf{B} + \text{rot}\mathbf{E} = 0, \qquad \text{div}\mathbf{B} = 0. \tag{6}$$

ENERGY-MOMENTUM TENSOR

Using the general procedure, we obtain the canonical conservative energy-momentum tensor [13]:

$$T_{\mu\nu} = -F_{\mu\alpha}F_{\nu\alpha}\left(1 - \frac{1}{2}\theta_{\gamma\beta}F_{\gamma\beta}\right) + \frac{1}{4}\theta_{\mu\alpha}F_{\nu\alpha}F_{\rho\beta}^2$$
$$-\theta_{\mu\beta}F_{\gamma\nu}F_{\rho\beta}F_{\gamma\rho} - \left(F_{\mu\alpha}F_{\nu\gamma} + F_{\nu\alpha}F_{\mu\gamma}\right)\theta_{\alpha\beta}F_{\gamma\beta} - \delta_{\mu\nu}\mathcal{L}. \tag{7}$$

The canonical energy-momentum tensor (7) is non-symmetric, but for classical electrodynamics at $\theta \to 0$, it becomes symmetric. The trace of the energy-momentum tensor (7) is not equal to zero; that indicates the trace anomaly at the classical level. This is the consequence of the violation of the dilatation symmetry: $x'_\mu = \lambda x_\mu$ for NC spaces.

Using the general procedure [14], and varying the action corresponding to the Lagrangian (1) on the metric tensor, we find the symmetric energy-momentum tensor:

$$T_{\mu\nu}^{sym} = T_{\mu\nu} + \frac{1}{4}\theta_{\nu\alpha}F_{\mu\alpha}F_{\rho\beta}^2 - \theta_{\nu\beta}F_{\gamma\mu}F_{\rho\beta}F_{\gamma\rho}. \tag{8}$$

Here the conservative tensor $T_{\mu\nu}$ is defined by Eq. (7). The components of the stress tensor tensor, found from Eq. (8), are given by

$$T_{44}^{sym} = \frac{\mathbf{E}^2 + \mathbf{B}^2}{2}\left[1 + (\boldsymbol{\theta}\cdot\mathbf{B})\right] - (\mathbf{E}\cdot\mathbf{B})(\boldsymbol{\theta}\cdot\mathbf{E}). \tag{9}$$

$$T_{m4}^{sym} = -i\left\{\left[1 + (\boldsymbol{\theta}\cdot\mathbf{B})\right](\mathbf{E}\times\mathbf{B}) + \frac{1}{2}\left(\mathbf{B}^2 - \mathbf{E}^2\right)(\mathbf{E}\times\boldsymbol{\theta})\right\}_m. \tag{10}$$

$$T_{mn}^{sym} = E_m E_n + B_m B_n - \frac{1}{2}\delta_{mn}\left(\mathbf{E}^2 + \mathbf{B}^2\right) + (\boldsymbol{\theta}\cdot\mathbf{B})\left(3E_m E_n + B_m B_n\right)$$

$$+ \frac{1}{2}\left(\mathbf{E}^2 + \mathbf{B}^2\right)\left[B_m\theta_n + \theta_m B_n - 3\delta_{mn}(\boldsymbol{\theta}\cdot\mathbf{B})\right] \quad (11)$$

$$- (\mathbf{E}\cdot\mathbf{B})\left[\theta_m E_n + \theta_n E_m - \delta_{mn}(\boldsymbol{\theta}\cdot\mathbf{E})\right] - (\boldsymbol{\theta}\cdot\mathbf{E})(E_m B_n + E_n B_m)$$

$$- (\mathbf{E}\times\boldsymbol{\theta})_m(\mathbf{B}\times\mathbf{E})_n - (\mathbf{E}\times\boldsymbol{\theta})_n(\mathbf{B}\times\mathbf{E})_m.$$

Eqs. (9)-(11) lead to the trace (anomaly) of the symmetric energy-momentum tensor:

$$T_{\mu\mu}^{sym} = 2(\boldsymbol{\theta}\cdot\mathbf{B})\left(\mathbf{E}^2 - \mathbf{B}^2\right) - 4(\boldsymbol{\theta}\cdot\mathbf{E})(\mathbf{E}\cdot\mathbf{B}). \quad (12)$$

From Eqs. (7), (12), we find the relation $T_{\mu\mu}^{sym} = 2T_{\mu\mu}$. The trace anomaly can contribute to the cosmological constant, and as a result, this trace anomaly should be taken into account by consideration of the inflation theory.

The trace anomaly vanishes in the case of the plane electromagnetic waves as two Lorentz invariants $\mathbf{E}^2 - \mathbf{B}^2$, $(\mathbf{E}\cdot\mathbf{B})$ are equal to zero, as well as for classical electrodynamics at $\theta = 0$.

DIRAC'S QUANTIZATION

In accordance with the general procedure [15] of gauge covariant quantization, we find from Eq. (1) (in leading order of θ)) the momenta:

$$\pi_i = \frac{\partial\mathscr{L}}{\partial(\partial_0 A_i)} = -E_i[1 + (\boldsymbol{\theta}\cdot\mathbf{B})] + (\boldsymbol{\theta}\cdot\mathbf{E})B_i + (\mathbf{E}\cdot\mathbf{B})\theta_i, \quad (13)$$

$$\pi_0 = \frac{\partial\mathscr{L}}{\partial(\partial_0 A_0)} = 0.$$

A primary constraint follows from Eq. (13):

$$\varphi_1(x) = \pi_0, \quad \varphi_1(x) \approx 0. \quad (14)$$

Here Dirac's symbol $\approx$ is used for equations which hold only weakly. Taking into account the equality $\pi_i = -D_i$, we find the Poisson brackets

$$\{A_i(\mathbf{x},t), D_j(\mathbf{y},t)\} = -\delta_{ij}\delta(\mathbf{x}-\mathbf{y}), \quad (15)$$

$$\{B_i(\mathbf{x},t), D_j(\mathbf{y},t)\} = \varepsilon_{ijk}\partial_k\delta(\mathbf{x}-\mathbf{y}). \quad (16)$$

In the quantum theory, we should make the replacement $\{B,D\} \to -i[B,D]$, where $[B,D] = BD - DB$ is the commutator. Using Eqs. (3), (13), and the relation $\mathscr{H} = \pi_\mu \partial_0 A_\mu - \mathscr{L}$, we arrive at the density of the Hamiltonian:

$$\mathscr{H} = \frac{1}{2}\left(\mathbf{E}^2 + \mathbf{B}^2\right)[1 + (\boldsymbol{\theta}\cdot\mathbf{B})] - (\boldsymbol{\theta}\cdot\mathbf{E})(\mathbf{E}\cdot\mathbf{B}) - \pi_m \partial_m A_0. \quad (17)$$

As the primary constraint (14) have to be a constant of motion, we find the condition

$$\partial_0 \pi_0 = \{\pi_0, H\} = -\partial_m \pi_m = 0, \tag{18}$$

where $H = \int d^3x \mathcal{H}$ is the Hamiltonian. Eq. (18) leads to the secondary constraint:

$$\varphi_2(x) = \partial_m \pi_m, \qquad \varphi_2(x) \approx 0. \tag{19}$$

This is simply the Gauss law (see Eq. (4)). It is easy to find $\partial_0 \varphi_2 = \{\varphi_2, H\} \equiv 0$, i.e. no additional constraints. There are no second class constrains here because $\{\varphi_1, \varphi_2\} = 0$, as similar to classical electrodynamics [15]. In accordance with the general approach [15], to have the total density of Hamiltonian, we add to Eq. (17) the Lagrange multiplier terms:

$$\mathcal{H}_T = \mathcal{H} + v(x)\pi_0 + u(x)\partial_m \pi_m. \tag{20}$$

The density energy, obtained from Eq. (17), is $\mathcal{E} \equiv T_{44}^{sym}$ (see Eq. (9)). The electric field, with the accuracy of $\mathcal{O}(\theta^2)$, is given by

$$E_i = -\pi_i[1-(\boldsymbol{\theta}\cdot\mathbf{B})] - (\boldsymbol{\theta}\cdot\boldsymbol{\pi})B_i - (\boldsymbol{\pi}\cdot\mathbf{B})\theta_i, \tag{21}$$

so that the total density of Hamiltonian becomes

$$\mathcal{H}_T = \frac{\pi^2 + \mathbf{B}^2}{2} + (\boldsymbol{\theta}\cdot\mathbf{B})\frac{\mathbf{B}^2 - \pi^2}{2} + (\boldsymbol{\theta}\cdot\boldsymbol{\pi})(\boldsymbol{\pi}\cdot\mathbf{B})$$
$$+ v(x)\pi_0 + (u(x) + A_0)\partial_m \pi_m, \tag{22}$$

where $\mathbf{B} = \text{rot}\mathbf{A}$.

Using Eq. (22), we obtain the Hamiltonian equations

$$\partial_0 A_i = \{A_i, H\} = \frac{\delta H}{\delta \pi_i}$$
$$= \pi_i[1-(\boldsymbol{\theta}\cdot\mathbf{B})] + (\mathbf{B}\cdot\boldsymbol{\pi})\theta_i + (\boldsymbol{\pi}\cdot\boldsymbol{\theta})B_i - \partial_i A_0 - \partial_i u(x), \tag{23}$$

$$\partial_0 \pi_i = \{\pi_i, H\} = -\frac{\delta H}{\delta A_i} = \partial_n\{[(\partial_n A_i) - (\partial_i A_n)][1 + (\boldsymbol{\theta}\cdot\mathbf{B})]\}$$
$$+ \varepsilon_{iab}\partial_b \left[\theta_a \frac{\mathbf{B}^2 - \pi^2}{2} + \pi_a(\boldsymbol{\pi}\cdot\boldsymbol{\theta})\right], \tag{24}$$

$$\partial_0 A_0 = \{A_0, H\} = \frac{\delta H}{\delta \pi_0} = v(x), \quad \partial_0 \pi_0 = \{\pi_0, H\} = -\frac{\delta H}{\delta A_0} = -\partial_m \pi_m, \tag{25}$$

which are gauge-equivalent to the Euler-Lagrange equations. Taking into account the definition (5), we find that Eq. (24) coincides with the first equation in (4), and Eq. (23) is the gauge covariant form of Eq. (21). Gauss's law is the secondary constraint in this Hamiltonian formalism and is the second equation in (4). The first class constraints generate gauge transformations. In accordance with Eq. (25), the A_0 is arbitrary function and Eqs. (25) represent the time evolution of non-physical fields. The variables π_0, $\partial_m \pi_m$ equal zero as constraints.

THE COULOMB GAUGE

Using the gauge freedom here the radiation gauge will be considered. It should be noted that the gauge fixing approach is beyond the Dirac's approach. Using the gauge freedom (described by functions $v(x)$, $u(x)$), we impose new constraints:

$$\varphi_3(x) = A_0 \approx 0, \qquad \varphi_4(x) = \partial_m A_m \approx 0. \tag{26}$$

Then equations (see also [16])

$$\{\varphi_1(\mathbf{x},t), \varphi_3(\mathbf{y},t)\} = -\delta(\mathbf{x}-\mathbf{y}), \quad \{\varphi_2(\mathbf{x},t), \varphi_4(\mathbf{y},t)\} = \Delta_x \delta(\mathbf{x}-\mathbf{y}), \tag{27}$$

hold, where $\Delta_x \equiv \partial^2/(\partial x_m)^2$. Defining "coordinates" Q_i and conjugated momenta P_i ($i=1,2$):

$$Q_i = (A_0, \partial_m A_m), \qquad P_i = (\pi_0, -\Delta_x^{-1}\partial_m \pi_m), \tag{28}$$

we find

$$\{Q_i(\mathbf{x},t), P_j(\mathbf{y},t)\} = \delta_{ij}\delta(\mathbf{x}-\mathbf{y}), \quad \Delta_x^{-1} = -\frac{1}{4\pi|\mathbf{x}|}, \quad \Delta_x \frac{1}{4\pi|\mathbf{x}|} = -\delta(\mathbf{x}), \tag{29}$$

Canonical variables Q_i, P_i, Eq. (29) are not the physical degrees of freedom and must be eliminated. Using the matrix of Poisson brackets [16] $C_{ij} = \{\varphi_i(\mathbf{x},t), \varphi_j(\mathbf{y},t)\}$, and the definition of the Dirac brackets [16], [17], we obtain

$$\{\pi_0(\mathbf{x},t), A_0(\mathbf{y},t)\}^* = \{\pi_0(\mathbf{x},t), A_i(\mathbf{y},t)\}^* = \{\pi_i(\mathbf{x},t), A_0(\mathbf{y},t)\}^* = 0, \tag{30}$$

$$\{\pi_i(\mathbf{x},t), A_j(\mathbf{y},t)\}^* = -\delta_{ij}\delta(\mathbf{x}-\mathbf{y}) + \frac{\partial^2}{\partial x_i \partial y_j}\frac{1}{4\pi|\mathbf{x}-\mathbf{y}|} \quad (i,j=1,2,3), \tag{31}$$

$$\{\pi_\mu(\mathbf{x},t), \pi_\nu(\mathbf{y},t)\}^* = \{A_\mu(\mathbf{x},t), A_\nu(\mathbf{y},t)\}^* = 0 \quad (\mu,\nu=1,2,3,4). \tag{32}$$

Taking into consideration the Fourier transformation of the Coulomb potential, Eq. (31) takes the form

$$\{\pi_i(\mathbf{k}), A_j(\mathbf{q})\}^* = -(2\pi)^3 \delta(\mathbf{k+q})\left(\delta_{ij} - \frac{k_i k_j}{\mathbf{k}^2}\right). \tag{33}$$

The projection operator in the right side of Eq. (33) extracts the physical transverse components of vectors.

Following the general method, we can set all second class constraints strongly to zero, and as a result, only transverse components of the vector potential A_μ and momentum π_μ are physical independent variables. Pairs of operators (28) are absent in the reduced physical phase space and the physical Hamiltonian becomes $H^{ph} = \int d^3x \mathscr{E}$. Equations of motion obtained from this Hamiltonian coincide with Eqs. (23)-(25) at $u(x) = v(x) = 0$ (the Poisson brackets are replaced by the Dirac brackets). In the quantum theory, we should substitute Dirac's brackets by the quantum commutators according to the prescription $\{.,.\}^* \to -i[.,.]$. The fields $\mathbf{E}, \mathbf{B}, \mathbf{D}, \mathbf{H}$ are invariants of the gauge transformations and are measurable quantities (observables).

CONCLUSION

We have just considered the effective non-linear electrodynamics, which is equivalent to Maxwell's theory on NC spaces due to the Seiberg-Witten map. The canonical conservative, symmetric non-conservative energy-momentum tensors, and their non-zero traces were found. We have shown that there is a trace anomaly which is related with the violation of the dilation symmetry at the classical level. The trace anomaly is absent in the case of the plane electromagnetic waves. At high energy (on short distances), when space-time might be non-commutative, the trace anomaly can contribute to the cosmological constant, and as a result, effect cosmology and physics of early universe.

Dirac's quantization of the effective non-linear electromagnetic theory is similar to the quantization of classical electrodynamics, and includes first class constraints. The gauge fixing approach, on the basis of Coulomb's gauge, leads to second class constraints and transition from the Poisson brackets to the Dirac brackets. For consideration of the vacuum state one needs to construct the Fock representation or the wave functionals. The normalization conditions are formulated here in terms of functional integrals with the corresponding gauge conditions. This may involve, however, the introducing ghosts and negative norm states, which is beyond the Dirac approach. The gauge covariant Dirac quantization does not violate the Lorentz invariance and locality in space, and has an advantage compared to the reduced phase space approach (see [17]).

The quantization of the Maxwell theory on NC spaces within BRST-scheme was investigated in [4], [5], [19].

REFERENCES

1. Seiberg, N., and Witten, E., *JHEP* **9909**, 032 (1999); arXiv:hep-th/9908142.
2. Snyder, H., *Phys. Rev.* **71**, 38-41 (1947); **72**, 68-72 (1947).
3. Connes, A., *Noncommutative Geometry*, Academic Press, 1994.
4. Bichl, A., Grimstrup, J., Popp, L., Schweda, M., and Wulkenhaar, R., arXiv:hep-th/0102044.
5. Fruhwirth, I., Grimstrup, J.M., Morsli, Z., Popp, L., Schweda, M., arXiv:hep-th/0202092.
6. Jackiw, R., *Nucl. Phys. Proc. Suppl.* **108**, 30-36 (2002); arXiv:hep-th/0110057.
7. Guralnic, Z., Jackiw, R., Pi, S.Y., and Polychronakos, A.P., *Phys. Lett.* **B 517**, 450-456 (2001); arXiv:hep-th/0106044.
8. Gomis, J., and Mehen, T., *Nucl. Phys.* **B591**, 265-276 (2000); arXiv:hep-th/0005129.
9. Aharony, O., Gomis, J., and Mehen, T., *JHEP* **0009**, 023 (2000); arXiv:hep-th/0006236.
10. Riad, I.F., and Sheikh-Jabbari, M.M., *JHEP* **0008**, 045 (2000); arXiv:hep-th/0008132.
11. Sheikh-Jabbari, M.M., *Phys. Rev. Lett.* **84**, 5265-5268 (2000); arXiv:hep-th/0001167.
12. Hayakawa, M., *Phys. Lett.* **B478**, 394-400 (2000); arXiv:hep-th/9912094; arXiv:hep-th/9912167.
13. Kruglov, S.I., *Annales Fond. Broglie* (in press); arXiv:hep-th/0110059.
14. Landau, L.D. and Lifshits, E.M., *The Classical Theory of Fields*, Pergamon Press, 1975.
15. Dirac, P.A.M., *Lectures on Quantum Mechanics*, Yeshiva University, New York, 1964.
16. Hanson, A., Regge, T., Teitelboim, C., *Constrained Hamiltonian Systems*, Accademia Nationale Dei Lincei, Roma, 1976.
17. Henneaux, M., and Teitelboim, C., *Quantization of Gauge Systems*, Princeton University Press, Princeton, New Jesey, 1992.
18. Matschull, H.J., arXiv:quant-ph/9606031.
19. Wulkenhaar, R., *JHEP* **0203**, 024 (2002); arXiv:hep-th/0112248.

Variational Two Fermion Wave Equations in QED

Andrei G. Terekidi and Jurij W. Darewych

Department of Physics and Astronomy, York University, Toronto, M3J 1P3, Canada

Abstract. We consider a reformulation of QED in which covariant Green functions are used to solve for the electromagnetic field in terms of the fermion fields. A simple Fock-space variational trial state is used to derive relativistic two-fermion wave equations variationally. We require the trial state to be an eigenstate of the square of the total relativistic angular momentum operator, its projection and parity. It is shown that, in the case of different masses of the particle and the antiparticle, a mixing of single and triplet states takes place. For small coupling constants the fine structure is calculated analytically up to fourth order for all states and mass ratios.

INTRODUCTION

The Hamiltonian formalism of QFT is based on the rest-frame eigenvalue equation

$$\hat{H} | \psi \rangle = M | \psi \rangle, \qquad (1)$$

where the invariant mass M includes the masses of the constituent particles, and their kinetic and potential energies. There are very few problem for which exact solution of this equation can be obtained. In practice, it is necessary to use approximation methods, such as the widely-used covariant perturbation theory or lattice methods. However, in this work, we shall use the variational approach to describe two fermion bound states in QED. This method is based on the variational principle

$$\delta \langle \psi | \hat{H} - E | \psi \rangle_{t=0} = 0, \qquad (2)$$

which is equivalent to the exact equation (1) for completely general variations of $|\psi\rangle$. Variational solutions are only as good as the trial states that are used. Thus, it is important that the trial states possess as many of the features of the exact solution as possible.

THE CLASSICAL LAGRANGIAN AND THE REFORMULATED HAMILTONIAN FOR QED

To apply the variational method for QED we start from the classical Lagrangian density of two fermion fields interacting electromagnetically

$$\begin{aligned}\mathscr{L} = & \ \overline{\psi}(x)\left(i\gamma^\mu \partial_\mu - m_1 - q\gamma^\mu A_\mu(x)\right)\psi(x) + \overline{\phi}(x)\left(i\gamma^\mu \partial_\mu - m_2 - Q\gamma^\mu A_\mu(x)\right)\phi(x) \\ & - 1/4(\partial_\alpha A_\beta(x) - \partial_\beta A_\alpha(x))(\partial^\alpha A^\beta(x) - \partial^\beta A^\alpha(x)).\end{aligned}$$

The corresponding Euler-Lagrange equations of motion are the coupled Dirac-Maxwell equations

$$(i\gamma^\mu \partial_\mu - m_1)\psi(x) = q\gamma^\mu A_\mu(x)\psi(x), \qquad (3)$$
$$(i\gamma^\mu \partial_\mu - m_2)\phi(x) = Q\gamma^\mu A_\mu(x)\phi(x), \qquad (4)$$

$$\partial_\mu \partial^\mu A^\nu(x) - \partial^\nu \partial_\mu A^\mu(x) = j^\nu(x), \qquad (5)$$
$$j^\nu(x) = q\overline{\psi}(x)\gamma^\nu \psi(x) + Q\overline{\phi}(x)\gamma^\nu \phi(x), \qquad (6)$$

where $j^\nu(x)$ is the source current of the Maxwell equation and is made up of two fermion particle currents. It has been pointed out in [1] that various models in QFT, including QED, can be reformulated, using mediating-field Green functions, into a form that is particularly convenient for variational calculations. We shall implement such an approach to QED in this work.

Equations (3)-(6) can be decoupled in part by using the well-known formal solution [2, 3] of the Maxwell equation (5), namely

$$A_\mu(x) = A_\mu^0(x) + \int d^4x' D_{\mu\nu}(x-x') j^\nu(x'), \qquad (7)$$

where $D_{\mu\nu}(x-x')$ is a Green function (or photon propagator in QFT terminology), defined by

$$\partial_\alpha \partial^\alpha D_{\mu\nu}(x-x') - \partial_\mu \partial^\alpha D_{\alpha\nu}(x-x') = g_{\mu\nu}\delta^4(x-x'), \qquad (8)$$

and $A_\mu^0(x)$ is a solution of the homogeneous (or "free field") equation (5) with $j^\mu(x) = 0$. We recall, in passing, that equation (8) does not define the covariant Green function $D_{\mu\nu}(x-x')$ uniquely, but boundary conditions based on physical consideration pin down the form of the Green function. Substitution of the formal solution (7) into equations (3)-(4) yields the "partly reduced" nonlinear, coupled Dirac equations for two fermion fields,

$$(i\gamma^\mu \partial_\mu - m_1)\psi(x) = q\gamma^\mu \left(A_\mu^0(x) + \int d^4x' D_{\mu\nu}(x-x') j^\nu(x')\right)\psi(x), \qquad (9)$$
$$(i\gamma^\mu \partial_\mu - m_2)\phi(x) = Q\gamma^\mu \left(A_\mu^0(x) + \int d^4x' D_{\mu\nu}(x-x') j^\nu(x')\right)\phi(x). \qquad (10)$$

The partly-reduced equations (9), (10) are derivable from the stationary action principle with the Lagrangian density

$$\mathcal{L}_R = \overline{\psi}(x)\left(i\gamma^\mu \partial_\mu - m_1 - q\gamma_\mu A_0^\mu(x)\right)\psi(x) + \overline{\phi}(x)\left(i\gamma^\mu \partial_\mu - m_2 - Q\gamma_\mu A_0^\mu(x)\right)\phi(x)$$
$$-\frac{1}{2}\int d^4x' j^\mu(x') D_{\mu\nu}(x-x') j^\nu(x) \qquad (11)$$

One can proceed to do conventional covariant perturbation theory using the reformulated QED Lagrangian (11). The interaction part of (11) has a somewhat modified structure from that of the usual formulation of QED. Thus, there are two interaction terms. The last term of (11) is a "current-current" interaction which contains the photon propagator sandwiched between the fermionic currents. As such, it corresponds to Feynman diagrams without external photon lines. The term containing A_0^μ corresponds to diagrams that cannot be generated by the "current-current" interaction term, such as those involving external (physical) photon lines. The Hamiltonian density corresponding to (11) is

$$\mathcal{H}_R = \psi^\dagger(x)\left(-i\vec{\alpha}\cdot\nabla+m_1\beta\right)\psi(x)+\phi^\dagger(x)\left(-i\vec{\alpha}\cdot\nabla+m_2\beta\right)\phi(x)$$
$$+\frac{1}{2}\int d^4x'\, j^\mu(x')D_{\mu\nu}(x-x')j^\nu(x), \tag{12}$$

where we suppressed the Hamiltonian density for the free $A_0^\mu(x)$ field. At this stage we suppose that the system is quantized, and the fields obey the usual equal-time commutation rules.

VARIATIONAL TWO-BODY EQUATIONS AND FINE STRUCTURE

For a system like μ^+e^-, the simplest Fock-space trial state that can be written down in the rest frame is

$$|\mu^+e^-\rangle = \sum_{s_1 s_2}\int d^3\mathbf{p}\, F_{s_1 s_2}(\mathbf{p})\, a_{s_1}^\dagger(\mathbf{p})\, B_{s_2}^\dagger(-\mathbf{p}), \tag{13}$$

where $F_{s_1 s_2}$ are four adjustable functions. The variational principle (2), with normal-ordered Hamiltonian (12) and trial state (13), leads to the following equation

$$\sum_{s_1 s_2}\int d^3\mathbf{p}\, (\omega_p+\Omega_p-E)\, F_{s_1 s_2}(\mathbf{p})\, \delta F_{s_1 s_2}^*(\mathbf{p}) \tag{14}$$

$$-\frac{m_1 m_2 c^4}{(2\pi)^3}\sum_{\sigma_1\sigma_2 s_1 s_2}\int \frac{d^3 p\, d^3 q}{\sqrt{\omega_p\omega_q\Omega_p\Omega_q}} F_{\sigma_1\sigma_2}(\mathbf{q})\,(-i)\,\mathcal{M}_{s_1 s_2 \sigma_1 \sigma_2}(\mathbf{p},\mathbf{q})\, \delta F_{s_1 s_2}^*(\mathbf{p}) = 0,$$

where $\omega_p = \sqrt{\mathbf{p}^2 c^2+m_1^2 c^4}$, $\Omega_p = \sqrt{\mathbf{p}^2 c^2+m_2^2 c^4}$, and $\mathcal{M}_{s_1 s_2 \sigma_1 \sigma_2}(\mathbf{p},\mathbf{q})$ is the invariant matrix element corresponding to one-photon exchange:

$$\mathcal{M}_{s_1 s_2 \sigma_1 \sigma_2}^{ope}(\mathbf{p},\mathbf{q}) = \bar{u}_{pm_1 s_1}(-iq\gamma^\mu)u_{qm_1\sigma_1}iD_{\mu\nu}(q-p)\bar{v}_{-qm_2\sigma_2}(-iQ\gamma^\nu)v_{-pm_2 s_2}.$$

For a system like positronium e^+e^- we obtain an additional "virtual-annihilation" term. Note that the $\mathcal{M}$-matrix appears naturally in this formalism. Only "tree-level" diagrams arise with the simple trial state (13).

For states of given J^P we require that the trial state (13) must be the eigenstate of the square of the total angular momentum operator (in relativistic reduction), its projection, and parity, namely

$$\begin{bmatrix}\hat{\mathbf{J}}^2\\ \hat{J}_3\\ \hat{\mathcal{P}}\end{bmatrix}|\mu^+e^-\rangle = \begin{bmatrix}J(J+1)\\ m_J\\ P\end{bmatrix}|\mu^+e^-\rangle. \tag{15}$$

We can write the functions $F_{s_1 s_2}(\mathbf{p})$ in the general form

$$F_{s_1 s_2}(\mathbf{p}) = \sum_{\ell_{s_1 s_2} m_{s_1 s_2}} f_{s_1 s_2}^{\ell_{s_1 s_2} m_{s_1 s_2}}(p) Y_{\ell_{s_1 s_2}}^{m_{s_1 s_2}}(\hat{\mathbf{p}}), \tag{16}$$

where $p = |\mathbf{p}|$, $Y_{\ell_{s_1 s_2}}^{m_{s_1 s_2}}(\hat{\mathbf{p}})$ are the usual spherical harmonics and $f_{s_1 s_2}^{\ell_{s_1 s_2} m_{s_1 s_2}}(p)$ are momentum-space radial functions. Substitution of (16) into (13) and then into (15) leads to two categories of states.

Category 1 $(\ell = J)$

The singlet states $\ell = J$ $(J \geq 0)$, $P = (-1)^{J+1}$.

In this case $\ell_{s_1 s_2} \equiv \ell = J$, $m_{11} = m_{22} = 0$ and $m_{12} = m_{21} = m_J$. The nonzero components of $F_{s_1 s_2}(\mathbf{p})$ are $F_{\uparrow\downarrow}(\mathbf{p}) \equiv F_{12}(\mathbf{p})$, $F_{\downarrow\uparrow}(\mathbf{p}) \equiv F_{21}(\mathbf{p})$, and they have the form

$$F_{s_1 s_2}(\mathbf{p}) = f_{s_1 s_2}^{J m_{s_1 s_2}}(p) Y_J^{m_J}(\hat{\mathbf{p}}) = C_{J m_J}^{(sgl) m_{s_1 s_2}} f^J(p) Y_J^{m_J}(\hat{\mathbf{p}}), \tag{17}$$

The corresponding state (13) has the form

$$|sgl\rangle = \sum_{s_1 s_2} C_{J m_J}^{(sgl) m_{s_1 s_2}} \int d^3 \mathbf{p}\, f^J(p) Y_J^{m_J}(\hat{\mathbf{p}}) a_{s_1}^\dagger(\mathbf{p}) B_{s_2}^\dagger(-\mathbf{p}) |0\rangle. \tag{18}$$

The triplet states $\ell = J$ $(J > 0)$, $P = (-1)^{J+1}$.

In this case $\ell_{s_1 s_2} \equiv \ell = J$, $m_{11} = m_J - 1$, $m_{12} = m_{21} = m_J$, $m_{22} = m_J + 1$, and

$$F_{s_1 s_2}(\mathbf{p}) = f_{s_1 s_2}^{J m_{s_1 s_2}}(p) Y_J^{m_J}(\hat{\mathbf{p}}) = C_{J m_J}^{(tr) m_{s_1 s_2}} f^J(p) Y_J^{m_{s_1 s_2}}(\hat{\mathbf{p}}). \tag{19}$$

The corresponding state is

$$|tr\rangle = \sum_{s_1 s_2} C_{J m_J}^{(tr) m_{s_1 s_2}} \int d^3 \mathbf{p}\, f^J(p) Y_J^{m_{s_1 s_2}}(\hat{\mathbf{p}}) a_{s_1}^\dagger(\mathbf{p}) B_{s_2}^\dagger(-\mathbf{p}) |0\rangle. \tag{20}$$

Here $C_{J m_J}^{(sgl) m_{s_1 s_2}}$ and $C_{J m_J}^{(tr) m_{s_1 s_2}}$ are the Clebsch-Gordan (C-G) coefficients for spin $S = 0$ and $S = 1$ respectively. However the states (18) and (20) do not diagonalize the expectation value of the Hamiltonian (12). The diagonalization can be achieved by the following linear combinations (corresponding to "quasi-singlet" and "quasi-triplet" states)

$$|\text{``sgl''}\rangle = \sqrt{\frac{1+\xi}{2}} |sgl\rangle - \sqrt{\frac{1-\xi}{2}} |tr\rangle,$$

$$|\text{``tr''}\rangle = \sqrt{\frac{1-\xi}{2}} |sgl\rangle + \sqrt{\frac{1+\xi}{2}} |tr\rangle,$$

where $\xi = \left[4\left((m_1 - m_2)/(m_1 + m_2)\right)^2 J(J+1) + 1 \right]^{-1/2}$. These states are realized for $J > 0$. The states with $J = 0$ (including the ground state) are purely singlet states (18).

It is easy to see from the form of ξ that for the positronium case ($m_1 = m_2$) the quasi states become the true singlet and triplet states.

Category 2 ($\ell = J \mp 1$)

In this case $\ell_{s_1 s_2} \equiv \ell = J \mp 1$, $P = (-1)^J$, $m_{11} = m_J - 1$, $m_{12} = m_{21} = m_J$, $m_{22} = m_J + 1$, and

$$F_{s_1 s_2}(\mathbf{p}) = C^{m_{s_1 s_2}}_{J-1 m_J} f^{J-1}(p) Y^{m_{s_1 s_2}}_{J-1}(\hat{\mathbf{p}}) + C^{m_{s_1 s_2}}_{J+1 m_J} f^{J+1}(p) Y^{m_{s_1 s_2}}_{J+1}(\hat{\mathbf{p}}), \tag{21}$$

where $C^{m_{s_1 s_2}}_{J-1 m_J}$, $C^{m_{s_1 s_2}}_{J+1 m_J}$ are the C-G coefficients for spin $S = 1$.

We now return to equation (14) and replace the functions $F_{s_1 s_2}(\mathbf{p})$ by expressions (17) and (19). After completing the variational procedure we get the following relativistic two fermion radial equations in momentum space:

The quasi singlet and quasi triplet states $\ell = J$

$$(\omega_p + \Omega_p - E) f^J(p) = \frac{m_1 m_2 c^4}{(2\pi)^3} \int \frac{q^2 dq}{\sqrt{\omega_p \omega_q \Omega_p \Omega_q}} \mathcal{K}^{(\text{"sgl"},\text{"tr"})}(p,q) f^J(q), \tag{22}$$

where $\mathcal{K}^{(\text{"sgl"},\text{"tr"})}$ are the kernels

$$\mathcal{K}^{(\text{"sgl"},\text{"tr"})} = \sum_{s_1 s_2 \sigma_1 \sigma_2 m_J} \int d\hat{\mathbf{p}} \, d\hat{\mathbf{q}} \, C^{s_1 s_2 \sigma_1 \sigma_2}_{J m_J} (-i) \mathcal{M}_{s_1 s_2 \sigma_1 \sigma_2}(\mathbf{p},\mathbf{q}) Y^{m_J *}_J(\hat{\mathbf{p}}) Y^{m_J}_J(\hat{\mathbf{q}}).$$

The kernels are defined by invariant $\mathcal{M}$-matrices and the coefficients $C^{s_1 s_2 \sigma_1 \sigma_2}_{J m_J}$, which can be expressed through the C-G coefficients and the parameter ξ.

The triplet states $\ell = J \mp 1$

Substitution (21) into (14) leads to coupled equations for $f^{J-1}(p)$ and $f^{J+1}(p)$, which are conveniently written in matrix form,

$$(\omega_p + \Omega_p - E) \mathbb{F}(p) = \frac{m_1 m_2 c^4}{(2\pi)^3} \int \frac{dq \, q^2}{\sqrt{\omega_p \omega_q \Omega_p \Omega_q}} \mathbb{K}(p,q) \mathbb{F}(p), \tag{23}$$

where

$$\mathbb{F}(p) = \begin{bmatrix} f^{J-1}(p) \\ f^{J+1}(p) \end{bmatrix}, \quad \mathbb{K}(p,q) = \begin{bmatrix} \mathcal{K}_{11}(p,q) & \mathcal{K}_{12}(p,q) \\ \mathcal{K}_{21}(p,q) & \mathcal{K}_{22}(p,q) \end{bmatrix}$$

The form of the kernels $\mathcal{K}_{ij}$ is similar to those of the the previous cases.

In the nonrelativistic limit, equations (22) and (23) give the usual time- independent two particle radial Schrödinger equation in momentum space.

We used non-relativistic hydrogen-like radial functions to evaluate the expectation value of the energy E in equations (22) and (23), and thus to calculate perturbatively the relativistic energy corrections with accuracy $\sim \alpha^4$. The results are presented in the form $\Delta \varepsilon = E - \alpha^2 \mu c^2 / 2n^2 - Mc^2$, where $\alpha = qQ/c$ ($\hbar = 1$), $\mu = m_1 m_2 / (m_1 + m_2)$, $M = m_1 + m_2$.

For the *singlet state* $\ell = J = 0$ (which includes the ground state) we obtain

$$\Delta \varepsilon = -\frac{\alpha^4 \mu c^2}{n^3} \left[\frac{1}{2} + \frac{2\mu}{M} - \frac{1}{8n} \left(3 - \frac{\mu}{M} \right) \right], \tag{24}$$

while for the *quasi-singlet and quasi-triplet states* $\ell = J$ $(J > 0)$ the result is

$$\Delta \varepsilon_J^{(\mp)} = -\frac{\alpha^4 \mu c^2}{n^3} \left[\frac{1}{2J+1} \left(1 + \frac{1 \mp \xi^{-1}}{4J(J+1)} \right) - \frac{1}{8n} \left(3 - \frac{\mu}{M} \right) \right], \qquad (25)$$

where upper and lower signs correspond to quasi-singlet and quasi-triplet states respectively.

For the *triplet states* $\ell = J - 1$ $(J \geq 1)$ we have

$$\Delta \varepsilon_{J-1}^{tr} = -\frac{\alpha^4 \mu c^2}{n^3} \left[\frac{1}{2J-1} \left(1 - \frac{1}{2J} - \frac{2\mu}{M} \frac{1}{2J+1} \right) - \frac{1}{8n} \left(3 - \frac{\mu}{M} \right) \right], \qquad (26)$$

and for the *triplet states* $\ell = J + 1$ $(J \geq 0)$

$$\Delta \varepsilon_{J+1}^{tr} = -\frac{\alpha^4 \mu c^2}{n^3} \left[\frac{1}{2J+3} \left(1 + \frac{1}{2(J+1)} + \frac{2\mu}{M} \frac{1}{2J+1} \right) - \frac{1}{8n} \left(3 - \frac{\mu}{M} \right) \right]. \qquad (27)$$

Our formulas (24)-(27) for the relativistic energy corrections are valid to $O(\alpha^4)$ for arbitrary masses and for all nJ^P states. They reduce to the correct results for two well-known particular cases: When one of the masses is infinite, we get the Dirac equation result. If the masses are equal, we get the well known results for positronium (minus the virtual-annihilation term, which we have not written out here).

Our results for arbitrary masses agree with the quasipotential Bethe-Salpeter-based calculations of J. Connell [4], and the calculations of H. Hersbach [5], based on a relativistic formalism due to de Groot and Ruijgrok.

REFERENCES

1. Darewych J. W., *Annales Fond. L. de Broglie (Paris)* **23**, 15 (1998); Darewych J. W., *Causality and Locality in Modern Physics*, G. Hunter et al. (eds.), Kluwer, Dordrecht, pp. 333-344, (1998).
2. Jackson J. D., *Classical Electrodynamics*, John Wiley, New York, (1975).
3. Barut A. O., *Electrodynamics and Classical Theory of Fields and Particles*, Dover, New York, (1980).
4. Connell John H., *Phys. Rev. D* **43**, pp. 1393-1402 (1991).
5. Hersbach H., *Phys. Rev. A* **46**, pp. 3657-3670 (1992).

Breaking of Supersymmetry in a U(1) Model with Stueckelberg Fields

S.V.Kuzmin* and D.G.C.McKeon[†]

Department of Physics and Astronomy, University of Western Ontario, London,Canada N6A 3K7
[†]*Department of Applied Mathematics, University of Western Ontario, London,Canada N6A 5B7*

Abstract. We consider a superfield model in which interactions exist between a chiral matter field, a real U(1) vector superfield and a novel auxiliary chiral field that has many features of the standard Stueckelberg field. This extra field affords the opportunity of having F-term breaking of supersymmetry as well as providing a Stueckelberg mass to the vector field. An unconventional gauge choice leads to a model in terms of component fields that has both an attractive mass spectrum and interesting interactions. The main obstacle for having a realistic mass spectrum in the usual broken supersymmetric QED is the presence of a goldstino (massless fermion) and generation of a mass for the vector field through the Higgs mechanism. In our model fermionic degrees of freedom give rise to two (well separated) massive Dirac fields and an arbitrarily small vector mass generated by two contributions that compensate for each other: a Stueckelberg and Higgs mechanism.

INTRODUCTION

Almost immediately after discovery of supersymmetry (SUSY), supersymmetric quantum electrodynamics (QED) with combined SUSY and U(1) invariance was proposed [1]. In exact supersymmetric models, fermions and bosons have degenerate masses. Therefore, in order to realize supersymmetry in particle physics, we must consider models in which supersymmetry is broken. The spontaneous breaking of SUSY QED was considered by Fayet and Iliopoulos [2], but the resulting model has no phenomenological interest, because of its non-physical mass spectrum having a massive vector and a massless spinor (Goldstino).

The mechanism for spontaneous breaking of SUSY QED in the standard approach of ref. [2] is the usual Higgs mechanism [3] for creating a mass for the vector field while preserving the initial masslessness of its Fermionic partner. Here we explore a different way of breaking SUSY that might be considered as being the "opposite" of the usual mechanism. We will start from a massive gauge superfield by using a SUSY generalization of the Stueckelberg approach [4]. The supersymmetric partner of the massive vector field is now a massive Fermion and spontaneous symmetry breaking does not convert the massive Fermion into a massless one (i.e. - no Goldstino). We will also show the possibility of constructing a Lagrangian that allows for an "anti-Higgs" mechanism of spontaneous symmetry breaking: the non-zero VEV of scalar fields creates a tachyonic contribution to the vector mass which compensates for the bare Stueckelberg mass.

THE MODEL

The Wess-Zumino action for SUSY QED [1] with a Fayet-Iliopoulos term [2], written in superfield form [5] is

$$I_{WZ} = \int d^4x d^2\theta d^2\bar{\theta} \left[\Phi_1 e^{+eV}\bar{\Phi}_1 + \Phi_2 e^{-eV}\bar{\Phi}_2\right] \\ + q\int d^4x d^2\theta \Phi_1\Phi_2 + q\int d^4x d^2\bar{\theta}\bar{\Phi}_1\bar{\Phi}_2 \\ + \frac{1}{4}\int d^4x d^2\theta W^\alpha W_\alpha + \frac{1}{4}\int d^4x d^2\bar{\theta}\bar{W}_\alpha \bar{W}^\alpha + \kappa\int d^4x d^2\theta d^2\bar{\theta} V \quad (1)$$

The two chiral matter superfields Φ_1 and Φ_2 have equal and opposite charges, V is a real gauge superfield, κ is the Fayet-Iliopoulos (FI) constant and W_α is a chiral spinor superfield $W_\alpha = -\frac{1}{4}\bar{D}^2 D_\alpha V$. I_{WZ} is obviously both SUSY invariant and invariant under the $U(1)$ transformation $V \to V + i(\Lambda - \bar{\Lambda})$, $\Phi_1 \to \Phi_1 e^{-ie\Lambda}$ and $\Phi_2 \to \Phi_2 e^{+ie\Lambda}$ where Λ is a chiral gauge superfield.

The action of Eq.(1) contains all renormalizable gauge invariant interactions that can be built using gauge and matter superfields. When generalizing this model, we add new fields in a way consistent with $U(1)$ symmetry [4]. We introduce a chiral Stueckelberg superfield S [6] so that the transformation

$$S \to S + m\Lambda \quad (2)$$

leaves the action invariant. (Here m is a Stueckelberg mass.) The invariant combination $V - i\frac{1}{m}(S - \bar{S})$ leads to a FI term that is manifestly gauge invariant (not just invariant up to a surface term). It also gives rise to a $U(1)$ invariant mass term in the action

$$m^2\left(V - i\frac{1}{m}(S - \bar{S})\right)^2. \quad (3)$$

S can also be coupled to the matter fields Φ_1 and Φ_2 in a $U(1)$ invariant fashion

$$\Phi_1 e^{i\frac{e}{m}S} + h.c \quad \text{and} \quad \Phi_2 e^{-i\frac{e}{m}S} + h.c.. \quad (4)$$

leading to so-called F-breaking terms. These extra interactions make SUSY QED richer, but they cannot eliminate the vector mass as the sign of the kinetic terms for the scalar fields appearing in both matter (1) and Stueckelberg superfields (3) are restricted, leading to positive contributions to the vector mass from both the Stueckelberg mass term and from spontaneous symmetry breaking.

Only interactions that do not have a fixed overall sign can possibly compensate for the vector mass arising from (1) and (3). These can be generated by

$$\Phi_1 e^{eV} e^{i\frac{e}{m}\bar{S}} + h.c. \quad \text{and} \quad \Phi_2 e^{-eV} e^{-i\frac{e}{m}\bar{S}} + h.c.. \quad (5)$$

The standard analysis of the model of Eq.(1) is based on a special gauge choice, the Wess-Zumino gauge [1], in which ($C = M = \chi = 0$). This gauge eliminates the non-polynomial character of the Lagrangian. With the extra interactions of (4-5), neither the Wess-Zumino gauge nor the gauge ($S = 0$) will eliminate non-polynomial interactions arising in Eqs.(1,4,5). The Wess-Zumino action is non-polynomial only in the $\theta, \bar{\theta}$ - independent components of V (the field C) as any power series in θ or $\bar{\theta}$ terminates. The conclusion that the component field Lagrangian will be polynomial if and only if we choose $C = 0$, can be avoided. In order to eliminate e^{eC} we call any component field appearing in Φ_1 f_1, and note following structure

$$e^{+eC} \times \hat{P}(C, \chi, \bar{\chi}, M, \bar{M}, V_\mu, \lambda, \bar{\lambda}, D, \partial_\mu, \partial^2) f_1 \bar{f}_1 \tag{6}$$

arising from the term $\Phi_1 e^{+eV} \bar{\Phi}_1$, where $\hat{P}$ is a polynomial. The simple field redefinition

$$e^{+e\frac{C}{2}} f_1 = \tilde{f}_1, \quad e^{e\frac{C}{2}} \partial_\mu f_1 = \partial_\mu \tilde{f}_1 - \frac{e}{2} \tilde{f}_1 \partial_\mu C, \quad etc. \tag{7}$$

will eliminate non-polynomiality in (6) without the need for any gauge fixing.

For Φ_2 equations similar to (6-7) hold with the replacements $1 \to 2$ and $e \to -e$. The mass term in (1) is invariant with respect to the redefinition of (7). We see therefore that the Wess-Zumino action can be written in a polynomial form without gauge fixing so that $C = 0$. In terms of component fields C appears only in a combination with D which permits its elimination

$$\left(D + \frac{1}{4} \partial^2 C \right) \equiv \tilde{D}. \tag{8}$$

With this, we have a polynomial Lagrangian without the field C (before gauge fixing). The component field Lagrangian for the WZ QED model in both the Wess-Zumino gauge ($C = M = \chi = 0$) and after the field redefinitions (7,8) combined with the gauge conditions $M = \chi = 0$, are identical. The consequences of these two distinct approaches are hidden in case of WZ QED because ImA_g disappears from the transformations of the new fields. With extra fields in theory, this hidden gauge freedom reappears in new interactions and is used to fix these extra fields.

The interaction terms of Eqs.(4,5) can potentially give rise to non-polynomial interactions. A gauge that preserves the field redefinition (7) and non-polynomiality is

$$ReA_s = 0 \quad and \quad ImA_s = -\frac{m}{2} C. \tag{9}$$

Our gauge is now completely specified. With the usual gauge fixing, only the second power of the vector superfield survives. In our gauge, up to the fourth power of $V - i\frac{1}{m}(S - \bar{S})$ survives. If $n = 1$, we obtain the FI term; $n = 2$ is the Stueckelberg mass term. For $n = 3$ we have

$$\kappa_3 (\bar{F}_s \psi_s \psi_s + F_s \bar{\psi}_s \bar{\psi}_s + m W_\mu \psi_s \sigma^\mu \bar{\psi}_s) \tag{10}$$

and $n = 4$ results in a four-Fermi interaction

$$\kappa_4 \psi_s \psi_s \bar{\psi}_s \bar{\psi}_s \tag{11}$$

Eq.(11) describes a non-renormalizable interaction, but can be used as a "compensator" if such an interaction is produced after elimination of auxiliary fields.

The Lagrangian we are considering, now consists of four parts. First, the Lagrangian of Eq. (1) is identical to the Lagrangian for SUSY QED in the usual Wess-Zumino gauge,

$$\begin{aligned}L_{WZ} =& -\tfrac{1}{4}F_{\mu\nu}F^{\mu\nu} - i\tfrac{1}{2}\lambda\sigma^\mu\partial_\mu\bar\lambda + i\tfrac{1}{2}\partial_\mu\lambda\sigma^\mu\bar\lambda + 2D^2 \\ &+ \{-\tfrac{1}{4}A_1\partial^2\bar A_1 - \tfrac{1}{4}\bar A_1\partial^2 A_1 + \tfrac{1}{2}\partial_\mu A_1\partial^\mu\bar A_1 - i\tfrac{1}{4}\psi_1\sigma^\mu\partial_\mu\bar\psi_1 + i\tfrac{1}{4}\partial_\mu\psi_1\sigma^\mu\bar\psi_1 \\ &+ F_1\bar F_1 + eA_1\bar A_1 D + \tfrac{1}{4}e^2 A_1\bar A_1 W_\mu W^\mu - \tfrac{1}{2}eA_1\bar\psi_1\bar\lambda - \tfrac{1}{2}e\bar A_1\psi_1\lambda \\ &+ ie\tfrac{1}{2}W_\mu(\bar A_1\partial^\mu A_1 - A_1\partial^\mu\bar A_1) + \tfrac{1}{4}eW_\mu\psi_1\sigma^\mu\bar\psi_1\} + \{1\to 2, e\to -e\} \\ &+ q(A_1 F_2 + \bar A_1\bar F_2 + A_2 F_1 + \bar A_2\bar F_1 - \tfrac{1}{2}\psi_1\psi_2 - \tfrac{1}{2}\bar\psi_1\bar\psi_2) + \kappa D \end{aligned} \qquad (12)$$

Next, there is a Stueckelberg mass term

$$\begin{aligned}L_{St} =& m^2\tfrac{1}{2}\int d^2\theta d^2\bar\theta\left(V - i\tfrac{1}{m}(S-\bar S)\right)^2 = \\ & F_s\bar F_s - \tfrac{1}{2}m\psi_s\lambda - \tfrac{1}{2}m\bar\psi_s\bar\lambda - i\tfrac{1}{4}\psi_s\sigma^\mu\partial_\mu\bar\psi_s + i\tfrac{1}{4}\partial_\mu\psi_s\sigma^\mu\bar\psi_s + \tfrac{1}{4}m^2 W_\mu W^\mu\end{aligned} \qquad (13)$$

Thirdly, there is the interaction of the Stueckelberg superfield with the second matter superfield Φ_2

$$p\int d^2\theta\Phi_2 e^{-i\frac{e}{m}S} + h.c. = p\left(F_2 + \frac{e}{m}A_2 F_s - \frac{1}{2}\frac{e}{m}\psi_2\psi_s - \frac{1}{4}\frac{e^2}{m^2}A_2\psi_s\psi_s + h.c.\right) \qquad (14)$$

Finally, there is the interaction between Φ_1, V and S

$$\begin{aligned}g\int d^2\theta d^2\bar\theta\Phi_1 e^{eV} e^{i\frac{e}{m}\bar S} + h.c. =& g\left(eA_1 D + i\tfrac{1}{2}e\partial_\nu A_1 W^\nu + \tfrac{1}{4}e^2 A_1 W_\nu W^\nu\right. \\ &+ \tfrac{1}{2}\tfrac{e^2}{m}\bar\psi_s\bar\lambda A_1 - \tfrac{1}{2}e\psi_1\lambda - i\tfrac{1}{4}\tfrac{e}{m}\psi_1\sigma^\mu\partial_\mu\bar\psi_s \\ &+ \tfrac{1}{4}\tfrac{e^2}{m}\psi_1\sigma^\nu\bar\psi_s W_\nu + i\tfrac{1}{4}\tfrac{e}{m}\partial_\mu\psi_1\sigma^\mu\bar\psi_s + \tfrac{e}{m}F_1\bar F_s - \tfrac{1}{4}\tfrac{e^2}{m^2}F_1\bar\psi_s\bar\psi_s + h.c.\Big) \end{aligned} \qquad (15)$$

(The substitutions $F_s \to iF_s$ and $\psi_s \to i\psi_s$ have been made). In Eqs. (14) and (15) we use only one of the matter fields in order to simplify calculations. The interaction of Eqs.(10,11) could also occur.

Upon eliminating the auxiliary fields, we obtain the scalar potential U, some Yukawa contributions and a four-Fermi interaction which can be cancelled by adding a term of the form of Eq. (11).

Finding the extremum of the potential U is cumbersome. To simplify the analysis, we impose $p = qg$. This reduces the amount of work required to analyze U, but does not destroy the attractive features of our model. Now the form of U is

$$U = \frac{1}{8}\left(\kappa + eA_1^2 + eB_1^2 - eA_2^2 - eB_2^2 + 2egA_1\right)^2 + q^2(A_1 + g)^2 + q^2(B_1^2 + A_2^2 + B_2^2) \qquad (16)$$

where now A_i and B_i are the real and imaginary parts of the complex scalar fields. Its minimum is determined by

$$\frac{\partial U}{\partial A_i} = \frac{\partial U}{\partial B_i} = 0 \qquad (17)$$

If $2q^2 > \left|\tfrac{e}{2}(\kappa - eg^2)\right|$, then $B_{10} = A_{20} = B_{20} = 0$ and $A_{10} = -g$ is the only possible solution and U has the minimum value $U(-g, 0, 0, 0) = \tfrac{1}{8}(\kappa - eg^2)^2 > 0$ and consequently we have both SUSY and $U(1)$ breaking. The mass matrix for the scalar fields is automatically diagonal and gives the following masses for the scalar fields

$$m_{A_1}^2 = m_{B_1}^2 = \frac{e}{4}(\kappa - eg^2) + q^2, \quad m_{A_2}^2 = m_{B_2}^2 = -\frac{e}{4}(\kappa - eg^2) + q^2 \qquad (18)$$

Note that we can have mass splitting for scalar fields even without the FI-term ($\kappa = 0$). The FI mechanism is not necessary for symmetry breaking in our model.

In matrix notation, the kinetic and mass parts of the spinor sector of our model (with the relation $p = gq$ and after rescaling $\lambda \to \lambda/\sqrt{2}$) are

$$L^{kin} = -i\frac{1}{4}\mathbf{v}^T \mathbf{K} \sigma^\mu \partial_\mu \bar{\mathbf{v}} + h.c., \quad L^{mass} = -\frac{1}{4}\mathbf{v}^T \mathbf{U} \mathbf{v} + h.c. \qquad (19)$$

where $\mathbf{v}^T = (\psi_1, \psi_2, \lambda, \psi_s)$ and

$$\mathbf{K} = \begin{vmatrix} 1 & 0 & 0 & a \\ 0 & 1 & 0 & 0 \\ 0 & 0 & 1 & 0 \\ a & 0 & 0 & 1 \end{vmatrix}, \quad \mathbf{U} = \begin{vmatrix} 0 & b & 0 & 0 \\ b & 0 & 0 & ab \\ 0 & 0 & 0 & c \\ 0 & ab & c & 0 \end{vmatrix} \qquad (20)$$

with

$$a = \frac{ge}{m}, \quad b = q, \quad c = \frac{m}{\sqrt{2}}(1+a^2). \qquad (21)$$

We now set $\mathbf{v} = \mathbf{tw}$ with $\mathbf{t}^T \mathbf{K} \mathbf{t} = \mathbf{I}$ and $\mathbf{t}^T \mathbf{U} \mathbf{t} = \mathbf{M}$. Here $\mathbf{I}$ is the identity matrix and $\mathbf{M}$ is a diagonal mass matrix, whose elements are the roots of the equation $|\mathbf{U} - x\mathbf{K}| = 0$ which has two pairs of solutions, $x_{1,2} = \pm m_+, x_{3,4} = \pm m_-$ (fermionic masses), where

$$m_\pm^2 = \frac{c^2 + b^2(1-a^2) \pm \left[(c^2 + b^2(1-a^2))^2 - 4(1-a^2)b^2 c^2\right]^{\frac{1}{2}}}{2(1-a^2)}. \qquad (22)$$

The condition for the vector field to be almost massless is

$$m_W^2 = \frac{1}{2}(m^2 - e^2 g^2) = \frac{1}{2}m^2(1-a^2) \to 0^+ \qquad (23)$$

If $a \to 1^-$, then $m_\pm^2 > 0$ and all solutions to Eq. (22) are real, justifying use the transformation $\mathbf{t}$ on $\mathbf{v}$. Moreover, in this limit, there can be a large mass splitting between two fermions.

$$m_+^2 = \frac{c^2}{(1-a^2)} = \frac{m^2(1+a^2)^2}{2(1-a^2)}, \quad m_-^2 = b^2 = q^2 \qquad (24)$$

If we choose our parameters so that

$$q^2 >> \frac{m^2(1+a^2)^2}{2(1-a^2)}, \quad m^2 \simeq (1-a^2) << 1 \qquad (25)$$

we have the following mass spectrum for the model

$$m_W \cong 0^+ << m_+ << m_{A_1} = m_{B_1} < m_- < m_{A_2} = m_{B_2}. \qquad (26)$$

DISCUSSION

It had been shown [7] that in a very general class of SUSY theories with supersymmetry breaking (spontaneously or possibly in certain explicit ways) the following mass formula holds at tree level

$$\sum_J (-1)^{2J} (2J+1) m_J^2 = 0 \tag{27}$$

where m_J is the mass associated with a field of spin J. This relation between masses often serves as a proof that spontaneous breaking of supersymmetry does not lead to an acceptable particle spectrum and that explicit, so-called soft breaking [8], is needed in order to modify (27). In our model this relation does not hold (see eqs.(18,22,23)).

The mass formula (27) was derived in ref.[7] in general form only for arbitrary interaction of chiral multiplets but for gauge theories a model-independent proof does not exist and the demonstration of (27) heavily based on a particular model, namely that of ref. [2]. This model leads to the vanishing of the determinant of the fermionic mass matrix which implies the existence of a massless fermion, which is the Goldstino, in the theory. In our model of eqs.(12-15), the determinant of fermionic mass matrix (20) is not zero. In our case this is not a necessary condition for Fermions to be massless as cross terms occur in the kinetic part and both of the matrices have to be diagonalized simultaneously. Possibly, some relations exist between the validity of eq.(27) and the presence of a Goldstino in gauge models, but in our model it appears as these conditions are not satisfied.

We have not attempted to find a phenomenological application of this particular (very simple) model. It is rather meant to be a demonstration of a new mechanism for breaking SUSY that has the attractive feature of having no massless Goldstino and a vector with arbitrarily small mass. We believe this deserves further investigation. The most crucial question is to consider the possibility of a total elimination of the vector mass and the renormalizability of this model. This will be reported separately.

We acknowledge the support provided by NSERC of Canada.

REFERENCES

1. Wess, J., and Zumino, B., *Nucl.Phys.* **B78**, 1 (1974).
2. Fayet, P., and Iliopoulos, J., *Phys.Lett.* **B51**, 461 (1974).
3. Higgs, P.W., *Phys.Lett.* **12**, 132 (1964); *Phys.Rev.Lett.* **13**, 508 (1964); *Phys.Rev.* **145**, 1156 (1966).
4. Stueckelberg, E.C.G., *Helv.Phys.Acta* **11**, 225, 299 (1938); **30**, 209 (1957).
5. Salam. A., and Strathdee, J., *Nucl.Phys.* **B76**, 477 (1974).
6. Delbourgo, R., *J.Phys.* **G1**, 800 (1975).
7. Ferrara, S., Girardello, L., and Palumbo, F., *Phys.Rev.* **D20**, 403 (1979).
8. Girardello, L., and Grisaru, M.T., *Nucl.Phys.* **194B**, 65 (1982).

Self Energy of Chiral Leptons in a Background Hypermagnetic Field

J. Cannellos*, E. J. Ferrer† and V. de la Incera†

* Physics Department, SUNY at Buffalo, Buffalo, New York 14260, USA
† Physics Department, SUNY-Fredonia, Fredonia, New York 14063, USA

Abstract. We consider the effects of a constant hypermagnetic field on the finite-temperature dispersion relations of leptons in the chiral phase of the electroweak model. In the strong field approximation, for which fermions are mainly confined to the lowest Landau level (*LLL*), we find that chiral-lepton propagation becomes maximally anisotropic and aligned to the direction of the external field, and that the external hypermagnetic field counteracts the appearance of the thermal "mass" that would be created in its absence.

HYPERMAGNETIC FIELDS IN THE *EW* MODEL

The origin of large-scale magnetic fields found in galaxies, galactic halos and clusters of galaxies [1] remains an open question, but compelling arguments favor their primordial nature. Strong primordial magnetic fields on the order of 10^{24} G at the electroweak (*EW*) scale can be generated during the *EW*-phase transition [2]. However, only its $U(1)$ gauge component, the hypermagnetic field, could exist in the symmetric *EW* phase, since its non-Abelian component would decay by acquiring an infrared magnetic mass $\sim g^2 T$ at high temperatures [3].[1] In this work, we consider the effects of strong primordial hypermagnetic fields on the propagation of chiral leptons in the hot plasma that existed before the *EW*-phase transition. With this aim in mind, we derive the self-energies of chiral leptons in the one-loop approximation, and then use them to solve the dispersion relations for the left and right-handed leptons.

ONE-LOOP LEPTON SELF ENERGIES

In the presence of an external hypermagnetic field along the $\mathscr{OZ}$-axis, the chiral fermion propagators, $S_{l_{R(L)}}$, obey the following equations:

$$\delta^4(x-y) = \left[\left(\gamma^\mu \Pi_\mu^{R(L)} - \Sigma_{l_{R(L)}}\right) S_{l_{R(L)}}\right](x-y), \tag{1}$$

---

[1] Abelian and non-Abelian *electric* fields will be Debye screened by thermal effects, producing a short-range decay for both fields [4].

where chiral-lepton[2] self energies are $\Sigma_{l_{R(L)}}(x,y)$, $i\Pi_\mu^{R(L)}(x) = \partial_\mu + ig^{R(L)}B_\mu^{ext}(x)$, with $B_\mu^{ext}(x) = Hx_1 g_{\mu 2}$ the external hyperpotential, and where right (R) and left (L)-handed fermion $U(1)$ coupling constants are defined as

$$g^R \equiv g', \quad \text{and} \quad g^L \equiv g'/2. \tag{2}$$

The one-loop chiral Schwinger-Dyson equations for Σ are

$$\Sigma_{l_L}(x,y) = \Delta_{l_L}^{(B,l_L)}(x,y) + \Delta_{l_L}^{(W^{3,\pm},l_L)}(x,y), \quad \text{and} \quad \Sigma_{e_R}(x,v) = \Delta_{e_R}^{(B,e_R)}(x,y), \tag{3}$$

where the $U(1)$ and the neutral and charged $SU(2)$ gauge-boson fields are, respectively, B, W^3 and $W^\pm$. The one-loop bubble graphs are[3]

$$\Delta_{l_L}^{(B,l_L)}(x,y) \equiv i\frac{g'^2}{4} R\gamma^\mu D_{\mu\nu}^{(B)}(x-y) S_{l_L}(x,y) \gamma^\nu L,$$

$$\Delta_{l_L}^{(W^3,l_L)}(x,y) \equiv i\frac{g^2}{4} R\gamma^\mu D_{\mu\nu}^{(W^3)}(x-y) S_{l_L}(x,y) \gamma^\nu L,$$

$$\Delta_{l_L}^{(W^\pm,l_L)}(x,y) \equiv i\frac{g^2}{2} R\gamma^\mu D_{\mu\nu}^{(W^\pm)}(x-y) S_{l_L}(x,y) \gamma^\nu L,$$

$$\Delta_{e_R}^{(B,e_R)}(x,y) \equiv ig'^2 L\gamma^\mu D_{\mu\nu}^{(B)}(x-y) S_{e_R}(x,y) \gamma^\nu R. \tag{4}$$

RITUS-TRANSFORMATION TO MOMENTUM SPACE

Since neither gauge fields B nor W interact with the hypermagnetic field, their propagators can be diagonalized in momentum space by Fourier transformation. All gauge-boson propagators have generic form,

$$D_{\mu\nu}(x-y) = \int \frac{d^4q}{(2\pi)^4} \frac{e^{iq\cdot(x-y)}}{q^2 - i\varepsilon} \left(g_{\mu\nu} - (1-\xi)\frac{q_\mu q_\nu}{q^2} \right). \tag{5}$$

where ξ is the proper gauge-fixing parameter.

Fermion propagators, $S_{l_{R(L)}}$, however, depend on the hypermagnetic field and cannot be Fourier transformed. Instead, we use Ritus' method of E_p functions [5], which are the eigenfunctions of the hypercharged particles in the background field, to diagonalize the fermion Green's function and self energy[4]. We extend Ritus' method for electrons in a

---

[2] In symmetric EW theory, the left-handed leptons are indistinguishable, and indicated generically by $l_L \in \{v_L, e_L\}$. The right-handed lepton singlet, $l_R = e_R$.

[3] We neglect graphs with scalar-bosons, ϕ, i.e., $\Delta_{l_L}^{(\phi,e_R)}(x,y) \equiv iG_e^2 D_\phi(x,y) G_{e_R}(x,y)$ and $\Delta_{e_R}^{(\phi,l_L)}(x,y) \equiv iG_e^2 G_{l_L}(x,y) D_\phi(x,y)$, since $G_e^2 \ll g^2, g'^2$. Moreover, we do not consider the contribution of tadpole diagrams, since the early universe, unlike the dense stellar medium, is almost charge symmetric ($\mu = 0$), with a particle-antiparticle asymmetry of $\sim 10^{-10} - 10^{-9}$.

[4] Ritus' method was originally developed for spin-1/2 charged particles [5] and it has been recently extended to spin-1 charged particles, i.e., to the W boson [6].

constant electromagnetic field to chiral fermions in a constant hypermagnetic field in the $\mathscr{OZ}$-direction, with gauge potential, $B_\mu^{ext} = (0,0,Hx_1,0)$. The chiral-lepton propagators are given by

$$S_{l_{R(L)}}(x,y) = \sum_{l''} \int \frac{dk_0 dk_2 dk_3}{(2\pi)^4} E_k^{R(L)}(x) \frac{1}{\gamma \cdot \bar{k}_{l_{R(L)}}} \bar{E}_k^{R(L)}(y), \quad \text{where} \tag{6}$$

$$E_k^{R(L)}(x) = e^{i(k_0 x^0 + k_2 x^2 + k_3 x^3)} \sum_\sigma N^{R(L)}(n(\sigma)) D_n\left[\rho^{R(L)}(x_1)\right] \Delta(\sigma). \tag{7}$$

Above, $\rho^{R(L)}(x_1) = \sqrt{2|g^{R(L)}H|}\left(x_1 - k_2/g^{R(L)}H\right)$, normalization factors are $N^{R(L)}(n(\sigma)) = \left(4\pi g^{R(L)}H\right)^{1/4}/\sqrt{n(\sigma)!}$, spin-projection matrices are $\Delta(\sigma) = diag(\delta_{\sigma 1}, \delta_{\sigma -1}, \delta_{\sigma 1}, \delta_{\sigma -1})$, $\sigma = \pm 1$, the adjoint matrix is $\bar{E}_k^{R(L)} = \gamma^0 (E_k^{R(L)})^\dagger \gamma^0$, and $D_n(\rho)$ is the parabolic-cylinder function of index n. The diagonalized propagators in Eq. (6) are, then, functions of the eigenvalues,

$$\left(\bar{k}_{l_{R(L)}}\right)_\mu = (k_0, 0, -sgn(g'H)\sqrt{2|g^{R(L)}H|l''}, k_3), \quad l'' = 0, 1, 2, \ldots, \tag{8}$$

where Landau levels, l'', $n = 0, 1, 2, \ldots$, and $\sigma = \pm 1$, are related as follows:

$$n = n(l'', \sigma) \equiv l'' + sgn(g'H)\frac{\sigma}{2} - \frac{1}{2}. \tag{9}$$

In momentum space, where $\delta^{(4)}(p-p') \equiv \delta_{ll'}\delta(p_0 - p'_0)\delta(p_2 - p'_2)\delta(p_3 - p'_3)$, the Schwinger-Dyson equations for $\Sigma_{l_{R(L)}}$ at one-loop and in the Feynman gauge, $\xi = 1$, are

$$(2\pi)^4 \delta^{(4)}(\bar{p} - \bar{p}')\Sigma_{l_{R(L)}}(\bar{p}) = iG_{l_{R(L)}}^2 \int d^4 x d^4 x' \sum_{l''=0}^\infty \int \frac{d^3 k}{(2\pi)^4} \int \frac{d^4 q}{(2\pi)^4} \frac{e^{iq\cdot(x-x')}}{q^2}$$

$$\times \bar{E}_p(g^{R(L)}, x) \gamma^\mu E_k(g^{R(L)}, x) \left(\frac{1}{\gamma \cdot \bar{k}}\right) \bar{E}_k(g^{R(L)}, x') \gamma_\mu E_{p'}(g^{R(L)}, x') R(L),$$

with $\quad G_{l_L}^2 = [\left(\frac{g'}{2}\right)^2 + 3\left(\frac{g}{2}\right)^2]$, and $\quad G_{eR}^2 = (g')^2$. \hfill (10)

Σ AT $T \neq 0$ IN THE LLL

After performing integrations and spin sums, Eq. (10) becomes

$$\Sigma_{l_{R(L)}}(\bar{p}) = -2iG_{l_{R(L)}}^2 \sum_{l''=0}^\infty \int \frac{d^4 q}{(2\pi)^4} \frac{e^{-\frac{1}{2|g'H|}q_\perp^2}}{q^2} \frac{1}{\bar{\kappa}^2} \tag{11}$$

$$\times \left\{ \delta_{l,l''}\left(\gamma^\perp \bar{\kappa}_\perp\right) + \left(\delta_{l,l''-sgn(g'H)}\Delta(1) + \delta_{l,l''+sgn(g'H)}\Delta(-1)\right)\left(\gamma^\parallel \bar{\kappa}_\parallel\right)\right\} R(L),$$

where, by the integrals over momentum δ-functions, the internal-lepton eigenvalues, $\left(\overline{k}_{l_{R(L)}}\right)_\mu$, have become

$$\overline{\kappa}_\mu = \left(p_0 - q_0,\ 0,\ -sgn\left(g'H\right)\sqrt{2\left|g^{R(L)}\right|l''},\ p_3 - q_3\right). \tag{12}$$

In the EW-symmetric phase, one expects that the field to temperature ratio satisfies $H/T^2 \sim 2$ [8], and the Landau level gap is $\Delta l \simeq |g'H| \sim O(T^2)$. The thermal energy is, then, of the same order as the energy of the Landau-level gap, making small the probability of thermally inducing occupation of any but the lower Landau levels. A qualitative approximate equivalence then exists with the strong-field approximation, $H/T^2 \gg 1$, for which the lowest Landau level (LLL) only survives. In the LLL, $l = 0$, the internal transverse-momentum term, $\delta_{l,l''}\left(\gamma^\perp \overline{\kappa}_\perp\right)$, vanishes. However, longitudinal-momentum term, $\overline{\kappa}_\parallel$, survives when $l'' = \pm sgn(g'H)$. By sign choice, $g'H = |g'H|$, and after Wick-rotation to Euclidean space,

$$\Sigma_{l_{R(L)}}(\bar{p}) = I_{l_{R(L)}}(\bar{p})(-i\gamma_4 \pm \gamma_3)R(L), \quad \text{where} \tag{13}$$

$$I_{l_{R(L)}}(\bar{p}) = 2G_{l_{R(L)}}^2 \int \frac{d^4q}{(2\pi)^4} \frac{e^{-\frac{1}{2|g^{R(L)}H|}q_\perp^2}}{q_4^2 + q_\perp^2 + q_3^2} \frac{i(\bar{p}_4 - q_4) - (\bar{p}_3 - q_3)}{(\bar{p}_4 - q_4)^2 + 2\left|g^{R(L)}H\right| + (\bar{p}_3 - q_3)^2}. \tag{14}$$

The internal-lepton transverse-momentum term, $2\left|g^{R(L)}H\right|$, above acts as a field-dependent infrared cutoff so that momentum integrals are finite. We separately evaluate the vacuum (vac) and the thermal (T) parts of $I_{l_{R(L)}}(\bar{p})$, in the infrared region, $\left|\bar{p}_\parallel\right| \ll \sqrt{g'H}$, using the Matsubara (imaginary-time) formalism. In Minkowski space, we obtain

$$I_{l_{R(L)}}^{(vac)}(\bar{p}) = \frac{G_{l_{R(L)}}^2}{2\pi}\left[\frac{a}{4\pi}\right](\bar{p}_0 \pm \bar{p}_3), \quad I_{l_{R(L)}}^{(T)}(\bar{p}) = \frac{G_{l_{R(L)}}^2}{2\pi}\left[\frac{T^2}{2\left|g^{R(L)}H\right|}\right](\bar{p}_0 \pm \bar{p}_3). \tag{15}$$

Upon combining $I_{l_{R(L)}}^{(vac)}(\bar{p})$ and $I_{l_{R(L)}}^{(T)}(\bar{p})$, the total self energy, $\Sigma_{R(L)}(\bar{p})$, is[5]

$$\Sigma_{R(L)}(\bar{p}) = A_{R(L)}(\bar{p}_0 \pm \bar{p}_3)\left(\gamma^0 \pm \gamma^3\right)R(L), \quad \text{where,}$$

$$A_{R(L)} = \frac{G_{l_{R(L)}}^2}{2\pi}\left[\frac{a}{4\pi} + \frac{T^2}{2\left|g^{R(L)}H\right|}\right] \quad (a \simeq 0.855). \tag{16}$$

---

[5] In covariant form, Eq. (16) is $\Sigma_{R(L)}(\bar{p}) = A_{R(L)}\left[\bar{\not{p}}_\parallel \pm \bar{p}^\mu \widehat{H}^*_{\mu\nu}\gamma^\nu\right]R(L)$, where $\widehat{H}^*_{\mu\nu}$ is the dimensionless hypermagnetic field dual tensor.

SELF-ENERGY COVARIANT STRUCTURE

The presence of a thermal bath of fermions and a background field will introduce a special frame that breaks Lorentz symmetry. The resulting general structure of the fermion self-energy, $\Sigma_{l_{R(L)}}$, takes the covariant form,

$$\Sigma_{l_{R(L)}}(\bar{p}) = \left[a_{1R(L)}\vec{p}_{\parallel} + a_{2R(L)}\vec{p}_{\perp} + b_{R(L)}\rlap{/}{u} + c_{R(L)}\widehat{\rlap{/}{H}} \right] R(L). \quad (17)$$

Above, $u_\mu = (1,0,0,0)$ is the four-velocity of the center of mass of the hypermagnetized medium (such that in the rest frame $u_\mu H^{\mu\nu} = 0$), the notation $\widehat{H}_\mu = H_\mu / |H_\mu|$ was introduced, with the hypermagnetic field in covariant form given by $H_\mu = \frac{1}{2}\varepsilon_{\mu\nu\rho\lambda}u^\nu H^{\rho\lambda}$, and a_1, a_2, b, and c are Lorentz-scalar coefficients which depend on the momenta, temperature and hypermagnetic field. Identification of Eqs. (16) and (17) reveals that the structure with coefficient, a_2, is absent in the *LLL*, and that coefficients, b and c, for the right and the left-handed lepton self energies are, respectively,

$$b_R = A_R(p \cdot H), \ b_L = -A_L(p \cdot H), \ c_R = -A_R(p \cdot u), \ c_L = A_L(p \cdot u). \quad (18)$$

DISPERSION RELATIONS

The fermion dispersion equation with radiative corrections is $\det[\bar{p} \cdot \gamma + \Sigma] = 0$. Substitution of Eq. (16) for either Σ_{l_R} or Σ_{l_L} gives

$$\bar{p}_0^2 = \bar{p}_3^2. \quad (19)$$

Thus, in the strong-field regime, chiral leptons behave as massless particles (with index of refraction, $n = 1$) propagating only along the field lines. By comparison, in the broken-symmetry *EW* phase, the effect of high magnetic fields on neutrino propagation permits less anisotropy [6], since in that case neutrinos can propagate in all directions. This different behavior is due to the different values of the neutrino charge in the different phases, that is, thanks to its non-zero hypercharge, it is minimally coupled to the hypermagnetic field, whereas its electrical neutrality allows it to couple to the magnetic field only through radiative corrections.

CONCLUSIONS

We have shown that the temperature-dependent poles in the chiral-fermions' one-loop Green's functions found by Weldon [8] at zero field, i.e., giving rise to effective masses, $M_T = G_{l_{R(L)}}^2 T^2/8$, disappear in a sufficiently strong hypermagnetic field [9]. Moreover, the dispersion relations show that chiral leptons, and in particular, neutrinos, have maximal anisotropic propagation, with alignment along the direction of the external field. We conclude that, if such large fields existed before the *EW*-phase transition, so great an anisotropy would be expected to have left a trace in the as yet undetected present neutrino cosmic background. Conversely, such a footprint, if detectable, would be sufficient evidence for the existence of strong primordial hypermagnetic fields.

REFERENCES

1. Sofue, Y., Fujimoto, M., and Wielebinski, R., *Ann. Rev. Astron. Astrophys.* **24,** 459 (1986); Kronberg, P.P., *Rep. Prog. Phys.* **57,** 325 (1994); Beck, R., et. al., *Ann. Rev. Astron. Astrophys.* **34,** 153 (1996).
2. Vachaspati, T., *Phys. Lett.* **B265,** 258 (1991); Kibble, T.W., and Vilenkin, A., *Phys. Rev.* **D52,** 679 (1995); Baym, G., Bodeker, D., and McLerran, L., *Phys. Rev.* **D53,** 662 (1996); Joyce, M., and Shaposhnikov, M., *Phys. Rev. Lett.* **79,** 1193 (1997).
3. Linde, A.D., *Rep. Prog. Phys.* **42,** 389 (1979); *Phys. Lett.* **B96,** 178 (1980); Kajantie, K., Laine, M., Rummukainen, K., and Shaposhnikov, M., *Nucl. Phys.* **B493,** 413 (1997).
4. Fradkin, E.S., *Proceedings Lebedev Phys. Inst.* **29,** 7 (1965), Eng. Transl., (Consultant Bureau, New York 1967); Rebhan, A.K., *Phys. Rev.* **D48,** R3967 (1993); *Nucl. Phys.* **B430,** 319 (1994); Braaten, E., and Nieto, A., *Phys. Rev. Lett.* **73,** 2402 (1994); Baier, R., and Kalashnikov, O.K., *Phys. Lett.* **B328,** 450 (1994).
5. Ritus, V.I., *Ann. of Phys. (NY)* **69,** 555 (1972); *ZhETF* **75,** 1560 (1978); *ZhETF* **76,** 383 (1979); in *Issues in Intense-Field Quantum Electrodynamics*, ed. V. L. Ginzburg (Nova Science, Commack 1987).
6. Elizalde, E., Ferrer, E.J., and de la Incera, V., *Annals of Phys. (NY)* **295,** 33 (2002).
7. Giovannini, M., and Shaposhnikov, M., *Phys. Rev.* **D57,** 2186 (1998); Elmfors, P., Enqvist, K., and Kainulainen, K., *Phys. Lett.* **B440,** 269 (1998).
8. Weldon, H.A., *Phys. Rev.* **D26,** 2789 (1982).
9. Cannellos, J., Ferrer, E.J., and de la Incera, V., hep-ph/0204126, to appear in *Phys. Lett. B*.

Magnetic behaviour of SO(5) superconductors

R. MacKenzie

Laboratoire-René-J.-A.-Lévesque, Université de Montréal
C.P. 6128, Succ. Centre-ville, Montréal, Qc H3C 3J7, Canada

Abstract. The distinction between type I and type II superconductivity is re-examined in the context of the SO(5) model recently put forth by Zhang. Whereas in conventional superconductivity only one parameter (the Ginzburg-Landau parameter κ) characterizes the model, in the SO(5) model there are two essential parameters. These can be chosen to be κ and another parameter, β, related to the doping. There is a more complicated relation between κ and the behaviour of a superconductor in a magnetic field. In particular, one can find type I superconductivity even when κ is large, for appropriate values of β.

INTRODUCTION

In this talk, recent work on magnetic properties of the SO(5) model of high-temperature superconductivity (HTSC) will be presented. After reviewing the case of conventional superconductivity and some relevant facts of HTSC, the SO(5) model will be introduced. The behaviour of a system described by this model when placed in a magnetic field will be analysed. The main conclusion is that in strongly underdoped superconductors, the critical value of the Ginzburg-Landau (GL) parameter κ can be much larger than the conventional value. Thus, a large value of κ can be associated with a type I superconductor. The application of these ideas to HTSC will be discussed briefly. This work forms the bulk of Refs. [1, 2].

PRELIMINARIES

Conventional superconductivity

In this section we briefly review some features of conventional superconductivity (SC). This material is well-known, and can be found in greater detail in almost any introductory SC textbook, such as Tinkham [3].

The phase transition in a conventional superconductor can be described at low energies by an effective theory, known as a Ginzburg-Landau (GL) theory, written in terms of the SC order parameter ϕ (a complex field representing the Cooper pair amplitude) and the electromagnetic field. This GL theory can be expressed in terms of a Helmholtz free energy, which takes the following form:

$$F = \int d\mathbf{x} \left\{ f_n - \frac{a_1^2}{2}|\phi|^2 + \frac{b}{4}|\phi|^4 + \frac{1}{2m^*}\left|\left(-i\hbar\nabla - \frac{e^*\mathbf{A}}{c}\right)\phi\right|^2 + \frac{\mathbf{h}^2}{8\pi} \right\}.$$

Here, f_n is a constant, $\mathbf{h} = \nabla \times \mathbf{A}$ is the microscopic magnetic field and a_1, b are parameters. The minimum of the potential is $|\phi|^2 = a_1^2/b \equiv v^2$.

There are two characteristic length scales in this model: the coherence length $\xi = (\hbar^2/m^*a_1^2)^{1/2}$ and the magnetic field penetration depth $\lambda = (m^*c^2/4\pi e^{*2}v^2)^{1/2}$. These are, roughly, the Compton wavelengths of the scalar and electromagnetic fields, respectively. By scaling out all dimensionful quantities, one finds that the behaviour of a SC described by the above free energy is determined by one dimensionless parameter, the GL parameter $\kappa = \lambda/\xi$. Some typical values for this parameter appear in Table 1.

TABLE 1. Typical parameter values for various categories of superconductors.

	λ (Å)	ξ (Å)	κ
Simple metal	300	1000	.3
Alloy	2000	50	40
High-T_c	2000	20	100

There are two very different classes of (conventional) SCs, depending on the value of κ. If $\kappa < 1/\sqrt{2}$, the material is said to be type I, while if $\kappa > 1/\sqrt{2}$, it is said to be type II. The behaviour when a magnetic field is applied to a superconductor differs greatly for these two classes.

There are several ways to see this. One way is to consider a configuration where a magnetic field equal to the so-called thermodynamic critical field H_c^0 is applied to the superconductor.[1] The sign of the energy of a surface separating SC and normal regions is the telling quantity. If it is positive, the system would prefer to minimize the amount of surface for a given magnetic flux; this is achieved if the flux penetrates in a macroscopic region. In contrast, if the surface energy is negative, the flux will form a lattice of flux tubes of the minimum allowable flux. One can calculate numerically the surface energy of a boundary between SC and normal regions. Even quantitatively, one can argue that for small κ the surface energy is positive, while for large κ it is negative (see Figure 1).

An alternative way to determine the distinction between type I and type II superconductors is to consider the energetics of vortices of varying winding number. Since the magnetic flux of a vortex is proportional to its winding number, the energy per unit winding number indicates whether it is energetically favourable for a given amount of flux to penetrate in many unit-winding-number vortices or in one large vortex. The former will

---

[1] If a weak magnetic field is applied to a superconductor, the field is expelled; if a strong field is applied, SC is destroyed and the field penetrates the material. The critical field is the transitional value, i.e., that where the (Gibbs) free energy of the normal phase in the magnetic field is equal to that of the SC phase in the field's absence.

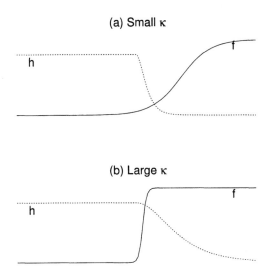

FIGURE 1. Field profiles at SC/normal surface for (a) Type I and (b) Type II superconductors. f is a rescaled ϕ. The slow variation of f when $\kappa \ll 1$ gives rise to a positive contribution to the surface energy; the slow variation of h has the opposite effect in the opposite limit. (Figure from "Magnetic Properties of SO(5) Superconductivity" by M. Juneau, R. MacKenzie and M.-A. Vachon, in Annals of Physics, Volume 298, 421, copyright 2002, Elsevier Science (USA), reproduced by permission of the publisher.)

be the case if the energy per unit winding number is of positive slope, while the latter will be true if it is of negative slope (see [1]).

The magnetization curves of type I and type II superconductors also highlight their different behaviour in a magnetic field (see Figure 2). As the magnetic field is increased in a type I superconductor, the field is completely expelled until H_c^0 is reached, at which point SC is destroyed macroscopically. In contrast, in a type II superconductor, at a lower field H_{c1}^0 the field starts to penetrate the superconductor in vortices; when the upper critical field H_{c2}^0 is reached, SC is finally destroyed. These critical fields vary as a function of κ; one finds $H_{c2}^0 = \sqrt{2}\kappa H_c^0$. The transition point between type I and type II SC occurs when these two critical fields are equal, which occurs at $\kappa = \kappa_c = 1/\sqrt{2}$.

High-temperature superconductivity

In this section, a couple of relevant facts of HTSC are presented. The key observation which leads to the SO(5) model of HTSC is that these materials exhibit two very different phases at low temperature, depending on the degree of doping. At sufficiently high doping, one sees SC, while at lower values of the doping (including the undoped case) the materials are antiferromagnetic (AF), as shown in Figure 3.

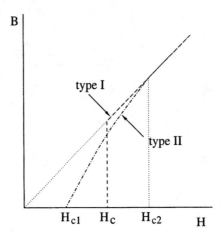

FIGURE 2. B vs. H for type I and type II superconductors.

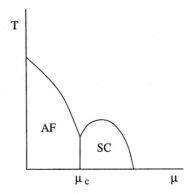

FIGURE 3. Approximate phase diagram for high-temperature superconductors and for SO(5) superconductivity.

An important property of HTSCs is that, as mentioned above (see Table 1), they are "highly type II", i.e., κ is very much larger than its critical value. (Values in the vicinity of 100 are typical; the lowest value we have seen reported is 17.)

However, as we shall see presently, in the SO(5) model, κ alone does not determine the magnetic behaviour (i.e., the type) of a superconductor. Indeed, it is possible that, in spite of having a very large κ, HTSCs (if described by the SO(5) model) might exhibit type I behaviour under certain conditions.

SO(5) SUPERCONDUCTIVITY

Motivation; Ginzburg-Landau model

As mentioned above, the presence of both AF and SC in HTSCs suggests the possibility of a sort of unification of these phenomena, both of which involve spontaneous symmetry breaking. This possibility was put forth by Zhang [4], who wrote down a model in terms of a five-component real order parameter. The five components are the real and imaginary components of the SC order parameter ϕ and the three components of the AF order parameter η. A GL theory which has an approximate SO(5) rotational symmetry can then be written down; the free energy is

$$F = \int d\mathbf{x} \left\{ \frac{\mathbf{h}^2}{8\pi} + \frac{\hbar^2}{2m^*} \left| \left(\nabla + \frac{ie^*\mathbf{A}}{\hbar c} \right) \phi \right|^2 + \frac{\hbar^2}{2m^*} (\nabla \eta)^2 + V(\phi, \eta) \right\},$$

where the potential is

$$V(\phi, \eta) = -\frac{a_1^2}{2}\phi^2 - \frac{a_2^2}{2}\eta^2 + \frac{b}{4}(\phi^2 + \eta^2)^2.$$

There are now three relevant length scales: λ, ξ and ξ' (the characteristic length of the η field). Rescaling now reduces the number of essential parameters to two, which can be taken to be κ and $\beta \equiv (a_2^2/a_1^2)$. The latter is related to the doping; $\beta = 1$ at the AF-SC boundary, and $\beta < 1$ in the SC phase.

The potential for $\beta < 1$ is shown in Figure 4. There are two important features of the potential. First, it is minimized at a nonzero value of ϕ and $\eta = 0$, so the ground state is indeed SC. Second, if ϕ is somehow forced to be zero, then η will be nonzero.

In fact, there are two situations when ϕ is indeed forced to be zero: firstly, in the core of a vortex [4, 5, 6, 1], and secondly, if the superconductor is placed in a sufficiently strong magnetic field.

Magnetic properties

This "induced antiferromagnetism" can have a dramatic effect on the critical fields, and on the type of superconductor described by the model. The effect on the thermodynamic critical field H_c arises because it is found by comparing the Gibbs free energies of SC and AF (rather than normal) states. The other critical fields $H_{c1,2}$ are affected because vortex energetics are affected by the AF core.

Both H_c and H_{c2} can be calculated analytically [2]:

$$H_c = H_c^0 \sqrt{1 - \beta^2};$$
$$H_{c2} = H_{c2}^0 (1 - \beta) = \sqrt{2}\kappa H_c^0 (1 - \beta).$$

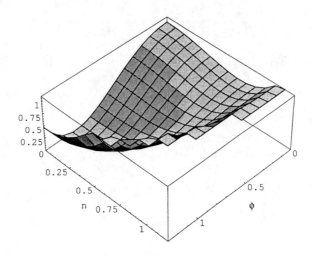

FIGURE 4. SO(5) potential as a function of ϕ and η.

As in the conventional case, equality of these critical fields indicates the boundary of type I/II behaviour. We can thus obtain a curve in the β-κ plane which represents the boundary separating type I and type II behaviour (Figure 5). Analytically, this curve is given by

$$\kappa_c(\beta) = \frac{1}{\sqrt{2}}\sqrt{\frac{1+\beta}{1-\beta}}.$$

This expression is confirmed (numerically) by analysis of surface energy and vortex energetics.

Application to high-temperature superconductivity

In HTSC, as mentioned above, κ is typically of order 100; thus, HTSCs are considered highly type II. However, as the previous section demonstrates, in the SO(5) model β also plays a role in the nature of the superconductor. In particular, for any $\kappa > 1/\sqrt{2}$, if β is sufficiently large, the material is type I. We can invert the above expression for $\kappa_c(\beta)$ to obtain the following expression for $\beta_c(\kappa)$, valid if $\kappa \gg 1$ $\beta_c(\kappa) \simeq 1 - \kappa^{-2}$. For example, if $\kappa = 100$, the material is a type I superconductor if $\beta < 0.9999$ while it is type II if $\beta > 0.9999$. (Since $\beta < 1$ in the SC state, there is only a minute range of β corresponding to type II.) If $\kappa = 17$ (the smallest value reported for a HTSC), $\beta > 0.996$ for the material to be type II – still a small window, but much greater than that for $\kappa = 100$.

The parameter β is related to measurable quantities: $\beta = 1 - 8m^*\hbar^{-2}\xi(\mu^2 - \mu_c^2)$, where μ, μ_c are the chemical potential and the critical chemical potential (that at the SC-AF boundary) and ξ is the charge susceptibility.

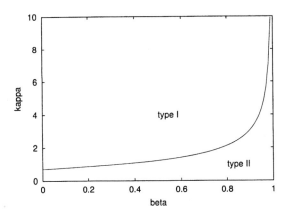

FIGURE 5. Curve delineating type I vs type II superconductivity in the κ-β plane. (Figure from "Magnetic Properties of SO(5) Superconductivity" by M. Juneau, R. MacKenzie and M.-A. Vachon, in Annals of Physics, Volume 298, 421, copyright 2002, Elsevier Science (USA), reproduced by permission of the publisher.)

Thus, we have the possibility of a fairly dramatic test of the SO(5) model; however, the experimental situation in the underdoped region appears rather delicate. For example, the appearance of inhomogeneities (stripe formation, phase separation) could mask the appearance of type I behaviour. Nonetheless, since the degree to which a superconductor is type II is reduced as the doping is reduced, one could hope to see signs of a reduction in the rigidity of the vortex lattice.

ACKNOWLEDGMENTS

This work was supported by the Natural Science and Engineering Research Council of Canada.

REFERENCES

1. Juneau, M., MacKenzie, R., Vachon, M.-A., and Cline, J., *Phys. Rev. B (Rapid Communications)*, **65**, 140512 (2002).
2. Juneau, M., MacKenzie, R., and Vachon, M.-A., *Ann. Phys.*, **298**, 421 (2002).
3. Tinkham, M., *Introduction to superconductivity*, 2nd edn., McGraw-Hill, 1996, ISBN 007064876.
4. Zhang, S.-C., *Science*, **275**, 1089 (1997).
5. Arovas, D., Berlinsky, A., Kallin, C., and Zhang, S.-C., *Phys. Rev. Lett.*, **79**, 2871 (1997).
6. Alama, S., Berlinsky, A., Bronsard, L., and Giorgi, T., *Phys. Rev. B*, **60**, 6901 (1999).

The Cosmological Constant Problem and Nonlocal Quantum Gravity

J.W. Moffat

Department of Physics, University of Toronto, Toronto, Ontario M5S 1A7, Canada

Abstract. A nonlocal quantum gravity theory is presented which is finite and unitary to all orders of perturbation theory. Vertex form factors in Feynman diagrams involving gravitons suppress graviton and matter vacuum fluctuation loops by introducing a low-energy gravitational scale, $\Lambda_{Gvac} < 2.4 \times 10^{-3}$ eV. Gravitons coupled to non-vacuum matter loops and matter tree graphs are controlled by a vertex form factor with the energy scale, $\Lambda_{GM} < 1 - 10$ TeV. A satellite Eötvös experiment is proposed to test a violation of the equivalence principle for coupling of gravitons to pure vacuum energy compared to matter.

INTRODUCTION

It is generally agreed that the cosmological constant problem is one of the most severe problems facing modern particle and gravitational physics. It is believed that its solution could significantly alter our understanding of particle physics and cosmology [1].

There is now mounting observational evidence [2] that the universe is accelerating and that there exists some form of dark energy. One possible explanation for the accelerating expansion of the universe is a small cosmological constant corresponding to a vacuum energy density, $\rho_{vac} \sim (2.4 \times 10^{-3}\,\text{eV})^4$.

There have been many attempts to solve the cosmological constant problem (CCP). Weinberg's theorem [3] disallows all adjustment models involving extra fields such as a dynamical scalar field. Higher-dimensional models of the brane-bulk type with finite volume extra-dimensions do not avoid fine-tuning [4].

Superstring theory (M-theory) has not yet provided a solution to the CCP. This could be due to the problem of understanding how to introduce supersymmetry breaking into string theory models, although there may be some deeper reason for the failure.

In the following, I will describe a possible resolution of the CCP, based on a model of a nonlocal quantum gravity theory and field theory that suppresses the coupling of gravity to vacuum energy density. The theory can be tested by performing Eötvös experiments on Casimir vacuum energy in satellites.

Gravitational Coupling to Vacuum Energy

We can define an effective cosmological constant

$$\lambda_{\text{eff}} = \lambda_0 + \lambda_{\text{vac}}, \tag{1}$$

where λ_0 is the "bare" cosmological constant in Einstein's classical field equations, and λ_{vac} is the contribution that arises from the vacuum density $\lambda_{vac} = 8\pi G \rho_{vac}$.

Already at the standard model electroweak scale $\sim 10^2$ GeV, a calculation of the vacuum density ρ_{vac}, based on local quantum field theory, results in a discrepancy of order 10^{55} with the observational bound

$$\rho_{vac} \leq 10^{-47} (\text{GeV})^4. \tag{2}$$

This results in a severe fine-tuning problem of order 10^{55}, since the virtual quantum fluctuations giving rise to λ_{vac} must cancel λ_0 to an unbelievable degree of accuracy. This is the "particle physics" source of the cosmological constant problem.

Nonlocal Quantum Gravity

Let us consider a model of nonlocal gravity with the action $S = S_g + S_M$, where $(\kappa^2 = 32\pi G)$ [1]:

$$S_g = -\frac{2}{\kappa^2} \int d^4x \sqrt{-g} \left\{ R[g, \mathscr{G}^{-1}] + 2\lambda_0 \right\} \tag{3}$$

and S_M is the matter action, which for the simple case of a scalar field ϕ is given by

$$S_M = \frac{1}{2} \int d^4x \sqrt{-g} \left(g^{\mu\nu} \mathscr{G}^{-1} \nabla_\mu \phi \mathscr{F}^{-1} \nabla_\nu \phi - m^2 \phi \mathscr{F}^{-1} \phi \right). \tag{4}$$

Here, $\mathscr{G}$ and $\mathscr{F}$ are nonlocal regularizing, *entire* functions and ∇_μ is the covariant derivative with respect to the metric $g_{\mu\nu}$. As an example, we can choose the covariant functions

$$\mathscr{G}(x) = \exp\left[-\mathscr{D}(x)/\Lambda_G^2\right],$$

$$\mathscr{F}(x) = \exp\left[-(\mathscr{D}(x) + m^2)/\Lambda_M^2\right], \tag{5}$$

where $\mathscr{D} \equiv \nabla_\mu \nabla^\mu$, and Λ_G and Λ_M are gravitational and matter energy scales, respectively [5, 6].

We expand $g_{\mu\nu}$ about flat Minkowski spacetime: $g_{\mu\nu} = \eta_{\mu\nu} + \kappa h_{\mu\nu}$. The propagators for the graviton and the ϕ field in a fixed gauge are given by

$$\bar{D}^\phi(p) = \frac{\mathscr{G}(p)\mathscr{F}(p)}{p^2 - \bar{m}^2 + i\varepsilon}, \tag{6}$$

---

[1] The present version of a nonlocal quantum gravity and field theory model differs in detail from earlier published work [5, 6]. A paper is in preparation in which more complete details of the model will be provided.

$$\bar{D}^G_{\mu\nu\rho\sigma}(p) = \frac{(\eta_{\mu\rho}\eta_{\nu\sigma} + \eta_{\mu\sigma}\eta_{\nu\rho} - \eta_{\mu\nu}\eta_{\rho\sigma})\mathscr{G}(p)}{p^2 + i\varepsilon}, \tag{7}$$

where $\bar{m}^2 = m^2\mathscr{G}(p)$.

Unitarity is maintained for the S-matrix, because $\mathscr{G}$ and $\mathscr{F}$ are *entire* functions of p^2, preserving the Cutkosky rules.

Gauge invariance can be maintained by satisfying certain constraint equations for $\mathscr{G}$ and $\mathscr{F}$ in every order of perturbation theory. This guarantees that $\nabla_\nu T^{\mu\nu} = 0$.

Resolution of the CCP

In flat Minkowski spacetime, the sum of all *disconnected* vacuum diagrams $C = \sum_n M_n^{(0)}$ is a constant factor in the scattering S-matrix $S' = SC$. Since the S-matrix is unitary $|S'|^2 = 1$, then we must conclude that $|C|^2 = 1$, and all the disconnected vacuum graphs can be ignored. This result is also known to follow from the Wick ordering of the field operators.

Due to the equivalence principle *gravity couples to all forms of energy*, including the vacuum energy density ρ_{vac}, so we can no longer ignore these virtual quantum fluctuations in the presence of a non-zero gravitational field. Quantum corrections to λ_0 come from loops formed from massive standard model (SM) states, coupled to external graviton lines at essentially zero momentum.

Consider the dominant contributions to the vacuum density arising from the graviton-standard model loop corrections. We shall adopt a model consisting of a photon loop coupled to gravitons, which will contribute to the vacuum polarization loop coorection to the bare cosmological constant λ_0. The covariant photon action is [7]:

$$S_A = -\frac{1}{4}\sqrt{-g}g^{\mu\nu}g^{\alpha\beta}\mathscr{G}^{-1}F_{\mu\alpha}\mathscr{F}^{-1}F_{\nu\beta}, \tag{8}$$

with

$$F_{\mu\alpha} = \partial_\mu A_\alpha - \partial_\alpha A_\mu. \tag{9}$$

The lowest order correction to the graviton-photon vacuum loop will have the form (in Euclidean momentum space):

$$\Pi^{Gvac}_{\mu\nu\rho\sigma}(p) = \kappa^2 \int \frac{d^4q}{(2\pi)^4} V_{\mu\nu\lambda\alpha}(p,-q,-q-p)$$

$$\times \mathscr{F}^\gamma(q^2)D^\gamma_{\lambda\beta}(q^2)V_{\rho\sigma\beta\gamma}(-p,q,p-q)\mathscr{F}^\gamma((p-q)^2)D^\gamma_{\alpha\gamma}((p-q)^2)\mathscr{G}^{Gvac}(q^2), \tag{10}$$

where $V_{\mu\nu\rho\sigma}$ is the photon-photon-graviton vertex and in a fixed gauge:

$$D^\gamma_{\mu\nu} = -\frac{\delta_{\mu\nu}}{q^2} \tag{11}$$

is the free photon propagator. Additional contributions to $\Pi^{Gvac}_{\mu\nu\rho\sigma}$ come from tadpole graphs [7].

This leads to the vacuum polarization tensor

$$\Pi^{\text{Gvac}}_{\mu\nu\rho\sigma}(p) = \kappa^2 \int \frac{d^4q}{(2\pi)^4} \frac{1}{q^2[(q-p)^2]}$$

$$\times K_{\mu\nu\rho\sigma}(p,q) \exp\left\{-q^2/\Lambda_M^2 - [(q-p)^2]/\Lambda_M^2 - q^2/\Lambda_{\text{Gvac}}^2\right\}. \quad (12)$$

For $\Lambda_{\text{Gvac}} \ll \Lambda_M$, we observe that from power counting of the momenta in the loop integral, we get

$$\Pi^{\text{Gvac}}_{\mu\nu\rho\sigma}(p) \sim \kappa^2 \Lambda_{\text{Gvac}}^4 N_{\mu\nu\rho\sigma}(p^2)$$

$$\sim \frac{\Lambda_{\text{Gvac}}^4}{M_{\text{PL}}^2} N_{\mu\nu\rho\sigma}(p^2), \quad (13)$$

where $N(p^2)$ is a finite remaining part of $\Pi^{\text{Gvac}}(p)$ and $M_{\text{PL}} \sim 10^{19}$ GeV is the Planck mass.

We now have

$$\rho_{\text{vac}} \sim M_{\text{PL}}^2 \Pi^{\text{Gvac}}(p) \sim \Lambda_{\text{Gvac}}^4. \quad (14)$$

If we choose $\Lambda_{\text{Gvac}} \leq 10^{-3}$ eV, then the quantum correction to the bare cosmological constant λ_0 is suppressed sufficiently to satisfy the observational bound on λ, *and it is protected from large unstable radiative corrections.*

This provides a solution to the cosmological constant problem at the energy level of the standard model and possible higher energy extensions of the standard model. The universal fixed gravitational scale Λ_{Gvac} corresponds to the fundamental length $\ell_{\text{Gvac}} \leq 1$ mm at which virtual gravitational radiative corrections to the vacuum energy are cut off.

The gravitational form factor $\mathscr{G}$, *when coupled to non-vacuum SM gauge boson or matter loops*, will have the form in Euclidean momentum space

$$\mathscr{G}^{\text{GM}}(q^2) = \exp\left[-q^2/\Lambda_{\text{GM}}^2\right]. \quad (15)$$

If we choose $\Lambda_{GM} = \Lambda_M > 1 - 10$ TeV, then we will reproduce the standard model experimental results, including the running of the standard model coupling constants, and $\mathscr{G}^{GM}(q^2) = \mathscr{F}^M(q^2)$ becomes $\mathscr{G}^{GM}(0) = \mathscr{F}^M(q^2 = m^2) = 1$ on the mass shell. *This solution to the CCP leads to a violation of the WEP for coupling of gravitons to vacuum energy and matter.* This could be checked experimentally in a satellite Eötvös experiment on the Casimir vacuum energy [8].

We observe that the required suppression of the vacuum diagram loop contribution to the cosmological constant, associated with the vacuum energy momentum tensor at lowest order, demands a low gravitational energy scale $\Lambda_{\text{Gvac}} \leq 10^{-3}$ eV, which controls the coupling of gravitons to pure vacuum graviton and matter fluctuation loops.

In our finite, perturbative quantum gravity theory nonlocal gravity produces a long-distance infrared cut-off of the vacuum energy density through the low energy scale $\Lambda_{\text{Gvac}} < 10^{-3}$ eV [6][2]. Gravitons coupled to *non-vacuum* matter tree graphs and matter

---

[2] The energy scale, $\Lambda_G \sim 10^{-3}$ eV, has also been considered by R. Sundrum and G. Dvali, G. Gabadadze and M. Shifman [9].

loops are controlled by the energy scale: $\Lambda_{GM} = \Lambda_M > 1 - 20$ TeV

The rule is: When external graviton lines are removed from a matter loop, leaving behind *pure* matter fluctuation vacuum loops, then those initial graviton-vacuum loops are suppressed by the form factor $\mathscr{G}^{\text{Gvac}}(q^2)$ where q is the internal matter loop momentum and $\mathscr{G}^{\text{Gvac}}(q^2)$ is controlled by $\Lambda_{\text{Gvac}} \leq 10^{-3}$ eV. On the other hand, e.g. the proton first-order self-energy graph, coupled to a graviton is controlled by $\Lambda_{GM} = \Lambda_M > 1 - 20$ TeV *and does not lead to a measurable violation of the equivalence principle.*

The scales Λ_M and Λ_{Gvac} are determined in loop diagrams by the quantum non-localizable nature of the gravitons and standard model particles. The gravitons coupled to matter and matter loops have a nonlocal scale at $\Lambda_{GM} = \Lambda_M > 1 - 20$ TeV or a length scale $\ell_M < 10^{-16}$ cm, whereas the gravitons coupled to pure vacuum energy are localizable up to an energy scale $\Lambda_{\text{Gvac}} \sim 10^{-3}$ eV or down to a length scale $\ell_{\text{Gvac}} > 1$ mm.

The fundamental energy scales Λ_{Gvac} and $\Lambda_{GM} = \Lambda_M$ are determined by the underlying physical nature of the particles and fields and do not correspond to arbitrary cut-offs, which destroy the gauge invariance, Lorentz invariance and unitarity of the quantum gravity theory for energies $> \Lambda_{\text{Gvac}} \sim 10^{-3}$ eV. The underlying explanation of these physical scales must be sought in a more fundamental theory[3]

Conclusions

We have described a possible solution to the cosmological constant problem. The particle physics resolution requires that we construct a nonlocal quantum gravity theory, which has vertex form factors that are different for gravitons coupled to quantum *vacuum* fluctuations and matter. This predicts a measurable violation of the WEP for coupling to vacuum energy, but not to matter-graviton couplings or to *non-vacuum matter loops*. This leads to a suppression of all standard model vacuum loop contributions and, thereby, avoids a fine-tuning cancellation between the "bare" cosmological constant λ_0 and the vacuum contribution λ_{vac}. It retains the experimental agreement of the standard model and classical Einstein gravity. A satellite Eötvös experiment for Casimir vacuum energy could experimentally decide whether nature does allow a vacuum energy WEP violation.

Even though we can succeed in our nonlocal quantum gravity scenario to explain why λ_{eff} is small, without excessive fine tuning, we are still confronted with the "coincidence" problem associated with dark energy and the existence of a small, positive cosmological constant [1].

As a model of a future fundamental, nonlocal quantum gravity theory, it does provide clues as to the resolution of the "infamous" cosmological constant problem.

---

[3] It is interesting to note that if we choose $\Lambda_{\text{GM}} = \Lambda_M = 5$ TeV, then we obtain $\Lambda_{\text{Gvac}} = \Lambda_M^2/M_{\text{PL}} = 2.1 \times 10^{-3}$ eV.

ACKNOWLEDGMENTS

I thank Michael Clayton and George Gillies for helpful and stimulating discussions. This work was supported by the Natural Sciences and Engineering Research Council of Canada.

REFERENCES

1. Straumann, N., *On the Cosmological Constant Problem and the Astronomical Evidence for a Homogeneous Energy Density with Negative Pressure*, e-print astro-ph/0203330; Peebles, P.J.E., and Ratra, B., *The Cosmological Constant and Dark Energy*, e-print astro-ph/0207347; Ellwanger, U., *The Cosmological Constant*, e-print hep-ph/0203252.
2. Perlmutter, S., et al. *Ap. J.* **483**, 565 (1997); Riess, A.G., et al. *Astron. J.* **116**, 1009.
3. Weinberg, S., *Rev. Mod. Phys.* **61**, 1 (1989).
4. Cline, J., e-print hep-ph/0207155.
5. Moffat, J.W., *Phys. Rev.* **D41**, 1177 (1990); Evens, D., Moffat, J.W., Kleppe, G., and Woodard, R.P., *Phys. Rev.* **D43**, 499 (1991).
6. Moffat, J.W., e-print hep-ph/0102088.
7. Capper, D.M., Leibbrandt, G., Medrano, M.R., *Phys. Rev.* **D8**, 4320 (1973); Capper, D.M., Duff, M.J., and Halpern, L., *Phys. Rev.* **10**, 461 (1974).
8. Ross, D.K., *Il Nuovo Cim.* **114**, 1073 (1999).
9. Sundrum, R., *JHEP* **9907**, 001 (1999); Dvali, G., Gabadadze, G., and Shifman, M., e-print hep-th/0202174.
10. Sudarsky, D., *Class. Quant. Grav.* **12**, 579 (1995).

GENERAL RELATIVITY

Critical behaviour in spherically symmetric scalar field collapse in any dimension

M. Birukou*, V. Husain†, G. Kunstatter*, E. Vaz*, M. Olivier** and B. Preston*

*Dept. of Physics and Winnipeg Institute of Theoretical Physics
University of Winnipeg, Winnipeg, Manitoba Canada R3B 2E9.
†Dept. of Mathematics and Statistics
University of New Brunswick, Fredericton, N.B. Canada E3B 1S5.
**Dept. de Physique,
Universite Laval, Quebec, QC, Canada G1K 7P4.

Abstract. Gravitational collapse in general relativity is known to exhibit remarkable critical behaviour. Until recently this critical behaviour has only been analyzed numerically in special cases. We describe a formalism which allows the construction of code that calculates spherically symmetric scalar field collapse with spacetime dimension and cosmological constant as input parameters. This code correctly reproduces all previously known results, and sets the stage for a systematic analysis of critical behaviour as a function of spacetime dimension and cosmological constant. New results in this direction are presented.

Introduction

As first shown in the seminal work of Choptuik in 1993 [1], gravitational collapse in general relativity exhibits remarkable critical behaviour. In particular, given a one parameter family of initial data sets, one finds numerically for a variety of matter fields that as the parameter is tuned, a transition from reflection of infalling matter to black hole formation is observed. Two types of behaviour occur close to this transition point. Depending on the matter type, black holes form with zero or non-zero initial mass, and the matter field exhibits discrete or continuous self-similarity.

The basic formula that determines the horizon radius at the threshold of black hole formation is [1]

$$R_{BH} \sim (a - a_*)^\gamma, \qquad (1)$$

where a is any initial data parameter, and $a_* < a$ is its critical value. The critical value $a = a_*$ gives a naked singularity, and illustrates a violation of the cosmic censorship conjecture. The critical exponent γ is found (at least in even spacetime dimensions) to be "universal" for a fixed matter type in the sense that it is independent of the functional form and parameters in the initial matter profile. For scalar field collapse in 4d, Choptuik obtained a critical exponent of $\gamma = 0.36$.

Although a variety of matter fields have been studied [2, 3], most numerical studies of spherically symmetric collapse have been confined to four spacetime dimensions. There have been some extensions beyond spherical symmetry, as well as a semi-analytic perturbation theory understanding of the critical exponent. (Recent reviews may be found in Refs.[2, 3].) However, very little has been learned about the dependence of the critial exponent on spacetime dimension, or the effects of non-zero cosmological constant. The only exceptions are the massless minimally coupled scalar field by Garfinkle et. al. [7] in six spacetime dimensions with zero cosmological constant ($\gamma \sim 0.424$), and by Pretorius and Choptuik [8] ($\gamma \sim 1.2$), and Husain and Olivier[9] ($\gamma \sim 0.81$) in three dimensions with negative cosmological constant. Interestingly, the paper by Garfinkle et. al.[7] started with the Einstein action in *arbitrary* spacetime dimension. The authors of that paper stated that they: "... expect that some numerical code can be used to evolve the d-dimensional Einstein equations with d as an input parameter..." but they were unable to find such a code, so their subsequent analysis was restricted to six dimensions.

The purpose of present paper is to describe a formalism and numerical code that calculates spherically symmetric scalar field collapse using both the space-time dimension, d and the cosmological constant, Λ, as input parameters. This code, which reproduces all previously known results, sets the stage for a systematic analysis of the dependence of the critical exponent on d and Λ. In the following, we will first describe the general formalism, then provide some relevant details of the numerical procedure and finally present some new results for the critical exponent in four dimensions with non-zero cosmological constant, and five dimensions with and without cosmological constant.

Spherically symmetric gravity in d-dimensions

Our formalism starts with the action for Einstein gravity with cosmological constant and minimally coupled scalar field in d spacetime dimensions:

$$S^{(d)} = \frac{1}{16\pi G^{(d)}} \int d^d x \sqrt{-g^{(d)}} \left[R(g^{(d)}) - \Lambda \right] - \int d^d x |\partial \chi|^2. \tag{2}$$

To impose spherical symmetry, we write the d-dimensional metric $g_{\mu\nu}$ as

$$ds^2_{(d)} = \bar{g}_{\alpha\beta} dx^\alpha dx^\beta + r^2(x^\alpha) d\Omega_{(d-2)}, \tag{3}$$

where $d\Omega_{(d-2)}$ is the metric on S^{d-2} and $\alpha, \beta = 1, 2$. We now make the following field redefinitions:

$$\phi := \frac{n}{8(n-1)} \left(\frac{r}{l}\right)^n, \qquad g_{\alpha\beta} := \phi^{2(n-1)/n} \bar{g}_{\alpha\beta}, \tag{4}$$

where $n \equiv d - 2$. (Note that this requires us to restrict to $n > 1$.) These definitions yield a dimensionally reduced action of the form:

$$S = \frac{1}{2G} \int d^2 x \sqrt{-g} \left[\phi R(g) + V^{(n)}(\phi, \Lambda) \right] - \int d^2 x \sqrt{-g} \, H^{(n)}(\phi) |\partial \chi|^2 \tag{5}$$

where $l \equiv (G^{(d)})^{n/2}$, $G \equiv \pi n/(n-1)$ and we have defined

$$H^{(n)}(\phi) \equiv \frac{8(n-1)}{n} \phi \tag{6}$$

$$V^{(n)}(\phi, \Lambda) \equiv \frac{1}{n} \left(\frac{8(n-1)}{n} \right)^{\frac{1}{n}} \phi^{\frac{1}{n}} \left(-l^2 \Lambda + \frac{n^2}{8} \left(\frac{8(n-1)}{n} \right)^{\frac{n-2}{n}} \phi^{-2/n} \right), \tag{7}$$

An overall factor proportional to the unit n-sphere volume has been dropped. The action (5) is one that has been studied extensively [10]. For arbitrary $V(\phi)$ and $H(\phi)$, it corresponds to generic dilaton gravity theory coupled to a scalar field in two spacetime dimensions. In the present context, the dilaton ϕ is proportional to the area of the n-sphere a radial distance r from the origin.

The vacuum equations ($\chi = 0$) for this class of theories can be solved exactly. In the coordinates $x = l\phi$, the vacuum solution for the metric is

$$ds^2 = -(j(\phi) - 2GM)dt^2 + (j(\phi) - 2GM)^{-1}dx^2 \tag{8}$$

where M plays the role of mass, and $j(\phi) \equiv \int_0^\phi d\tilde{\phi}\, V(\tilde{\phi})$.

This solution applies for all $n > 1$, with

$$j^{(n)}(\phi) = \frac{1}{n}\left(\frac{8(n-1)}{n}\right)^{\frac{1}{n}}\left(-l^2\Lambda\left(\frac{n}{n+1}\right)\phi^{\frac{n+1}{n}} + \frac{n^3}{8(n-1)}\left(\frac{8(n-1)}{n}\right)^{\frac{n-2}{n}}\phi^{\frac{n-1}{n}}\right) \tag{9}$$

The metric (8) is singular at $\phi = 0$ even when $M = 0$. Up to numerical constants, $j(\phi) \to \phi^{1-1/n} \to 0$ as $\phi \to 0$. This not a physical singularity since the corresponding physical metric $\bar{g}$ is flat at the origin. However, the vanishing of $j(\phi)$ doesaffect the choice of boundary conditions in our numerical method.

We now go to double null coordinates, in which the metric is parametrized as

$$ds^2 = -2lg(u,v)\phi'(u,v)dudv \tag{10}$$

where the prime denotes partial differentiation with respect to the null coordinate v. (Recall that this is just the $u - v$ part of the physical metric.) In these coordinates, the relevant field equations are remarkably simple:

$$\dot{\phi}' = -\frac{l}{2}V^{(n)}(\phi)g\phi' \tag{11}$$

$$\frac{g'\phi'}{g\phi} = 8\pi(\chi')^2 \tag{12}$$

$$(\phi\chi')^\cdot + (\phi\dot{\chi})' = 0. \tag{13}$$

In the above, the dot refers to differentiation with respect to u, which is treated like the "time" coordinate for the purposes of the following numerical integration. Eqs. (11-13) are virtually identical in form to those studied in [9] in the context of 2+1 dimensional AdS gravity. Amazingly, the dependence on n and Λ appears only in the dilaton potential $V^{(n)}$ in [11]. It is this feature that allows us to consider arbitrary n and Λ within the same code.

The evolution equations may be put in a form more useful for numerical solution by replacing the scalar field χ, by a new field $h \equiv \chi + 2\phi\chi'/\phi'$.

This yields evolution equations of the form:

$$\dot{\phi} = -\tilde{g}/2 \tag{14}$$

$$\dot{h} = \frac{1}{2\phi}(h - \chi)(g\phi V - \tilde{g}), \tag{15}$$

where
$$\tilde{g} = \int_u^v (g\phi'V)d\tilde{v}. \qquad (16)$$

and χ is now to be considered a functional of h and g given by

$$\chi = \frac{1}{2\sqrt{\phi}} \int_u^v dv \left[\frac{h\phi'}{\sqrt{\phi}}\right] + \frac{K_3(u)}{\sqrt{\phi}} \qquad (17)$$

The integration constant $K_3(u)$ must be zero because the definition of h requires $h = \chi$ at $\phi = 0$. The function g is obtained by integrating the constraint (12):

$$g = K_1(u)\exp\left[4\pi \int_u^v dv \frac{\phi'}{\phi}(h-\chi)^2\right] \qquad (18)$$

where $K_1(u)$ is again an integration constant (i.e. independent of the "spatial" coordinate v).

Since we are considering a spherical, hollow shell of collapsing matter with no black hole initially in the interior, we must ensure that the spacetime at the center of the shell corresponds to flat Minkowski spacetime. In order to determine the correct boundary conditions on our field variables, we transform the vacuum ($M = 0$) metric (8) to double null coordinates (10) to obtain:

$$ds^2 = -j(\phi)dudv \qquad (19)$$

where $u = t - \phi^*$ and $v = t + \phi^*$, with the generalized "tortoise coordinate" ϕ^* defined by

$$\phi^* = l \int_0^\phi \frac{d\phi}{j(\phi)} \qquad (20)$$

With these definitions, the origin $\phi = 0$ corresponds to the surface $v = u$. Moreover, for the vacuum solution

$$\phi' \equiv \frac{\partial \phi}{\partial v} = \frac{1}{2}\frac{j(\phi)}{l} \qquad (21)$$

Comparing metric (19) to our general form (10) we see that $g = 1$ for the vacuum solution. This determines the integration constant $K_1(u) = 1$. Moreover, we find that

$$\dot{\phi} \equiv \frac{\partial \phi}{\partial u} \to -\frac{1}{2}\frac{j(\phi)}{l} \qquad (22)$$

which vanishes at $\phi = 0$ in agreement with the expression (16).

Numerical Method

The u coordinate is considered as "time," and a discretization of the coordinate v gives coupled ODEs by the prescription

$$h(u,v) \to h_i(u), \qquad \phi(u,v) \to \phi_i(u). \qquad (23)$$

where $i = 0, \cdots, N$ specifies the v grid. Initial data for these two functions are prescribed at $u = 0$, from which the functions $g(u,v), \tilde{g}(u,v)$ are constructed. Evolution in the 'time' variable u is performed using the 4th. order Runge-Kutta method.

Our initial data for the dilaton is $\phi(0,v) = v$ and we have considered two types of initial scalar field configurations in order to test for universality of the critical exponent: Gaussian and tanh functions

$$\chi_G(u=0,\phi) = a\phi \exp\left[-\left(\frac{\phi-\phi_0}{\sigma}\right)^2\right], \text{ and } \chi_T(u=0,\phi) = a\tanh(\phi). \quad (24)$$

In both cases we varied the amplitude a in order to measure the critical exponent.

For reasons given above, boundary conditions at fixed u are $\tilde{g}_k = 0$ and $g_k = 1$ where k is the index corresponding to the position of the origin, i.e. $\phi_k = 0$. All grid points $0 \leq i \leq k-1$ correspond to ingoing rays that have reached the origin and are dropped from the grid. These conditions are equivalent to $g|_{r=0} = g(u,u) = 1$.

In order to check for black hole formation we examine the function

$$ah \equiv g^{\alpha\beta}\partial_\alpha\phi\partial_\beta\phi = -\frac{\dot{\phi}}{lg}, \quad (25)$$

whose vanishing signals the formation of an apparent horizon.

The general procedure is then the following. For fixed amplitude, a, the code is run to evolve the initial data given above. At each time step (i.e. u-step), the function ah is checked to see if it vanishes anywhere besides the origin. If so, the run is terminated, and the value of ϕ at the vanishing point determines the area of the horizon and corresponding radius R_{bh} on black hole formation. In practice we cannot achieve $ah = 0$ precisely, but instead look for some optimal small value which is typically $\sim 10^{-4}$. In the subcritical case, it is expected that all the radial grid points reach zero without detection of an apparent horizon. This is the signal of pulse reflection. The results (a, R_{bh}) for many values of a are plotted and fitted to a relationship of the form [1] in order to determine the critical exponent γ.

Results and Prospects

The code was first tested by checking known results in four and six dimensions. It was then run for the five dimensional case with zero, positive and negative cosmological constant. We are currently doing runs to map out the dependence of the critical exponent on the two parameters n and Λ. The figures below show the scaling law for a couple of specific cases. The squares represent the points (a, R_{ah}) and the lines are the least squares fit to these points. All the graphs we obtained exhibit an oscillation about the least squares fit line. This is a known feature for zero cosmological constant, and is concomitant with discrete self-similarity of the critical solution. Our results for negative comological constant also show this feature, which indicates that the critical solution for this case also has discrete self-similarity.

The following is a list of results obtained to date:

$\underline{d=4, \Lambda=0}$: We find exponents $\gamma = 0.36$ and $\gamma = 0.35$ for Gaussian and tanh initial data respectively. These are in good agreement ($\sim 4\%$) with earlier studies [1, 2, 3], and represent a good test of our formalism and code.

$\underline{d=6, \Lambda=0}$: We find $\gamma = 0.44$ and $\gamma = 0.41$ for Gaussian and tanh initial data respectively. For comparison, the result in Ref. [7] is $\gamma = 0.424$. In six dimensions this is the

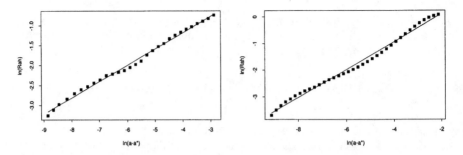

FIGURE 1. $5d, \Lambda = 0$: logarithmic plot of apparent horizon radius R_{ah} versus amplitude $(a - a_*)$; $\gamma = 0.41$ (left, from tanh data) and $\gamma = 0.52$ (right, from Gaussian data).

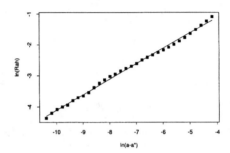

FIGURE 2. $5d, \Lambda = -1$, Gaussian data: logarithmic plot of apparent horizon radius R_{ah} versus amplitude $(a - a_*)$; $\gamma = 0.49$

first evidence for universality since Ref. [7] contains results only for a specific gaussian form of initial data, different from the one used here.

$d = 5, \Lambda = 0$: Fig. 1 gives the results for the tanh and Gaussian initial data; $\gamma = 0.41$ and $\gamma = 0.52$ respectively. This suggests that there is no universality in five dimensions. The reason for this is not clear to us and it would be worthwhile calculating the exponents using the subcritical approach suggested in Ref. [13], where the Ricci scalar at the origin is calculated near criticality from below. It is also worth noting that a similar lack of universality is manifested in the 3–dimensional AdS case using the supercritical apparent horizon method of computing γ [14].

$d = 5, \Lambda = -1$: Fig. 2 give the results for the Gaussian data; $\gamma = 0.49$.

$d = 5, \Lambda = +1$: This is an interesting case because of the presence of a cosmological horizon in addition to the potential apparent horizon. Fig. 3 shows graphs of the scalar field and apparent horizon functions h and ah in the left and right columns respectively, as functions of ϕ, prior to and at the onset of apparent horizon formation in the two successive rows. Note the location of the cosmological horizon in the right hand column

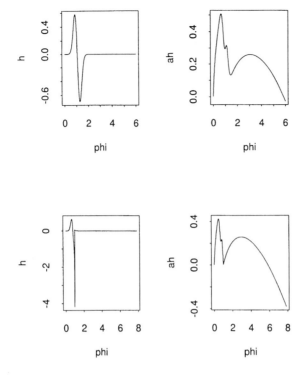

FIGURE 3. 5d, $\Lambda = +1$, Gaussian data: plots of the scalar field h and the ah function Eqn. (25) prior to (top row) and nearer apparent horizon formation. Note the expansion of the ϕ grid from 6 to 8 in the bottom two graphs.

near $\phi = 5.8$. We find that the function $\phi(u,v)$ evolved such that instead of the radial grid contracting as for the zero and negative Λ cases, it expanded as the scalar field moved towards the origin. This feature is visible in Fig. 6: the range of ϕ in the lower graphes has expanded to 8 from 6. In fact the closer is the onset of apparent horizon formation, the larger the range of the ϕ variable (and hence the radial grid). This prevented us from extracting accurate apparent horizon radii since the interesting features became confined to an ever shrinking part of the grid.

Given the successful tests of our results with earlier work, and the new results in five dimensions, we are in a position to probe other sectors of the (d,Λ) parameter space. Three issue are of particular interest: (i) Further verification and understanding of the apparent lack of universality in five dimensions. (ii) Probing of the negative Λ case to see if there is a dependence of γ on the initial radial position of the collapsing pulse. That this may be the case is suggested by the intuition that a collapsing pulse will "feel" the cosmological constant if it starts at a radial position comparable to the cosmological constant length scale, whereas if it starts out very close to the origin compared to this scale, the effect of Λ on the collapse should be negligible. (iii) Understanding the positive Λ case, particularly its expanding grid pecularity. These and other related issues are currently under investigation.

Acknowledgements: This work was supported in part by the Natural Sciences and Engineering Research Council of Canada. We thank J. Gegenberg for helpful conversations.

REFERENCES

1. M. Choptuik, Phys. Rev. Lett. **70**, 9 (1993).
2. C. Gundlach, Living Rev. Rel. 2, 4 (1999).
3. L. Lehner, Class. Quant. Grav. 18, R25-R86 (2001).
4. V. Husain, E. Martinez, and D. Nunez, Class. Quant. Grav. **13** 1183 (1996).
5. V. Husain, E. Martinez, and D. Nunez, Phys. Rev. **D50** 3783 (1994).
6. D. Garfinkle, Phys. Rev. **D51**, 5558 (1995).
7. D. Garfinkle, C. Cutler and G. Duncan, Phys. Rev. **D60**, 104007 (1999).
8. F. Pretorius and M. Choptuik, Phys. Rev. **D62** 124012 (2000).
9. V. Husain and M. Olivier, Class. Quant. Grav **18** L1-L10 (2001).
10. J. Gegenberg, D. Louis-Martinez and G. Kunstatter, Phys. Rev. **D51** 1781 (1995); D. Louis-Martinez and G. Kunstatter, Phys. Rev. **D52** 3494 (1995).
11. D. Goldwirth and T. Piran, Phys. Rev. **D36**, 3575 (1987).
12. R.M. Wald and V. Iyer, Phys. Rev. **D44**, 3719 (1991).
13. G. Garfinkle and C. Duncan, Phys. Rev. **D58** (1998) 064024.
14. F. Pretorius, personal communication.
15. L. Lehner, Int. J. Mod. Phys. **D9** 459 (2000).
16. M. Birukou, V. Husain, G. Kunstater, E. Vaz, M. Olivier, Phys. Rev. D in press, gr-qc/0201026.

Spacetime Ambiguities

Kayll Lake [1]

Department of Physics, Queen's University, Kingston, Ontario, Canada, K7L 3N6

Abstract. With the advent of the database GRDB, geometric objects can be subjected interactively to any algorithmic procedure. Given an entry in the database (*e.g* a metric), one might like to know, for example, if the entry is (say) necessarily a perfect fluid solution in general relativity. In this note I point out that degeneracies arise, so that reversing Einstein's equations leads to multiple inequivalent interpretations of a given "exact solution".

INTRODUCTION

The sheer size and complexity of even routine calculations in general relativity has hindered its development, integration and application. The advent of modern computer algebra systems offers the potential for fundamental advances in general relativity and related fields including: (i) enhancement of our understanding of spacetime, (ii) the relationship of spacetime and its integration to more modern concepts and (iii) the application of general relativity (and generalizations) to realistic astrophysical situations. The computer interface GRTensorJ allows the program GRTensor (grtensor.org) to be run interactively over the World-Wide-Web and has been used to develop a geometric "database" GRDB (grdb.org). GRDB is designed to be a database of geometric objects in the general area of differential geometry with no a priori restriction on the number of dimensions. It contains exact solutions of Einstein's field equations without being restricted to them. By design the database is highly interactive. The records (manifolds) are generally stored only in terms of a metric or a set of basis vectors along with constraints and referencing data. Other elements can be calculated and displayed as required either interactively or from an extensive menu. GRDB reflects the future of the client's side of symbolic computation: no learning curve, no software updates and a knowledge base instead of documentation [1].

The usual procedure for finding exact solutions of Einstein's equations involves writing down a phenomenological energy-momentum tensor in a set of coordinates so that the field equations can be integrated, often with the aide of simplifying assumptions. In view of the difficulty of solving Einstein's equations, inverse problems are important. An inverse problem of interest (not explicitly involving the classification problem for the Ricci tensor) may be stated thus: Given a spacetime M with metric $g_{\alpha\beta}$, what, if any, fluid flow could generate M via Einstein's equations? In simple terms, given a metric in GRDB, one might like to know if it is necessarily, for example, a perfect fluid. In

---

[1] Electronic Address: lake@astro.queensu.ca

this note I point out that degeneracies could arise, so that reversing Einstein's equations could lead to multiple inequivalent interpretations of a given "exact solution".

REVERSING EINSTEIN'S EQUATIONS

To start, we need a phenomenological energy-momentum tensor. Here we take this to be

$$T^\alpha_\beta = \rho u^\alpha u_\beta + p_1 n^\alpha n_\beta + p_2 \delta^\alpha_\beta + p_2(u^\alpha u_\beta - n^\alpha n_\beta) - 2\eta \sigma^\alpha_\beta, \qquad (1)$$

where the congruence of unit timelike vectors u^α represents the fluid flow, σ^α_β is the shear associated with u^α and η the phenomenological shear viscosity. To restrict u^α consider the congruence tangent to an open region R of a two dimensional hypersurface Σ in a spacetime M (so the field u^α is necessarily irrotational). Under this restriction it follows that the condition

$$G^\beta_\alpha u^\alpha n_\beta = 0, \qquad (2)$$

where n^α is in the tangent space of M orthogonal to u^α and G^β_α is the Einstein tensor of M, determines u^α uniquely [2]. With (2) imposed, the fluid can be considered "non-conducting". We now set up systems of equations to be solved simultaneously for the phenomenological parameters (ρ, p_1, p_2, η) in terms of quantities that follow algorithmically from u^α and the metric alone [3].

Scalars

To proceed in a manifestly covariant way we construct scalars from the set $(G^\beta_\alpha, u^\alpha, n_\alpha)$. Four scalars linear in η are: $G^\beta_\alpha u^\alpha n_\beta$ (used in condition (2)), $G^\beta_\alpha u^\alpha u_\beta$, $G^\beta_\alpha n^\alpha n_\beta$ and $G^\alpha_\alpha \equiv G$. With (1) it follows that

$$G = 8\pi(-\rho + p_1 + 2p_2), \qquad (3)$$

and

$$G^\beta_\alpha u^\alpha u_\beta \equiv G1 = 8\pi\rho. \qquad (4)$$

In all cases we take (4) as the definition of the energy density ρ. Further,

$$G^\beta_\alpha n^\alpha n_\beta \equiv G2 = 8\pi(p_1 - 2\eta\Delta) \qquad (5)$$

where

$$\Delta \equiv \sigma^\beta_\alpha n^\alpha n_\beta. \qquad (6)$$

---

[2] The form of u^α follows algorithmically from the metric via (2) and the unity of u^α subject to a time orientation on Σ.

[3] The decomposition (1) imposes restrictions on ds^2. Writing $ds_M^2 = ds_\Sigma^2 + ds^2$ and writing the coordinates x^1, x^2 on Σ here we take $ds^2 = w(x^1, x^2, x^3, x^4)((dx^3)^2 + (dx^4)^2)$. This is somewhat more general than type B warped product spaces and includes, for example, all spherical spaces.

Rearrangement of (3), (4) and (5) gives

$$48\pi\eta\Delta = G + G1 - 3G2 + 16\pi P \tag{7}$$

where $P \equiv p_1 - p_2$. If $\sigma_\alpha^\beta = 0$ or $\eta \equiv 0$ or $P = 0$ then (3), (4) and (5) form a complete set. For example, it follows that

$$G + G1 = 3G2 \tag{8}$$

is a necessary and sufficient condition for a perfect fluid if $\sigma_\alpha^\beta = 0$ or $\eta \equiv 0$ or $P = 0$. More generally, however, we need to include scalars quadratic in η. These are $G_\beta^\alpha G_\delta^\beta n_\alpha u^\delta$, $G_\beta^\alpha G_\delta^\beta u_\alpha u^\delta$, $G_\beta^\alpha G_\delta^\beta n_\alpha n^\delta$, and $G_\beta^\alpha G_\alpha^\beta$. Only the last two scalars are useful here since with (1) it follows that $G_\beta^\alpha G_\delta^\beta n_\alpha u^\delta = 0$ and that $G_\beta^\alpha G_\delta^\beta u_\alpha u^\delta = -(8\pi\rho)^2$. With (1) it follows that

$$G_\beta^\alpha G_\alpha^\beta \equiv G3 = (8\pi)^2(\rho^2 - 4p_1\eta\Delta + 4p_2\eta\Delta + p_1^2 + 2p_2^2 + 4\eta^2\sigma_2), \tag{9}$$

where

$$\sigma_2 \equiv \sigma_\beta^\alpha \sigma_\alpha^\beta \tag{10}$$

and

$$G_\beta^\alpha G_\delta^\beta n_\alpha n^\delta \equiv G4 = (8\pi)^2(p_1^2 - 4p_1\eta\Delta + 4\eta^2\sigma_3), \tag{11}$$

where

$$\sigma_3 \equiv \sigma_\beta^\alpha \sigma_\delta^\beta n_\alpha n^\delta. \tag{12}$$

Equations (9) and (11) are not independent of (3), (4) and (5) but rather merely generate syzygies. Solving we find that

$$p1 = \frac{G2}{8\pi} + 2\eta\Delta, \tag{13}$$

and

$$p2 = \frac{G + G1 - G2}{16\pi} - \eta\Delta, \tag{14}$$

where, with (11)

$$(16\pi)^2 \eta^2 (\Delta^2 - \sigma_3) = G2^2 - G4, \tag{15}$$

and with (9)

$$(16\pi)^2 \eta^2 (3\Delta^2 - 2\sigma_2) = G^2 + 3(G1^2 + G2^2) - 2G2(G + G1) + 2(GG1 - G3). \tag{16}$$

Now in comoving coordinates it easy to show that

$$\Delta^2 = \frac{2}{3}\sigma_2 = \sigma_3. \tag{17}$$

However, Δ, σ_2 and σ_3 are invariantly defined quantities (the flow u^α is uniquely defined but its coordinate representation is not) and so (17) holds in general. Use of G3 and G4 merely leads us to the syzygies

$$G4 = G2^2, \tag{18}$$

149

and
$$G3 = \frac{G^2 + 3(G1^2 + G2^2)}{2} + G(G1 - G2) - G1G2, \qquad (19)$$

while η remains arbitrary [4].

Generalized Walker condition

Writing the coordinates x^1, x^2 on Σ with M orientated via $u^1 > 0$ it follows from (1) and Einstein's equations that

$$\widetilde{G}_2^1 \widetilde{G}_1^2 - (\widetilde{G}_1^1 - \widetilde{G}_3^3)(\widetilde{G}_2^2 - \widetilde{G}_3^3) \equiv \widetilde{G5} = (8\pi)^2 P(\rho + p_2), \qquad (20)$$

where $\widetilde{G}_\beta^\alpha = G_\beta^\alpha + 16\pi\eta\sigma_\beta^\alpha$ [5]. Although (20) holds in every M, it is not manifestly covariant. Use of (20) leads to local syzygies while η remains arbitrary.

EXAMPLES

Few examples of exact fluid solutions of Einstein's equations are available in non-comoving coordinates. From the recent work of Senovilla and Vera [4], their solution (40)

$$ds^2 = -(dt)^2 + (dx)^2 + \frac{\cos^{1+2\nu}(\mu x)}{\cosh^{2\nu-1}(\mu t)}(dy)^2 + \frac{\cos^{1-2\nu}(\mu x)}{\cosh^{-2\nu-1}(\mu t)}(dz)^2 \qquad (21)$$

has $\sigma_\alpha^\beta \neq 0$ but is consistent with (1) only for $\eta \equiv 0$. Equation (8) is satisfied and so the metric (21) necessarily represents a perfect fluid. The Lemaître space consists of the metric [5]

$$ds^2 = -(dt)^2 + \frac{(R'(t,r))^2 (dr)^2}{1 + f(r)} + R(t,r)^2 d\Omega^2, \qquad (22)$$

along with the constraints

$$\dot{R}(t,r) = \sqrt{2 \frac{m(r)}{R(t,r)} + f(r)} \;, \ddot{R}(t,r) = -\frac{m(r)}{R^2(t,r)}, \qquad (23)$$

$$\ddot{R}'(t,r) = -\frac{m'(r)}{R^2(t,r)} + 2\frac{m(r)R'(t,r)}{R^3(t,r)} \qquad (24)$$

---

[4] There are no independent scalars quartic or higher in η [2].
[5] If $\sigma_\alpha^\beta = 0$ and $\rho + p_2 \neq 0$ then it follows from the classic work of Walker [3] that $G_2^1 G_1^2 = (G_1^1 - G_3^3)(G_2^2 - G_3^3)$ is a necessary and sufficient condition for a perfect fluid.

and
$$\dot{R}'(t,r) = \frac{2m'(r)R(r,t) - 2m(r)R'(r,t) + f'(r)R^2(r,t)}{2R(r,t)\sqrt{(2m(r) + f(r)R(r,t))R(r,t)}}, \qquad (25)$$

where $' \equiv \frac{\partial}{\partial r}$ and $\cdot \equiv \frac{\partial}{\partial t}$. The field u^α is necessarily comoving. Equation (8) is satisfied and so if $\eta \equiv 0$ it follows that the source is a perfect fluid, in fact simply dust (since $G2 = 0$) with

$$\rho = \frac{m'(r)}{4\pi R^2(r,t) R'(r,t)}. \qquad (26)$$

However, it follows that the metric (22) (with the given constraints) can also be interpreted as an imperfect fluid of the same ρ but with $p_1 = 2\eta\Delta$ and $p_2 = -\eta\Delta$ with η arbitrary.

ACKNOWLEDGMENTS

This work was supported by a grant from the Natural Sciences and Engineering Research Council of Canada. Portions of this work were made possible by use of *GRTensorII* [6]. It is a pleasure to thank Nicos Pelavas for comments. A more complete account of this work is in progress with Mustapha Ishak.

REFERENCES

1. Ishak, M., and Lake, K., *Class. Quantum Grav.* **19**, 505-514 (2002).
2. Pollney, D., Pelavas, N., Musgrave, P. and Lake, K., *Computer Physics Communications* **115**, 381-394 (1998).
3. Walker, A. G., *Quarterly Journal of Mathematics* **6**, 81-93 (1935).
4. Senovilla, J.M and Vera, R., *Class. Quant. Grav.* **15**, 1737-1758 (1998).
5. This is commonly called the "Tolman-Bondi" metric and can be found in any elementary general relativity text as the spherically symmetric solution for "dust".
6. This is a package which runs within Maple. It is entirely distinct from packages distributed with Maple and must be obtained independently. The GRTensorII software and documentation is distributed freely on the World-Wide-Web from the address `grtensor.org`

Imposing a 4-Dimensional Background on General Relativity

Edward M. Schaefer

2822 New Providence Ct., Falls Church, VA 22042

Abstract. It is postulated that there exists a flat background spacetime whose increments can be directly measured by an observer, and on which local curved spacetimes exist. To make this premise work it must also be postulated that large spatial distances are measured by an observer using parallax and that the solutions to the Einstein Field Equations may be distorted to have a flat background. The mathematics of deriving background metrics and doing distortions are discussed. The distortion of the Schwarzschild Solution and some of its effects are discussed, including $r <= 2m$ being made inaccessible.

INTRODUCTION

Although it is well accepted and verified, General Relativity (GR)[1, 2, 3] is not without its problems. Among them are the singularities of black holes[4], reachable positions at which the math of GR breaks down.

In this article, it is assumed that solving the Einstein Field Equations (EFE) is only part of the process of generating local metrics of spacetime, with the behavior of the black hole spacetimes indicating that this is not the final step in constructing metrics. To move beyond the EFE, it is also assumed that there exists a flat background spacetime to which the local spacetimes must conform, as described below. The resultant theory calls for the local metrics to be distortions of the EFE Solutions. The geodesic equations of the distorted metrics remove black holes by making their part of the manifold inaccessible.

POSTULATES AND THE OVERALL PREMISE

Postulate 1: There exists a background spacetime which is Minkowski in form, has intervals whose relative lengths are directly measured by an observer in the spacetime using their local clock and rod, and on which the local curved spacetimes exist.

Postulate 1 requires that it be possible to differentiate between locally measured intervals and those as measured by a single observer. For example, the temporal coordinate of the Schwarzschild Solution is a background time: It is based on the observations of a distant observer and its intervals can differ from those of the local clock. However, for spatial intervals there is a problem since the use of rods at rest[5, 6] forces all at-rest observers to agree on spatial lengths. This requires the next postulate:

Postulate 2: The spatial distance to a distant position is measured by the parallax of that position along a local baseline whose length is measured using the local rod.

The use of parallax to measure large distances is justified in [7]. Its use creates a gravitational length contraction effect, which complements gravitational time dilation and causes spatial distances to be increased for observers at lower potentials.

As shown later, the background spacetimes for curved EFE solutions usually are not flat (as required by Postulate 1). So it is also assumed that:

Postulate 3: EFE solutions correctly describe submanifolds of equal maximal clock rate but not the separations between those submanifolds.

Overall Premise: The local metrics of spacetime are distortions of the EFE solutions in the direction of changing maximal clock rate and/or parallactically measured rod length such that their background spacetime is flat.

MATHEMATICAL REPRESENTATION

Extended Tensor Syntax

To support this discussion, an extended tensor syntax is used: $^{type}_{source}X^{vector\ indicies}_{form\ indicies}$, where *type* is a code defining the type of values in a portion of a metric tensor, and *source* is the identifying character of a metric tensor from which another tensor arises. (Example: $^{t}_{g}ds^2$ is the temporal part of the spacetime interval squared for the metric $g_{\mu\nu}$).

The background spacetime and its metric

From Postulate 1: "There exists a background spacetime ...". This means that there exists a background spacetime interval B and a background metric $b_{\mu\nu}$ such that

$$dB^2 = b_{\mu\nu} dx^\mu dx^\nu, \qquad (1)$$

where x^μ are the same coordinates used with the corresponding local metric or EFE Solution ($g_{\mu\nu}$).

Also from Postulate 1: The background spacetime "has intervals ... [that are] measured ... using [one's] local clock and rod". Given that at a distant position at a lower potential in a gravitational field there is

- a gravitational time dilation effect which decreases the observed temporal intervals between events,
- a gravitation length contraction effect (due to Postulate 2) which is of the same magnitude as the gravitational time dilation and increases the observed spatial distance between events[7], and
- a maximal relative rate at which a clock at the distant position may be observed to run ($\overset{*}{\tau}$) which is also the magnitude of the gravitational time dilation and length contraction effects,

it follows that the background and local metrics are related by:

$$b_{\mu\nu} = {}^{t}g_{\mu\nu}/(\overset{*}{\tau})^2 + {}^{s}g_{\mu\nu}(\overset{*}{\tau})^2, \qquad (2)$$

where $^t g_{\mu\nu}$ is the time-like part of the local metric and $^s g_{\mu\nu}$ is the space-like part of the local metric. How metrics are split into time-like and space-like parts is discussed in [8], and briefly in the example for the accelerated box below.

Distortion

From the Overall Premise: "The local metrics of spacetime are distortions of the EFE solutions ...". Distortion is an operation that maintains the coordinates but changes the geometry. It is represented by a mixed rank-two tensor $N^\mu{}_\nu$ such that $h_{\mu\nu} = g_{\mu\alpha} N^\alpha{}_\nu$, where $g_{\mu\nu}$ is the original metric tensor and $h_{\mu\nu}$ is its distortion.

The distortion tensor is

$$N^\mu{}_\nu = \delta^\mu_\nu + S^\mu{}_\nu \tag{3}$$

where δ^μ_ν is the Kronecker delta and $S^\mu{}_\nu$ is the stretching tensor. $S^\mu{}_\nu$ in turn is

$$S^\mu{}_\nu = Q \, \hat{s}^\mu \otimes \hat{s}_\nu \tag{4}$$

where Q is the coefficient of stretching[1], $\hat{s}^\mu$ is a unit vector oriented in the direction of stretching, and $\hat{s}_\nu$ is a unit form oriented in the direction of stretching. Q, $\hat{s}^\mu$, and $\hat{s}_\nu$ may be functions of position.

Implementation of the Overall Premise

The Overall Premise states that the distortion is "in the direction of maximal clock time ...". For the stretching tensor definition given in (4), this indicates that

$$\hat{s}^\mu = \overset{*,\mu}{\tau} / \left| \overset{*,\mu}{\tau} \right| \quad \text{and} \quad \hat{s}_\nu = \overset{*}{\tau}_{,\nu} / \left| \overset{*}{\tau}_{,\nu} \right|, \tag{5}$$

where "," is the covariant derivative operation.

The Overall Premise also states that the distortion is "such that the background becomes flat". So given a distorted background metric $i_{\mu\nu}$ such that

$$i_{\mu\nu} = b_{\mu\alpha} N^\alpha{}_\nu = b_{\mu\alpha} \left(\delta^\alpha_\nu + Q \, \hat{s}^\alpha \otimes \hat{s}_\nu \right), \tag{6}$$

there exists at each position in the spacetime a value for Q such that

$$_i R_{\nu\mu\sigma o} = 0 \tag{7}$$

where $_i R_{\nu\mu\sigma o}$ is the Riemann tensor for $i_{\mu\nu}$. In this article, only stationary spacetimes are discussed. In these cases, the direction of changing maximal clock time and therefore the distortion itself cannot have a time-like component. With only the space-like part of

---

[1] Q is related to the factor of change in the incremental distance between coordinates in the direction of stretching (n) by $Q = n^2 - 1$.

the metric being affected, a distortion of a local metric or EFE Solution will distort the background in the same way. Therefore the local metrics for stationary spacetimes are

$$j_{\mu\nu} = g_{\mu\alpha} N^{\alpha}{}_{\nu} \qquad (8)$$

where $N^{\alpha}{}_{\nu}$ is the distortion used to flatten the background metric for the EFE Solution $g_{\mu\alpha}$, obtained by using (2), (5), and (6), and then solving for Q in (7).

DISTORTIONS OF EFE SOLUTIONS

The Accelerated Box

The tetrad spacetime interval for an observer in Minkowski spacetime who is being accelerated in the $+z$ direction at a rate of a is[9]:

$$ds^2 = g_{\mu\nu} dx^{\mu} dx^{\nu} = (1+az)^2 dt^2 - dx^2 - dy^2 - dz^2. \qquad (9)$$

In (9), $\overset{*}{\tau} = 1+az$, and this occurs when a particle is undergoing an incremental displacement of $d\xi^{\mu} = (dt, 0, 0, 0)$ [where the tensor indices are (t, x, y, z)]. As shown in [8], the temporal component of (9) is then

$$^t ds^2 = g_{\mu\nu} d\xi^{\mu} d\xi^{\nu} = (1+az)^2 dt^2, \qquad (10)$$

and the spatial component of (9) is

$$^s ds^2 = ds^2 - {^t ds^2} = -dx^2 - dy^2 - dz^2. \qquad (11)$$

Using (1) and (2) with (10) and (11) gives a background spacetime interval for (9) of

$$dB^2 = dt^2 - (1+az)^2 (dx^2 + dy^2 + dz^2). \qquad (12)$$

It turns out that the Riemann tensor for (12) is null ($_b R_{\mu\nu\sigma o} = 0$). This means that Minkowski spacetime is not affected, and that this theory maintains GR's correspondence with Special Relativity for spacetimes far from gravitational sources.

The Schwarzschild Solution

The Schwarzschild Solution for the external vacuum spacetime surrounding a spherically symmetric non-rotating massive object is[10, 11]:

$$_g ds^2 = (1 - 2m/r) dt^2 - (1 - 2m/r)^{-1} dr^2 - r^2 d\theta^2 - (r \sin\theta)^2 d\phi^2. \qquad (13)$$

For (13), $\overset{*}{\tau} = \sqrt{1 - 2m/r}$, $d\xi^{\mu} = (dt, 0, 0, 0)$, and the background spacetime interval is

$$dB^2 = dt^2 - dr^2 - r(r - 2m)[d\theta^2 + (\sin\theta)^2 d\phi^2]. \qquad (14)$$

The Riemann tensor for (14) is not null[2], and so (13) needs to be distorted to have a flat background. Since the $\overset{*}{\tau}$ for (13) is a function only of r, the stretching tensor for (14) obtained using (4) and (5) is $S^r{}_r = -Q(r)$ with all other terms being zero[3]. (3) then gives a distortion tensor of

$$N^r{}_r = 1 - Q(r), \quad N^t{}_t = N^\theta{}_\theta = N^\phi{}_\phi = 1, \quad \text{all other terms 0.} \tag{15}$$

(6) and (15) produce a distorted background for (14) of

$$dB^2 = i_{\mu\nu}dx^\mu dx^\nu = dt^2 - [1 - Q(r)]dr^2 - r(r - 2m)[d\theta^2 + (\sin\theta)^2 d\phi^2]. \tag{16}$$

The independent non-zero terms of the Riemann tensor for (16) ($_iR_{\mu\nu\sigma\sigma}$) are

$$_iR_{r\theta r\theta} = \tfrac{1}{2}\left[\tfrac{d}{dr}Q(r)\right](r - m)/[1 - Q(r)] - m^2/[r(r - 2m)], \tag{17}$$

$$_iR_{r\phi r\phi} = {}_iR_{r\theta r\theta}(\sin\theta)^2, \quad \text{and} \tag{18}$$

$$_iR_{\theta\phi\theta\phi} = (\sin\theta)^2\left[Q(r)r^2 - 2rQ(r)m + m^2\right]/[1 - Q(r)]. \tag{19}$$

(17), (18), and (19) all go to zero [as required by (7)] when

$$Q(r) = -m^2/[r(r - 2m)]. \tag{20}$$

(8), (15), and (20) then produce a local metric ($j_{\mu\nu}$) for (13) whose spacetime interval is

$$_jds^2 = (1 - 2m/r)dt^2 - \left[(r - m)^2/(r - 2m)^2\right]dr^2 - r^2d\theta^2 - (r\sin\theta)^2 d\phi^2. \tag{21}$$

(21) is referred to as the Distorted Schwarzschild Solution.

The Equations of Motion for the Distorted Schwarzschild Solution

Given that the metrics generated by (8) [such as (21)] are the true local metrics of spacetime, it follows that the geodesic principle applies with respect to them. Therefore

$$\ddot{x}^\mu + {}_j\Gamma^\mu{}_{\rho\sigma}\dot{x}^\rho \dot{x}^\sigma = 0 \tag{22}$$

where $\dot{x}^\mu = dx^\mu/d\tau$, $\ddot{x}^\mu = d^2x^\mu/d\tau^2$, and $_j\Gamma^\mu{}_{\rho\sigma}$ are the connections for $j_{\mu\nu}$.
The geodesic equations for (21) are

$$\ddot{t} + 2m\dot{t}\dot{r}/[r(r - 2m)] = 0, \tag{23}$$

$$\ddot{r} + (r - 2m)^2 m(\dot{t})^2/\left[(r - m)^2 r^2\right] - m(\dot{r})^2/[(r - 2m)(r - m)]$$
$$- (r - 2m)^2 r\left[(\dot{\theta})^2 + (\sin\theta)^2 (\dot{\phi})^2\right]/(r - m)^2 = 0, \tag{24}$$

$$\ddot{\theta} + 2\dot{r}\dot{\theta}/r - \sin\theta \cos\theta (\dot{\phi})^2 = 0, \quad \text{and} \tag{25}$$

$$\ddot{\phi} + 2\dot{r}\dot{\phi}/r + 2\dot{\theta}\dot{\phi}\cos\theta/\sin\theta = 0. \tag{26}$$

---

[2] as shown by (17), (18), and (19) when $Q(r) = 0$
[3] The negative sign of $S^r{}_r$ is due to the Lorentz signage being used.

For an object at spatial rest in the coordinate system, $\dot{r} = \dot{\theta} = \dot{\phi} = 0$, and $\dot{t} = \sqrt{r/(r-2m)}$. Substituting these values into (24) produces

$$\ddot{r} = -m(r-2m)/\left[r(r-m)^2\right]. \tag{27}$$

In (27), $\lim_{r \to 2m} \ddot{r} = 0$. So at $r = 2m$ there is no downward coordinate acceleration from spatial rest. Additionally, if $\dot{r} \neq 0$ then $\lim_{r \to 2m} \ddot{r} = \infty$. Due to these effects, the coordinate space of $r < 2m$ is inaccessible in the geometry of (21), and $2m$ is now the minimum permissible size of an uncharged object of geometrized mass m. Additional effects arising from the use of (21) instead of (13) are discussed in [12]. This includes a demonstration that in low gravitational fields the difference in the incremental distance between radial coordinates in (13) and (21) is $\frac{1}{2}m^2/r^2$.

CONCLUSION

In conclusion, a model is being presented here that prohibits the formation of black holes, and corresponds well to Einstein's GR in cases where this theory and Einstein's have been tested so far. More research is needed to verify its viability and find testable predictions that differ from those of Einstein's GR.

REFERENCES

1. Einstein, A., *Annalen der Physik*, **49** (1916), as translated in *The Principle of Relativity*[13] pp. 111–164.
2. Wald, R. M., *General Relativity*, The University of Chicago Press, Chicago 60637, 1984.
3. Ohanian, H. C., and Ruffini, R., *Gravitation and Spacetime*, W. W. Norton & Co., New York, NY, 1994, 2nd edn.
4. Wald, R. M., *General Relativity*, chap. 9, pp. 211–242, in [2] (1984).
5. Einstein, A., *Relativity*, Bonanza Books, New York, NY, 1961, chap. XXIV, pp. 83–86.
6. Schmidt, H.-J., How to measure spatial distances? (1995), available at http://mentor.lanl.gov as preprint gr-qc/9512006.
7. Schaefer, E. M., Why distance in relativity is measured using parallax (2002), available at http://users.erols.com/ems57/Distortion/distance.html.
8. Schaefer, E. M., On the splitting of spacetime metrics (2002), available at http://users.erols.com/ems57/Distortion/splitting.html.
9. Misner, C. W., Thorne, K. S., and Wheeler, J. A., *Gravitation*, W. H. Freeman and Co., San Francisco, CA, 1973, section 6.6, p. 173, eq. (6.18).
10. Wald, R. M., *General Relativity*, chap. 6, p. 124, in [2] (1984), eq. (6.1.44).
11. Ohanian, H. C., and Ruffini, R., *Gravitation and Spacetime*, chap. 7, pp. 393–396, in [3] (1994).
12. Schaefer, E. M., A study of the Distorted Schwarzschild Solution (2002), available at http://users.erols.com/ems57/Distortion/distorted-Schwarzschild-study.html.
13. Lorentz, H. A., Einstein, A., Minkowski, H., and Weyl, H., *The Principle of Relativity*, Dover Publications, Inc., 180 Varick Street, New York, NY 10014, 1952, translated by W. Perrett and G.B. Jeffery.

MATRIX MODELS

Two-Dimensional Quantum Gravity as String Bit Hamiltonian Models

B. Durhuus[a,b] and C.-W. H. Lee[c,1]

[a] Department of Mathematics, University of Copenhagen, Universitetsparken 5, DK-2100, Copenhagen, Denmark.
[b] MaPhySto, Centre for Mathematical Physics and Stochastics,[2] Denmark.
[c] Department of Pure Mathematics, Faculty of Mathematics, University of Waterloo, Waterloo, Ontario, N2L 3G1, Canada.

Abstract. We construct a number of string bit Hamiltonian models which describe the dynamics of two-dimensional open or closed quantum space-time without matter. They turn out to be special cases of a Lie algebraic Hamiltonian model. The transition amplitudes of these models in the continuum limit agree with known results from Lorentzian gravity and Liouville gravity in the proper time gauge.

INTRODUCTION

Quantization of gravity is a major open problem in theoretical physics. Perturbation theory, a hugely successful tool in the quantization of the electroweak field and the high energy regime of the strong field, fails miserably in quantum gravity since gravity is not renormalizable. (See, e.g., Ref.[1] for more details.) Nevertheless, other tools are available. In particular, the success of lattice quantum chromodynamics has inspired physicists to discretize space-time by triangulation and quantize it before taking the continuum limit. Indeed, dynamical triangulation allows for detailed analysis of two-dimensional quantum gravity coupled with matter fields. (See, e.g., Ref.[2].) Yet difficulties persist. For instance, four-dimensional results are hard to come by, and conceptual problems relating e.g. to the notion of time, of which the problem of Wick rotating euclidean models is a specific aspect, remain hard to clarify.

String bit models [3] offer an alternative Hamiltonian approach to discretized quantum gravity in which a time parameter is built into the model at the outset. Each equal-time slice is discretized as a finite number of *links* (string bits in original context of string theory). The links form a closed loop in a closed universe; in an open universe, the slice is composed of a left-boundary link, a right-boundary link and interior links in between. The Hamiltonian couples states deviating from each other by either splitting a link or joining several adjacent links together. Some string bit models display a continuum limit at which each link has an infinitesimal length and the number of links scales to infinity

---

[1] speaker.
[2] funded by a grant from the Danish National Research Foundation.

in such a way that the equal-time slices have finite sizes. As will be seen in the sequel, the transition amplitudes agree in such cases with known results from other approaches. Remarkably, the string bit models we shall consider are special cases of a more abstract class of models whose Hamiltonian is an element of the sl_2 subalgebra of the Virasoro algebra.

Further details on the subject matter of this article can be found in Ref.[4].

BUILDING BLOCKS OF STRING BIT MODELS

Let $\bar{q}_\mu$, $a^{\mu_1}_{\mu_2}$ and q^μ be a $1 \times N$ row vector, an $N \times N$ matrix and an $N \times 1$ column vector, respectively, of annihilation operators. The corresponding creation operators are $\bar{q}^{\dagger\mu}$, $a^{\dagger\mu_2}_{\mu_1}$ and $q^\dagger_\mu$, respectively. Their commutators vanish except

$$\left[\bar{q}_{\mu_1}, \bar{q}^{\dagger\mu_2}\right] = \delta^{\mu_2}_{\mu_1}, \quad \left[a^{\mu_1}_{\mu_2}, a^{\dagger\mu_3}_{\mu_4}\right] = \delta^{\mu_1}_{\mu_4}\delta^{\mu_3}_{\mu_2} \text{ and } \left[q^{\mu_1}, q^\dagger_{\mu_2}\right] = \delta^{\mu_1}_{\mu_2}.$$

We consider a Fock representation with a normalized vacuum vector $|\Omega\rangle$ which is annihilated by all annihilation operators.

OPEN UNIVERSE

A typical quantum state of an open universe takes the form

$$|n\rangle_{\frac{1}{2}} := \frac{1}{N^{(n+1)/2}} \bar{q}^{\dagger\mu_1} a^{\dagger\mu_2}_{\mu_1} a^{\dagger\mu_3}_{\mu_2} \cdots a^{\dagger\mu_{n+1}}_{\mu_n} q^\dagger_{\mu_{n+1}} |\Omega\rangle$$

$$= \frac{1}{N^{(n+1)/2}} \bar{q}^\dagger (a^\dagger)^n q^\dagger |\Omega\rangle,$$

where the limit $N \to \infty$ is understood here and in all subsequent formulas involving N. (We shall account for the subscript $1/2$ later.) Physically, this state represents a time slice with n interior links and two boundary links. All states of this form span a Hilbert space $\mathcal{T}_{1/2}$ with inner product

$$\langle m|n\rangle_{\frac{1}{2}} = \delta_{mn}.$$

Physical observables are linear combinations of operators, or their restrictions to $\mathcal{T}_{1/2}$, of the form

$$\gamma^k_l := \frac{1}{N^{(k+l-2)/2}} a^{\dagger\mu_2}_{\mu_1} a^{\dagger\mu_3}_{\mu_2} \cdots a^{\dagger\nu_l}_{\mu_k} a^{\nu_{l-1}}_{\nu_l} a^{\nu_{l-2}}_{\nu_{l-1}} \cdots a^{\mu_1}_{\nu_1}$$

$$= \frac{1}{N^{(k+l-2)/2}} \text{Tr}[(a^\dagger)^k a^l],$$

$$l^0_0 := \bar{q}^{\dagger\mu} \bar{q}_\mu = \bar{q}^\dagger \bar{q}', \text{ or}$$

$$r^0_0 := q^\dagger_\mu q^\mu = (q^\dagger)^t q,$$

where $(q^\dagger)^t$ and $\bar{q}^t$ stand for the transpose of the vectors $q^\dagger$ and $\bar{q}$, respectively. Let σ_l^k be the restriction of γ_l^k to $\mathscr{T}_{1/2}$. Then

$$\sigma_l^k |n\rangle_{\frac{1}{2}} = \begin{cases} 0 & \text{if } l > n, \\ (n-l+1)|n-l+k\rangle_{\frac{1}{2}} & \text{if } l \leq n, \end{cases}$$

$$l_0^0 |n\rangle_{\frac{1}{2}} = |n\rangle_{\frac{1}{2}}$$

and $r_0^0 = l_0^0$ as operators acting on $\mathscr{T}_{1/2}$. Physically, σ_l^k annihilates a universe with less than l interior links and replaces l neighboring interior links with k ones in a universe with more than $l-1$ interior links. l_0^0 (r_0^0) annihilates the left (right) boundary link and then recreates it. Thus l_0^0 and r_0^0 are effectively the identity operator on $\mathscr{T}_{1/2}$.

Following is the Hamiltonian of a string bit model for this open universe:

$$\begin{aligned} H_{\frac{1}{2}} &:= \operatorname{Tr}\left[a^\dagger a + \frac{\lambda}{\sqrt{N}}(a^\dagger)^2 a + \frac{\lambda}{\sqrt{N}} a^\dagger a^2\right] - \frac{1}{4}\bar{q}^\dagger \bar{q}^t - \frac{1}{4}(q^\dagger)^t q \\ &= \sigma_1^1 + \lambda\sigma_1^2 + \lambda\sigma_2^1 - \frac{1}{4}l_0^0 - \frac{1}{4}r_0^0. \end{aligned}$$

In this Hamiltonian, λ is a real parameter; σ_1^1 yields the interior spatial volume energy; σ_1^2 splits any interior link into two; σ_2^1 combines any two adjacent links to one; and l_0^0 and r_0^0 are the volume energies of the left and right boundaries, respectively. Note that the boundary links contribute *negatively* to the volume energy.

Before we discuss the transition amplitude of this model, let us turn our attention to the closed universe first.

CLOSED, HOMOGENEOUS UNIVERSE

A typical quantum state of a time slice with n links takes the form

$$|n\rangle_1 := \frac{1}{N^{n/2}} \operatorname{Tr}(a^\dagger)^n |\Omega\rangle .$$

Such quantum states span a Hilbert space $\mathscr{T}_1$ with inner product given by

$$\langle m|n\rangle_1 = n\delta_{mn}.$$

Let g_l^k be the restriction of γ_l^k to $\mathscr{T}_1$. Then

$$g_l^k|n\rangle_1 = \begin{cases} 0 & \text{if } l > n, \\ n|n-l+k\rangle_1 & \text{if } l \leq n, \end{cases}$$

The Hamiltonian of a string bit model describing this universe reads

$$\begin{aligned} H_1 &:= \operatorname{Tr}\left[a^\dagger a + \frac{\lambda}{\sqrt{N}}(a^\dagger)^2 a + \frac{\lambda}{\sqrt{N}} a^\dagger a^2\right] \\ &= g_1^1 + \lambda g_1^2 + \lambda g_2^1. \end{aligned}$$

Physically, g_1^1 is the volume energy, g_1^2 splits any link into two and g_2^1 combines any two adjacent links to one.

SL_2 GRAVITY MODEL

The above two string bit models turn out to be special cases of a Lie algebraic model. Consider the closed universe first. Make the identifications

$$g_1^1 \leftrightarrow L_0, \; g_2^1 \leftrightarrow L_1 \text{ and } g_1^2 \leftrightarrow L_{-1}.$$

As operators acting on the Hilbert space $\mathscr{T}_1$, L_0, L_1 and L_{-1} satisfy the commutation relations

$$[L_0, L_1] = -L_1, \; [L_0, L_{-1}] = L_{-1} \text{ and } [L_1, L_{-1}] = 2L_0.$$

This is nothing but the sl_2 subalgebra of the Virasoro algebra. Note that

$$g_2^1|1\rangle_1 = 0, \; g_1^1|1\rangle_1 = |1\rangle_1 \text{ and } \langle 1|1\rangle_1 = 1.$$

Hence $|1\rangle_1$ plays the role of a normalized highest weight vector. The highest weight is $h = 1$.

As for the open universe, make the identifications

$$\sigma_1^1 - \frac{1}{2}l_0^0 \leftrightarrow L_0, \; \sigma_2^1 \leftrightarrow L_1 \text{ and } \sigma_1^2 \leftrightarrow L_{-1}.$$

Again L_0, L_1 and L_{-1} satisfy the sl_2 subalgebra of the Virasoro algebra. Since

$$\sigma_2^1|1\rangle_{\frac{1}{2}} = 0, \; \left(\sigma_1^1 - \frac{1}{2}l_0^0\right)|1\rangle_{\frac{1}{2}} = \frac{1}{2}|1\rangle_{\frac{1}{2}} \text{ and } \langle 1|1\rangle_{\frac{1}{2}} = 1,$$

$|1\rangle_{1/2}$ plays the role of a normalised highest weight vector, but here the highest weight is $h = 1/2$.

Thus we propose to consider an sl_2 gravity model with the Hamiltonian

$$H := L_0 + \lambda L_{-1} + \lambda L_1,$$

in a highest weight representation with arbitrary positive highest weight h.

As is well known, the $h = 1$ highest weight representation is the symmetric tensor product of two copies of the $h = 1/2$ highest weight representation. Therefore, the cyclically symmetric closed universe may be seen as two open universes glued together. Indeed, the following identification holds true:

$$|n\rangle_1 \leftrightarrow \sum_{k=1}^{n} |k\rangle_{\frac{1}{2}} \otimes |n-k+1\rangle_{\frac{1}{2}}.$$

Likewise, multiple tensor products of the open universe lead to "p-seamed" models in the terminology of [5].

SOLUTION TO QUANTUM GRAVITY MODELS

Consider the sl_2 gravity model introduced in the preceding section. The highest weight vector $|1\rangle_h$ and the vectors

$$|n+1\rangle_h := \frac{1}{n!} L_{-1}^n |1\rangle_h$$

for $n = 1, 2, \ldots$, form an orthogonal basis of the Hilbert space. The unnormalized transition amplitude at the continuum limit is defined as

$$\tilde{G}_u(L,L';T) := \lim_{a \to 0} a^\alpha \langle \frac{L'}{a} | e^{-\frac{2TH}{a}} | \frac{L}{a} \rangle_h, \tag{1}$$

where

$$\lambda = -\frac{1}{2} e^{-\frac{1}{2}\Lambda a^2}$$

in H, and α is chosen in such a way that a non-trivial limit exists. Define the normalized transition amplitude $\tilde{G}(L,L';T)$ by normalizing $\langle L'/a|$ and $|L/a\rangle_h$ in Eq.(1). Then $\alpha = 1$ and it can be shown that[3]

$$\tilde{G}(L,L';T) = \frac{\sqrt{\Lambda}}{\sinh(T\sqrt{\Lambda})} e^{-\sqrt{\Lambda}(L+L')\coth(T\sqrt{\Lambda})} I_{2h-1}\left(\frac{2\sqrt{\Lambda LL'}}{\sinh(T\sqrt{\Lambda})}\right), \tag{2}$$

where I_{2h-1} is a modified Bessel function.

Choosing different values of h in Eq.(2) allows us to compare our model with others. If $h = 1/2$, we shall recover the open universe as discussed in the previous section, and the transition amplitude will be in perfect agreement with that of the Lorentzian gravity model in Ref.[5]; if $h = 1$, we shall recover the cyclically symmetric closed universe, and the transition amplitude will be exactly the same as that of a cyclically symmetric version of the Lorentzian gravity model in Ref.[6]; if h is a positive integer, our result will be identical with the two-dimensional Liouville quantum gravity model in the proper time gauge with $h - 1$ being interpreted as a winding number [7]; lastly, if h is a positive half-integer, we shall obtain a $(2h)$-seamed gravity model as discussed at the end of the previous section [5].

---

[3] Let us take this opportunity to correct some typos in Ref.[4]. The first two equations on p.214 should read

$$\sum_{n=1}^{\infty} a_n n^S p^n \frac{\Gamma(n-1+2h)}{(n-1)!\Gamma(2h)} = 0 \quad \text{and} \quad \sum_{n=1}^{\infty} a_n p^n \frac{\Gamma(n-1+2h)}{(n-1)!\Gamma(2h)} e^{zn} = 0,$$

respectively. The first two formulae on p.215 should read

$$\langle \phi_R | \phi_R \rangle_h = \sum_{n=1}^{\infty} \sum_{m=1}^{\infty} c_R(n) c_R m^R p^{n+m} \langle n|m \rangle_h = \sum_{n=1}^{\infty} c_R(n) c_R n^R p^{2n} \frac{\Gamma(n-1+2h)}{(n-1)!\Gamma(2h)}$$

and

$$c_0^{-2} \simeq \sum_{r=0}^{R} \frac{(-2\sqrt{\Lambda}a)^{R+r}\Gamma(2h)}{\Gamma(R+2h)\Gamma(r+2h)} \binom{R}{r} \sum_{n=1}^{\infty} n^{R+r} \frac{\Gamma(n-1+2h)}{(n-1)!} p^{2n},$$

respectively.

DISCUSSION AND OPEN PROBLEMS

Are there physical interpretations for other positive values of h? We have yet to know. In any case, it seems highly relevant to uncover how general the equivalance between string bit models and two-dimensional quantum gravity is. (See also the discussion in Ref.[8].) Given the agreement between the sl_2 gravity model and others, does sl_2 play in two-dimensional quantum gravity a key role which we have yet to unearth?

Among other obvious directions of research, partly related to the previously mentioned, is the question whether there are solvable string bit models describing quantum gravity with matter, and whether there exist similar higher dimensional Hamiltonian models of quantum gravity.

ACKNOWLEDGEMENTS

We thank J. Ambjørn, H. P. Jakobsen, R. Mann, E. Poisson and S. G. Rajeev for discussions and helpful suggestions at various stages of this work. C.-W. H. Lee would also like to thank MaPhySto for financial support in the early phase.

REFERENCES

1. M. E. Peskin and D. V. Schroeder, *An Introduction to Quantum Field Theory*, Perseus Books Group, 1995.
2. J. Ambjørn, B. Durhuus, T. Jónsson, *Quantum Geometry*, Cambridge Monographs on Mathematical Physics, Cambridge University Press, 1997.
3. R. Giles and C. B. Thorn, Phys. Rev. **D 16**, 366 (1977); C. B. Thorn, in: Proceedings of Sakharov Conference on Physics, Moscow (1991) pp.447–454 [hep-th/9405069].
4. B. Durhuus and C.-W. H. Lee, Nucl. Phys. **B 623**, 201 (2002) [hep-th/0108149].
5. P. Di Francesco, E. Guitter and C. Kristjansen, Nucl. Phys. **B 567**, 515 (2000) [hep-th/9907084].
6. J. Ambjørn and R. Loll, Nucl. Phys. **B 536**, 407 (1999) [hep-th/9805108].
7. R. Nakayama, Phys. Lett. **B 325**, 347 (1994) [hep-th/9312158].
8. M. Arnsdorf, Class. Quant. Grav. **19**, 1065 (2002) [gr-qc/0110026].

On quantization of matrix models

Artem Starodubtsev

*Department of Physics, University of Waterloo,
University ave W, Waterloo, ON, Canada N2L 3G1,
and
Perimeter Institute for Theoretical Physics,
35 King st N, Waterloo, ON, Canada N2J 2W9,
astarodu@astro.uwaterloo.ca*

Abstract. The issue of non-perturbative background independent quantization of matrix models is addressed. The analysis is carried out by considering a simple matrix model which is a matrix extension of ordinary mechanics reduced to 0 dimension. It is shown that this model has an ordinary mechanical system evolving in time as a classical solution. But in this treatment the action principle admits a natural modification which results in algebraic relations describing quantum theory. The origin of quantization is similar to that in Adler's generalized quantum dynamics. The problem with extension of this formalism to many degrees of freedom is solved by packing all the degrees of freedom into a single matrix. The possibility to apply this scheme to various matrix models is discussed.

1. INTRODUCTION

Matrix models were proposed as a non-perturbative definition of string theory some time ago. There are several versions of matrix models. Generally they are defined as reductions of certain Super Yang Mills theories to one dimension (BFSS matrix model [1]) or to zero dimension (IKKT matrix model [2]).

One of the most interesting properties of these models is the dynamical origin of spacetime. For example the action of IKKT model

$$S = -Tr(\frac{1}{4}[A_\mu, A_\nu][A^\mu, A^\nu] + fermions) \qquad (1)$$

contains no a priori spacetime structure[1] The later arises from classical solutions of the model (1) and it is associate to the distribution of the eigenvalues of the matrices A_μ [3]. Spacetime is said to be "generated dynamically" in these models. Therefore one may hope that such model could provide a background independent definition of string theory so that different string theories in various background spacetimes would

---

[1] The action (1) contains a fixed flat Minkowskian metric $\eta^{\mu\nu}$ contracting lower indices. However in [3] was considered a possibility that this model can also in principle describe curved spacetime. The metric $\eta^{\mu\nu}$ is to be understood in this case as a metric in tangent space and the frame field forming the metrics of the manifold are to originate from matrices. All the background information encoded in $\eta^{\mu\nu}$ is the dimension and the signature. Here it is worth mentioning also cubic matrix models [4] which do not depend on any background metric structure at all.

be different solutions of a single theory. However the commonly used quantization procedure is that one first picks up a classical solution to the matrix model and then quantizes small perturbations around it. It is clear that the resulting quantum theory is no longer background independent.

The key question which is addressed in this note is whether we can define a nonperturbative background independent quantization of matrix models. Given that the basic entries of the classical theory are operators a natural question arises: Can matrix models produce quantum theory without quantization in the usual sense? If so, how can it be realized?

One proposal for this was made by Smolin [5] which is that matrix models can be interpreted as a non-local hidden variables theories, the eigenvalues being quantum observables and the entries being hidden variables. Quantum mechanics for the eigenvalues is then reproduced in the ordinary statistical mechanical description of the model.

In this note we will study the possibility that in matrix model framework, where there is no a priory spacetime structure, classical and quantum theory can be brought much closer to each other than they are in the presence of a background spacetime. The later would imply a possible relation between origin of space-time and quantization. There are some indications on the existence of such a relation also pointed out in [5]. Many of the notions on the basis of which the distinction between classical and quantum theory is generally made rely on the existence of classical spacetime. For example classical system is recognized as having a definite smooth trajectory, something what quantum system doesn't have. But the very notion of smooth trajectory can not be defined without using the classical notion of spacetime and its differential structure. Also quantum theory is characterized by its generic non-locality. But the notion of locality also relies on classical spacetime. Given all the above it is very plausible that matrix models with the generalized notion of spasetime that they give rise to can provide framework which is general for classical and quantum theory.

In the present paper the problem of quantization of matrix models is studied on a simple example which is a matrix extension of mechanical system. In section 2 the model is defined. It is shown that this matrix model reproduces ordinary mechanics. In section 3 a natural modification of this model is proposed. The equations of motion of this modified model describe quantum mechanics. In section 4 the scheme is generalized to systems with many degrees of freedom. This is done by a specific way of encoding the degrees of freedom which is very natural for matrix models. In section 5 possible applications to the marix models of interest are discussed.

2. MATRIX MODEL FOR A MECHANICAL SYSTEM WITH ONE DEGREE OF FREEDOM

In this section we will define a matrix model which reproduces the ordinary mechanics of a system with one degree of freedom x described by the following action S_L in the Lagrangian form.

$$S_L[x] = \int dt \left\{ \frac{1}{2}\left(\frac{dx}{dt}\right)^2 - V(x) \right\} \tag{2}$$

Here V is an arbitrary function of x which represents the self-interaction of x. The equation of motion are obtained by requiring that the variation of the action (2) $\delta S_L = 0$:

$$\frac{d^2 x}{dt^2} + \partial_x V(x) = 0. \tag{3}$$

The matrix model for the system (2) can be obtained by replacing the configuration coordinate x and the time derivative $\frac{d}{dt}$ by unknown operators (X and D respectively) and the integration operation $\int dt$ by the trace operation. The action (2) then takes the following operator form:

$$S_{op}[D, X] = Tr\left\{\frac{1}{2}[D, X]^2 - V(X)\right\}, \tag{4}$$

where $[A, B] = AB - BA$ is the operator commutator. The action (4) can also be obtained by matrix extension of (2) reduced to zero demension.

Now the statement is that the matrix model defined by the action (4) has the mechanical system defined by the action (2) as its solution. To see this consider the variation of the action (4) with respect to D and X. For (2) to be functionally differentiable the Tr operation has to be invariant with respect to cyclic permutations. The later is true for finite dimensional operators but this property generally does not extend to infinite dimensional ones. Here we will simply require that the variations δX and δD are such that the cyclic property of Tr holds. It can be shown that this requirement corresponds to the variations fixed at the endpoints. From the requirement that the variation of (2) vanishes we get the following equations for X and D

$$[D, [D, X]] + \partial_X V(X) = 0, \tag{5}$$

$$[X, [D, X]] = 0. \tag{6}$$

Classically eq. (5) coincides with the newtonian equation of motion. Eq. (6) has no classical analogs.

From now on we will look for solutions of operator equations such as (5,6) in a representation where X is a multiplication operator. Thus we assume that the operator algebra can be realized on a space of functions $\psi(X)$. This means that any operator from our algebra can be represented by the following series (finite or infinite)

$$O = \sum_{i,j=0}^{\infty} o_{ij} X^i \left(\frac{d}{dX}\right)^j = \sum_{j=0}^{\infty} o_j(X) \left(\frac{d}{dX}\right)^j. \tag{7}$$

We can now use this form for D and substitute it into eq.(6). Then one can show that one can introduce a parameter t such that

$$X = x(t) \times, \quad D = \frac{d}{dt} + d_0(t). \tag{8}$$

Now if we substitute this into (5) we recover the classical Newtonian equation (3). Therefore the matrix model based on the action (4) describes classical dynamics of the system defined by the action (2).

3. QUANTIZATION WITHOUT QUANTIZATION

Now one can show that by a minor modification of the action (4) the equations of motion (5,6) can be turned into those of quantum mechanics. The equations of motion in this framework are derived from variation of the action not only with respect to X but also with respect to D. Therefore the action (4) admits a nontrivial modification by adding an extra term to the action which depends purely on D. Consider the simplest possible such term which is linear in D:

$$S_{op}[D,X] = Tr\left\{\frac{1}{2}[D,X]^2 - V(X) - iD\right\}. \tag{9}$$

Here the coefficient in front of D is taken to be imaginary to provide hermiticity of the extra term. Being interpreted as a derivative operator D has to be anti-hermitian. Because we don't have any Hilbert space structure yet the hermiticity relations are to be understood as formal *-relations.

By variation of the action (9) we obtain the following modified equations of motion

$$[D,[D,X]] + \partial_X V(X) = 0, \tag{10}$$

$$[[D,X],X] = iI, \tag{11}$$

where I is the identity operator. The equations (5) and (10) are the same, eqs. (6) and (11) differ by them of iI in the r.h.s.

We can now solve eqs. (10,11) for D in the form (7). The solution will have (or, in general, will be unitary equivalent to) the following form

$$D = \frac{i}{2}\frac{d^2}{dX^2} - iV(X) + ic. \tag{12}$$

This is precisely the relation between the evolution generator and the hamiltonian of the ordinary quantum mechanics. It may be interpreted as the Schrodinger equation when represented on states or the Heisenberg equation when applied to operators. The only remaining freedom is the constant c indicating the arbitrariness in defining the ground state energy of the system.

One may notice that eq. (11) is the usual quantum mechanical commutation relation. This is where the quantization comes from. The origin of quantum mechanical commutation relation (11) in this model is analogous to that in generalized quantum dynamics proposed by Adler [6]. There a matrix extension of ordinary mechanical system evolving in time was considered. Then the invariance with respect to unitary transformations was interpreted as a gauge symmetry. The commutation relation of the form (11) was then induced by a term linear in the corresponding gauge field in the action.

One may also point out that we can in fact add any function of D to the action (4). This function can then be expanded in powers of D. The term linear in D thus gives the standard commutation relation of quantum mechanics. It is interesting to note that the next correctin to the action which is quadratic in D results in commutation relations leading to the minimum length uncertainty relation considered by Kempf [7].

4. SYSTEMS WITH MANY DEGREES OF FREEDOM

The generalization of all the above to systems with N degrees of freedom is not straightforward. If we simply took one copy of the action (9) per one degree or freedom and considered a sum of them adding also some terms representing interaction between these degrees of freedom, then instead of having one copy of the equation (11) per each degree of freedom we would have a single equation

$$\sum_{j=1}^{N} [[D, X_j], X_j] = iN, \qquad (13)$$

for the whole system. This is not enough to recover the the complete set of quantum mechanical equations of the system considered. The same problem also arose in [6]. A solution to this problem based on thermodynamical considerations has been given by Adler and Millard [8]. It is based on the fact that the net non-commutativity for all the degrees of freedom given by (13) is distributed uniformly between different degrees of freedom in thermodynamical approximation. This is analogous to how each degree of freedom carry a $(1/2)kT$ portion of energy at equilibrium.

Here we will give another solution to this problem which is very natural in the context of matrix models and which is exact, i.e. valid also apart from thermodynamical equilibrium. It is based on a specific way of encoding of the degrees of freedom of the theory which is extensively used in various matrix models [1, 2, 4].

Let our system have N degrees of freedom. They can be represented by N infinite dimensional matrices commuting with each other $X_1, X_2, ..., X_N$, $[X_i, X_j] = 0$. Then we can construct the following operator

$$\mathscr{X} = diag(X_1, X_2, .., X_N). \qquad (14)$$

This is an infinite dimensional matrix divided into N infinite dimensional blocks located at diagonal. The interaction between X_I can be introduced via including non-block-diogonal matrices in the potential term of the acton. The kinematical term in the action for $\mathscr{X}$ will be the same as that in (9). The variation of it with respect to D results in $[[D, X_j], X_j] = iI$. More detailed analisys shows that the complete set of equations of the model has non-trivial solutions only when $[[D, X_j], X_k] = iI\delta_{jk}$. Thus the standard quantum-mechanical equations are recovered also in this case.

5. DISCUSSION

The first natural question to ask is whether this scheme can be applied to matrix models used in string theory. If we try to apply the result of this paper for example directly to IKKT model(1) we encounter the same problem as with mechanical system with many degrees of freedom. The presence of a sum over μ and ν in (1) means that the quantum commutation relations derived from it are not completely determined. The solution to this problem would be the packing of all the matrices A_μ into a single matrix. The resulting matrix model wouldn't even have a preexisting dimension. The IKKT model then could be derived from this model by a symmetry breaking.

Such models already exist in literature. These are the cubic or Chern-Simons matrix models [4]. They are explicitly background independent and may be relevant not only to string theory but also to loop quantum gravity given the relation between loop quantum gravity and topological field theory [9]. The extra term in the action inducing quantization turns out to be very similar to that used to introduce the space non-commutativity in the non-commutative Chern-Simons theory [10]. This may imply a possible relation between space-time non-commutativity in non-commutative geometry and quantum mechanical non-commutativity.

ACKNOWLEDGMENTS

I am grateful to Stephen Adler, Satoshi Iso, Achim Kempf, and especially Lee Smolin for useful discussions.

REFERENCES

1. T. Banks, W. Fishler, S. H. Shenker, L. Susskind, *M theory as a matrix model: a conjecture* hep-th/9610043; Phys. Rev. D55 (1997) 5112.
2. N. Ishibashi, H. Kawai, Y. Kitazawa and A. Tsuchiya, *A large N reduced model as superstring* hep-th/9612115; Nucl. Phys. B498 (1997) 467; M. Fukuma, H. Kawai, Y. Kitazawa and A. Tsuchiya, *String field theory from IIB matrix model* hep-th/9705128, Nucl. Phys. B (Proc. Suppl) 68 (1998) 153.
3. H. Aoki, S. Iso, H. Kawai, Y. Kitazawa, T. Tada *Space-Time Structures from IIB Matrix Model* hep-th/9802085 Prog.Theor.Phys. 99 (1998) 713-746; S. Iso, H. Kawai *Space-Time and Matter in IIB Matrix Model - gauge symmetry and diffeomorphism* hep-th/9903217 Int.J.Mod.Phys. A15 (2000) 651-666
4. L. Smolin, *The exceptional Jordan algebra and the matrix string*, hep-th/0104050; *M theory as a matrix extension of Chern-Simons theory*, hep-th/0002009, Nucl.Phys. B591 (2000) 227-242; *The cubic matrix model and the duality between strings and loops*, hep-th/0006137; T. Azuma, S, Iso, H. Kawai and Y. Ohwashi, ıSupermatrix models, hep-th/0102168; T. Azuma, *Investigations of Matrix Theory via Super Lie Algebras*, hep-th/0103003.
5. L. Smolin *Matrix models as non-local hidden variables theories* hep-th/0201031
6. S. L. Adler *Quaternionic Quantum Mechanics and Quantum Fields*, Oxford University Press 1995
7. A. Kempf, G. Mangano and R. B. Mann, *Hilbert space representation of the minimal length uncertainty relation*, Phys. Rev. D **52**, 1108 (1995) [arXiv:hep-th/9412167]. A. Kempf, *Nonpointlike Particles in Harmonic Oscillators*, J. Phys. A **30**, 2093 (1997) [arXiv:hep-th/9604045].
8. S. L. Adler and A. C. Millard *Generalized Quantum Dynamics as Pre-Quantum Mechanics* hep-th/9508076; Nucl. Phys. B 473, 199 (1996)
9. L. Smolin, *Linking Topological Quantum Field Theory and Nonperturbative Quantum Gravity*, J.Math.Phys. 36 (1995) 6417-6455, gr-qc/9505028.
10. A. Polychronakos, *Noncommutative Chern-Simons terms and the noncommutative vacuum*, hep-th/0010264, JHEP 0011 (2000) 008; J. Kluson, *Matrix model and Chern-Simons theory*, hep-th/0012184; L. Susskind, *The Quantum Hall Fluid and Non-Commutative Chern Simons Theory*, hep-th/0101029

Large N Matrix Models and Noncommutative Fisher Information

A. Agarwal, L. Akant, G. S. Krishnaswami and S. G. Rajeev[1]

Department of Physics and Astronomy, University of Rochester, Rochester, New York 14627

Abstract. We interpret the action for 0+1-dimensional large N matrix models in the context of noncommutative probability theory. The actions of both 0-dimensional and 0+1-dimensional matrix models contain universal terms, free entropy and free Fisher information respectively. Their minimization properties are essential for the solution of matrix models. We also give a geometric interpretation of the action principle of 0+1-dimensional matrix models.

NONCOMMUTATIVE PROBABILITY SPACES

A noncommutative probability[1] space consists of an algebra A together with a linear functional $\tau : A \to C$. One often requires τ to be real valued, cyclically symmetric: $\tau(XY) = \tau(YX)$ for all $X, Y \in A$ and normalized to unity: $\tau(1) = 1$. In general the algebra A is choosen to be an operator algebra e.g. C^*-algebra or W^*-algebra. Then the elements of A are regarded as operator valued random variables and the functional τ gives the expectation of these random variables. A special case arises when one picks r algebraically independent, self adjoint elements $X^1, ..., X^r; (X^i)^* = X^i$ of A and forms the free algebra $C[X^1, ..., X^r]$[2]. Then one obtains a new noncommutative probability space by restricting τ to this algebra. At this point it is useful to introduce the multi-index notation where $X^I = X^{i_1...i_n} = X^{i_1}...X^{i_n}$. This special case is best exemplified by 0-dimensional matrix models.

0-DIMENSIONAL MATRIX MODELS

In 0-dimensional matrix models the generators of the free algebra are given by $N \times N$ hermitean matrices $M^1, ..., M^r$ and the functional τ is given by the expectation values of $U(N)$ invariant traces $\frac{1}{N} Tr M^I$

$$\tau(M^I) = \lim_{N \to \infty} \frac{\int dM e^{-NTrV(M)} \frac{1}{N} Tr M^I}{\int dM e^{-NTrV(M)}} \quad (1)$$

---

[1] Presented by 2nd author akant@pas.rochester.edu at MRST 2002
[2] One can think of the free algebra as the algebra of polynomials in noncommuting variables $X^1, ..., X^r$.

here $V(M) = \gamma_I M^I$ is a polynomial in $M^1, ..., M^r$ and the measure of integration is

$$dM = \prod_i dM_{ii} \prod_{i<j} dReM_{ij} dImM_{ij} \qquad (2)$$

these expectation values are called the moments and they are denoted by $G^I = \tau(M^I)$. The assignment of the factors of N in the above formula guarantees a smooth large N limit. One of the most important features of the large N limit is the factorization property:

$$\langle \frac{1}{N} TrM^I \frac{1}{N} TrM^J \rangle = \langle \frac{1}{N} TrM^I \rangle \langle \frac{1}{N} TrM^J \rangle + O(\frac{1}{N^2}). \qquad (3)$$

Now consider the partition function $Z = \lim_{N \to \infty} \int dM e^{-NTrV(M)}$; its invariance under an infinitesimal change of variables of the form $M^i \to M^i + \varepsilon \delta^{ij} f_j(M)$ and the use of the large N factorization property yield the factorized Schwinger-Dyson equations (SDE):

$$\gamma_{J_1 i J_2} G^{J_1 I J_2} + \delta^I_{I_1 i I_2} G^{I_1} G^{I_2} = 0. \qquad (4)$$

These are infinitely many equations for the moments G^I. In [4] we derived an action principle for these equations where the action functional

$$F(G) = \chi(G) + \gamma_I G^I \qquad (5)$$

contains a universal term $\chi(G)$ which is independent of the original action $V(M)$. In [4] we identified this universal term as the free entropy of Dan Voiculescu [2]. The SDE are given by $L_i^I F(G) = 0$ where the Lie derivative L_i^I is defined as

$$L_i^I = G^{J_1 I J_2} \frac{\partial}{\partial G^{J_1 i J_2}} = G^{JI} \frac{\partial}{\partial G^{Ji}} \qquad (6)$$

with

$$L_i^I \chi(G) = \delta^I_{I_1 i I_2} G^{I_1} G^{I_2}. \qquad (7)$$

0+1-DIMENSIONAL MATRIX MODELS

Now we have a reformulation of 0-dimensional matrix models in terms of the moments alone. Next we want to develop a similar reformulation for $0+1$-dimensional matrix models. As a first step let us introduce an infinite dimensional space $\mathcal{V}$ with a coordinate chart $\{G^I\}$ whose coordinate functions are labeled by multi-indices I. The set of all moments of a given 0-dimensional matrix model corresponds to a point in this space and the totality of such points forms a subspace of $\mathcal{V}$. This subspace is characterized by the moment inequalities [4] and the cyclic symmetry of the coordinate functions $G^{Ii} = G^{iI}$. Now we will construct a dynamical theory whose configuration space is $\mathcal{V}$. We want to write down a Schrödinger operator on $\mathcal{V}$.

We start by defining a metric on $\mathcal{V}$. However, technically it is much easier to define the inverse of the metric by

$$\Omega^{IJ} = \Omega(dG^I, dG^J) = \delta^{ij} \delta^I_{I_1 i I_2} \delta^J_{J_1 j J_2} G^{I_2 I_1 J_2 J_1}. \qquad (8)$$

Now we can choose the kinetic part of the hamiltonian to be the Laplace-Beltrami operator corresponding to this metric

$$K = -\frac{1}{2}\Delta_\Omega = -\frac{1}{2}\frac{1}{\sqrt{\Omega}}\partial_I(\sqrt{\Omega}\Omega^{IJ}\partial_J) \tag{9}$$

here $\Omega = det\Omega_{IJ}$ and $\partial_I = \frac{\partial}{\partial G^I}$.

The coordinate representation of momentum operators on a configuration space with metric Ω_{IJ} is given by[6]

$$\pi_I = -i(\partial_I + \frac{1}{2}\partial_I \ln\sqrt{\Omega}) \tag{10}$$

now we can write the kinetic term as

$$K = \frac{1}{2}\pi_I \Omega^{IJ}\pi_J + \frac{1}{4}\partial_I(\Omega^{IJ}\partial_J \ln\sqrt{\Omega}) + \frac{1}{8}\partial_I(\ln\sqrt{\Omega})\Omega^{IJ}\partial_J(\ln\sqrt{\Omega}). \tag{11}$$

and the full hamiltonian can be written as

$$H = \frac{1}{2}\pi_I \Omega^{IJ}\pi_J + \frac{1}{8}\partial_I(\ln\sqrt{\Omega})\Omega^{IJ}\partial_J(\ln\sqrt{\Omega}) + V(G). \tag{12}$$

here $V(G) = \gamma_I G^I$ is the potential term. We recognize this as the Jevicki-Sakita-Yaffe hamiltonian [7, 8]

In general the second term in K vanishes [5] so we excluded that from the defintion of H. Note that the hamiltonian contains an effective potential generated by the Laplace-Beltrami operator

$$V_{eff}(G) = \frac{1}{8}\partial_I(\ln\sqrt{\Omega})\Omega^{IJ}\partial_J(\ln\sqrt{\Omega}). \tag{13}$$

In [4] we showed that the free entropy $\chi(G) = \ln J$ where J is the determinant of the change of variable $G^I = TrM^I$. The pull-back of the metric Ω_{IJ} under this change of variables can be calculated easily and it follows that $\chi(G) = -\ln\sqrt{\Omega}$. Using this and the cyclic symmetry of the moments we can express V_{eff} as

$$V_{eff} = \frac{1}{8}(\partial_{Ii}\chi)\delta^{ij}G^{IJ}(\partial_{Jj}\chi). \tag{14}$$

We know that in the large N limit the quantum fluctuations are suppressed by a factor of $\frac{1}{N^2}$ so in this limit the above hamiltonian can be treated as a classical hamiltonian and the ground state is given by the minimization of

$$U(G) = V_{eff}(G) + V(G). \tag{15}$$

Notice the similarity between $F(G)$ of 0-dimensional models and $U(G)$, they both contain universal terms, $\chi(G)$ and $V_{eff}(G)$ respectively. In [4] we showed that $\chi(G)$ is the free entropy and below we show that $V_{eff}(G)$ is the noncommutative analog of Fisher's information first introduced by D. Voiculescu [3, 5]. So we see that the dynamics of large N matrix models are governed by variational principles where the action functionals contain universal terms which have natural interpretations in the context of noncommutative probability theory.

NONCOMMUTATIVE FISHER INFORMATION

In this section we want to show that V_{eff} is proportional to the noncommutative Fisher information (or the free Fisher information).

The conjugate variable (or the noncommutative Hilbert transform) S_j of the generator X^j is defined to be the solution of

$$\tau(S_j X^I) = \tau(X^{I_1})\tau(X^{I_2})\delta^I_{I_1 j I_2} \qquad (16)$$

and the free information of $X^1, ..., X^n$ is defined as

$$\Phi(X^1, ..., X^n) = \tau(S_i \delta^{ij} S_j) \qquad (17)$$

but from (7) we know that

$$L^I_i \chi(G) = \frac{\partial \chi}{\partial G^{Ji}} G^{JI} = G^{I_1} G^{I_2} \delta^I_{I_1 i I_2} \qquad (18)$$

or equivalently

$$\tau(\frac{\partial \chi}{\partial G^{Ji}} X^J X^I) = \tau(X^{I_1})\tau(X^{I_2})\delta^I_{I_1 i I_2}. \qquad (19)$$

Comparing this with the definition of the conjugate variable we see that

$$S_i = \frac{\partial \chi}{\partial G^{Ji}} X^I \qquad (20)$$

and the free Fisher information can be written as

$$\Phi(X^1, ..., X^n) = \tau(\frac{\partial \chi}{\partial G^{Ii}} X^I \delta^{ij} \frac{\partial \chi}{\partial G^{Jj}} X^J) = \frac{\partial \chi}{\partial G^{Ii}} \delta^{ij} G^{IJ} \frac{\partial \chi}{\partial G^{Jj}}. \qquad (21)$$

So we have

$$V_{eff}(G) = \frac{1}{8}\Phi(G) \qquad (22)$$

which is the desired result.

REFERENCES

1. D. V. Voiculescu, K. J. Dykema and A. Nica *Free Random Variables*, American Mathematical Society Providence, USA (1992);
2. D.V. Voiculescu, Invent. Math. **118**, 411 (1994)
3. D.V. Voiculescu, Invent. Math. **132**, 189 (1998)
4. L. Akant, G. Krishnaswami, S.G. Rajeev, hep-th/0111263, to appear in Int. J. Mod. Phys. A
5. A. Agarwal, L. Akant, G.S. Krishnaswami, S.G. Rajeev, in preparation
6. B.S. de Witt, Rev. Mod. Phys. **29**, 377 (1957)
7. A.Jevicki, B.Sakita, Nucl. Phys. B **165** (1980)
8. L.G. Yaffe, Rev. Mod. Phys. **54**, 407 (1982)

PARTICLE PHENOMENOLOGY

Single Top Production and Extra Dimensions

Alakabha Datta

Lab Rene J.-A. Levesque, Universite de Montreal,
C.P 6128 succrsale centre-ville
QC, H3C 3J7, Canada

Abstract. In extra dimension theories, where the gauge bosons of the standard model propagate in the bulk of the extra dimensions, there are Kaluza-Klein excitations of the standard model gauge bosons that can couple to the standard model fermions. In this paper we study the effects of the excited Kaluza-Klein modes of the W on single top production at the Tevatron.

Single top production is a very useful probe of the electroweak properties of the top quark which is believed to be intimately connected to electroweak symmetry breaking. For instance, single top production offers the possibility of measuring the Cabibbo-Kobayashi-Maskawa (CKM) matrix element V_{tb}, which is constrained by the unitarity of the CKM matrix to be ~ 1, and the polarization of the top quark to probe the $V-A$ nature of the weak interaction [1]. Single top production occurs within the SM in three different channels, the s-channel W^* production, $q\bar{q}' \to W^* \to t\bar{b}$, the t-channel W-exchange mode, $bq \to tq'$ (sometimes referred to as W-gluon fusion) and through tW^- production. The theory of single top production in the Standard Model(SM) has been extensively studied in the literature [2]. On the experimental front the D0 [3] and CDF [4] collaborations have already set limits on both the s-channel and t-channel cross sections using data collected during run I of the Fermilab Tevatron, and one expects discovery of these modes at the current run II. Measurement of the production cross section and of single top quarks is also planned at the CERN Large Hadron Collider (LHC) [5].

The large mass of the top quark, comparable to the electroweak symmetry breaking scale, makes the top quark sector very sensitive to new physics [6] and effects of new physics in singletop production have been studied extensively [7, 8, 9]. In this talk based on [10], we consider effects that extra dimension theories can produce in single top production at the Tevatron. If, in such theories, the gauge fields of the Standard Model(SM) live in the bulk of the extra dimensions then they will have Kaluza-Klein(KK) excitations and the KK excited W will contribute to single top production. Direct searches for these KK excitations have put a lower bound on the mass of the first KK excitation to be around a TeV. To study the physics of the KK excited W we use a model which is based on a simple extension of the SM to 5 dimensions (5D SM) [11]. However, we do not assume that this model represents all the physics beyond the standard model. The 5D SM is probably a part of a more fundamental underlying theory. The piece of the effective Lagrangian, obtained after integrating over the fifth dimension, that is relevant to our calculation is given by

$$\mathscr{L}^{ch} = \sum_{a=1}^{2} \mathscr{L}_a^{ch} + \mathscr{L}_{new} \qquad (1)$$

with

$$\mathscr{L}_a^{ch} = \frac{1}{2} m_W^2 W_a \cdot W_a + \frac{1}{2} M_c^2 \sum_{n=1}^{\infty} n^2 W_a^{(n)} \cdot W_a^{(n)}$$
$$- g W_a \cdot J_a - g\sqrt{2} J_a^{KK} \cdot \sum_{n=1}^{\infty} W_a^{(n)}, \qquad (2)$$

where $m_W^2 = g^2 v^2 / 2$, the weak angle θ is defined by $e = g s_\theta = g' c_\theta$, while the currents are

$$J_{a\mu} = \sum_\psi \bar{\psi}_L \gamma_\mu \frac{\sigma_a}{2} \psi_L,$$
$$J_{a\mu}^{KK} = \sum_\psi \varepsilon^{\psi_L} \bar{\psi}_L \gamma_\mu \frac{\sigma_a}{2} \psi_L. \qquad (3)$$

Here ε^{ψ_L} takes the value 1(0) for the ψ_L living in the boundary(bulk). The mass of the nth excited KK state of the W is given by $nM_c = n/R$ where R is the compactification radius. The term $\mathscr{L}_{new}$ represents the additional new physics beyond the 5 dimensional standard model the structure of which remains unknown till the full underlying theory is understood. The coupling of KK excited W to the standard model is determined in terms of the Fermi coupling, G_F, up to corrections of $O(m_Z^2/M_c^2)$ [11] . For $M_c \sim$ TeV the $O(m_Z^2/M_c^2)$ effects are small for single top production and therefore we do not include these effects in our calculations. We have ignored the mixing of the W with W_{KK} which is also an $O(m_Z^2/M_c^2)$ effect. Thus, assuming the W_{KK} decays only to standard model particles, the predicted effect of W_{KK} on single top production depends, in addition to the SM parameters, only on the unknown mass of the W_{KK}.

The cross section for $p\bar{p} \to t\bar{b}X$ is given by

$$\sigma(p\bar{p} \to t\bar{b}X) = \int dx_1 dx_2 [u(x_1)\bar{d}(x_2) + u(x_2)\bar{d}(x_1)] \sigma(u\bar{d} \to t\bar{b}). \qquad (4)$$

Here $u(x_i)$, $\bar{d}(x_i)$ are the u and the $\bar{d}$ structure functions, x_1 and x_2 are the parton momentum fractions and the indices $i=1$ and $i=2$ refer to the proton and the antiproton. The cross section for the process

$$u(p_1) + \bar{d}(p_2) \to W^* \to \bar{b}(p_3) + t(p_4),$$

is given by

$$\sigma = \sigma_{SM} \left[1 + 4\frac{A}{D} + 4\frac{C}{D} \right],$$

$$A = (s-M_W^2)(s-M_{W_{KK}}^2) + M_W M_{W_{KK}} \Gamma_W \Gamma_{W_{KK}},$$
$$C = (s-M_W^2)^2 + (M_W \Gamma_W)^2,$$
$$D = (s-M_{W_{KK}}^2)^2 + (M_{W_{KK}} \Gamma_{W_{KK}})^2, \quad (5)$$

and

$$\sigma_{SM} = \frac{g^4}{384\pi} \frac{(2s+M_t^2)(s-M_t^2)^2}{s^2[(s-M_W^2)^2 + (M_W \Gamma_W)^2]}. \quad (6)$$

Here $s = x_1 x_2 S$ is the parton center of mass energy while S is the $p\bar{p}$ center of mass energy. To calculate the width of the W_{KK} we will assume that it decays only to the standard model particles. The W_{KK} will then have the same decays as the W boson but in addition it can also decay to a top-bottom pair which is kinematically forbidden for the W boson. The width of the W_{KK}, $\Gamma_{W_{KK}}$, is then given by

$$\Gamma_{W_{KK}} \approx \frac{2M_{W_{KK}}}{M_W} \Gamma_W + \frac{2M_{W_{KK}}}{3M_W} \Gamma_W \cdot X,$$
$$X = (1 - \frac{M_t^2}{M_{W_{KK}}^2})(1 - \frac{M_t^2}{2M_{W_{KK}}^2} - \frac{M_t^4}{M_{W_{KK}}^4}). \quad (7)$$

where Γ_W is the width of the W boson and we have neglected the mass of the b quark along with the masses of the lighter quarks and the leptons.

For the first excited KK W state around a TeV we calculate the change in the single top production cross section in the presence of W_{KK} excitations.[1] We have used the CTEQ [12] structure functions for our calculations and obtain a standard model cross section of 0.30 pb for the process $p\bar{p} \to t\bar{b}X$ at $\sqrt{S} = 2$TeV. We find that the presence of the first W_{KK} can lower the cross section for the s-channel process by as much as 25 % for $M_{W_{KK}} \sim 1$TeV. The inclusion of the higher KK resonance can than lower the cross section by more than 40%. This has an important implication for the measurement of V_{tb} using the s-channel mode at the Tevatron. It was pointed out in Ref[8] that there could be models where the presence of an additional W(denoted as W') could lead to a measurement of the cross section for the s-channel $p\bar{p} \to t\bar{b}X$ smaller than the standard model prediction. This could, as pointed out in Ref [8], lead one to conclude that $V_{tb} < 1$ which could then be wrongly interpreted as evidence for the existence of new generation(s) of fermions mixed with the third generation. This work provides a specific example of such a model but the key point is that if V_{tb} were less than unity, one would see a decrease of *both* the s-channel and the t-channel cross section unlike the KK excited W case whose effect would mainly be on the s-channel process and not so much on the the t-channel process. Furthermore, most realistic models with an extra W' do not predict a large decrease in the s-channel cross section[9]. Hence a large decrease of the s-channel cross section *without* a corresponding decrease in the t-channel cross section would indicate the presence of KK excited W bosons.

---

[1] We have not included the QCD and Yukawa corrections to the single top quark production rate. They will enhance the total rate, but not change the percentage of the correction of new physics to the cross section.

ACKNOWLEDGMENTS

This work was supported in part by Natural Sciences and Engineering Research Council of Canada.

REFERENCES

1. G. Mahlon and S. Parke, Phys. Rev. D **55**, 7249 (1997); G. Mahlon and S. Parke, Phys. Lett. B **476**, 323 (2000).
2. S. S. Willenbrock and D. A. Dicus, Phys. Rev. D **34**, 155 (1986); C. P. Yuan, Phys. Rev. D **41**, 42 (1990); S. Cortese and R. Petronzio, Phys. Lett. B **253**, 494 (1991); R. K. Ellis and S. Parke, Phys. Rev. D **46**, 3785 (1992); D. O. Carlson and C. P. Yuan, Phys. Lett. B **306**, 386 (1993); T. Stelzer and S. Willenbrock, Phys. Lett. B **357**, 125 (1995); A. P. Heinson, A. S. Belyaev and E. E. Boos, Phys. Rev. D **56**, 3114 (1997); T. Stelzer, Z. Sullivan and S. Willenbrock, Phys. Rev. D **58**, 094021 (1998); B. W. Harris, E. Laenen, L. Phaf, Z. Sullivan and S. Weinzierl, arXiv:hep-ph/0207055.
3. V. M. Abazov et al. [D0 Collaboration], Phys. Lett. B **517**, 282 (2001); B. Abbott et al. [D0 Collaboration], Phys. Rev. D **63**, 031101 (2001); A. P. Heinson [D0 Collaboration], Int. J. Mod. Phys. A **16S1A**, 386 (2001).
4. C. I. Ciobanu [CDF Collaboration], Int. J. Mod. Phys. A **16S1A**, 389 (2001); D. Acosta et al. [CDF Collaboration], Phys. Rev. D **65**, 091102 (2002); T. Kikuchi, S. K. Wolinski, L. Demortier, S. Kim, and P. Savard [CDF Collaboration], Int. J. Mod. Phys. A **16S1A**, 382 (2001).
5. M. Beneke et al., in *Proceedings of the Workshop on Standard Model Physics (and More) at the LHC*, eds. G. Altarelli, and M. L. Mangano (CERN, Geneva, 2000), p. 456, arXiv:hep-ph/0003033.
6. R. D. Peccei and X. Zhang, Nucl. Phys. B **337**, 269 (1990); R. D. Peccei, S. Peris and X. Zhang, Nucl. Phys. B **349**, 305 (1991).
7. D. O. Carlson, E. Malkawi, and C. P. Yuan, Phys. Lett. B **337**, 145 (1994); G. L. Kane, G. A. Ladinsky and C. P. Yuan, Phys. Rev. D **45**, 124 (1992); T. G. Rizzo, Phys. Rev. D **53**, 6218 (1996); T. Tait and C. P. Yuan, Phys. Rev. D **55**, 7300 (1997); A. Datta and X. Zhang, Phys. Rev. D **55**, 2530 (1997). K. Whisnant, J. M. Yang, B. L. Young and X. Zhang, Phys. Rev. D **56**, 467 (1997); E. Boos, L. Dudko and T. Ohl, Eur. Phys. J. C **11**, 473 (1999); T. Tait and C. P. Yuan, Phys. Rev. D **63**, 014018 (2001); D. Espriu and J. Manzano, Phys. Rev. D **65**, 073005 (2002); G. Lu, Y. Cao, J. Huang, J. Zhang and Z. Xiao, arXiv:hep-ph/9701406. P. Baringer, P. Jain, D. W. McKay and L. L. Smith, Phys. Rev. D **56**, 2914 (1997); C. X. Yue and G. R. Lu, Chin. Phys. Lett. **15**, 631 (1998); T. Han, M. Hosch, K. Whisnant, B. L. Young and X. Zhang, Phys. Rev. D **58**, 073008 (1998); D. Atwood, S. Bar-Shalom, G. Eilam and A. Soni, Phys. Rev. D **54**, 5412 (1996); S. Bar-Shalom, D. Atwood and A. Soni, Phys. Rev. D **57**, 1495 (1998); E. Christova, S. Fichtinger, S. Kraml and W. Majerotto, Phys. Rev. D **65**, 094002 (2002); C. S. Li, R. J. Oakes and J. M. Yang, Phys. Rev. D **55**, 1672 (1997); C. S. Li, R. J. Oakes and J. M. Yang, Phys. Rev. D **55**, 5780 (1997); C. S. Li, R. J. Oakes, J. M. Yang and H. Y. Zhou, Phys. Rev. D **57**, 2009 (1998); A. Datta, J. M. Yang, B. L. Young and X. Zhang, Phys. Rev. D **56**, 3107 (1997); R. J. Oakes, K. Whisnant, J. M. Yang, B. L. Young and X. Zhang, Phys. Rev. D **57**, 534 (1998); K. i. Hikasa, J. M. Yang and B. L. Young, Phys. Rev. D **60**, 114041 (1999); P. Chiappetta, A. Deandrea, E. Nagy, S. Negroni, G. Polesello and J. M. Virey, Phys. Rev. D **61**, 115008 (2000).
8. T. Tait and C. P. Yuan, arXiv:hep-ph/9710372.
9. E. H. Simmons, Phys. Rev. D **55**, 5494 (1997).
10. A. Datta, P. J. O'Donnell, Z. H. Lin, X. Zhang and T. Huang, Phys. Lett. B **483**, 203 (2000).
11. See A. Delgado, A. Pomarol and M. Quirós, JHEP **0001**, 030 (2000) and references therein.
12. H. L. Lai *et.al*, CTEQ Collaboration, hep-ph/9903282.

Constraints on the vector quark model from rare B decays

M. R. Ahmady*, M. Nagashima† and A. Sugamoto†

*Department of Physics, Mount Allison University, Sackville, NB E4L 1E6 Canada
†Department of Physics, Ochanomizu University, 1-1 Otsuka 2, Bunkyo-ku, Tokyo 112, Japan

Abstract. The addition of an extra generation of vector-like quarks to the Standard Model leads to the flavour changing neutral current at the tree level as well as the non-unitarity of the CKM quark mixing matrix. We use the latest experimental results on the rare B decays, $B \to X_s \ell^+ \ell^-$ ($\ell = e, \mu$), $B \to X_s \gamma$ and B_s°-$\bar{B}_s^\circ$ mixing to constrain the parameters of this model.

Rare B decays are excellent venues for observing the signals of new physics beyond the Standard Model (SM). Present and near future dedicated B experiments are expected to access new rare B decay channels and to improve the precision of those which have already been measured. Therefore, a lot of theoretical attentions are devoted to the effect of new physics on these processes. One such beyond the SM scenarios is the extension of the SM to contain an extra generation of isosinglet quarks[1]:

$$\psi^i \equiv \begin{pmatrix} u^\alpha \\ d^\alpha \end{pmatrix}, \alpha = 1...4, \qquad (1)$$

where

$$\begin{pmatrix} u^4 \\ d^4 \end{pmatrix} \equiv \begin{pmatrix} U \\ D \end{pmatrix}. \qquad (2)$$

Both left- and right-handed components of U and D are $SU(2)_L$ singlets, and therefore, the Dirac mass terms of vectorlike quarks respects the electroweak gauge symmetry. However, for ordinary quarks, gauge invariant Yukawa couplings to an isodoublet scalar Higgs field are responsible for the mass generation via spontaneous symmetry breaking. The coupling between the vectorlike and the ordinary quarks are provided by the extra gauge invariant Yukawa couplings of the following form:

$$-f_d^{i4} \bar{\psi}_L^i D_R \phi - f_u^{i4} \bar{\psi}_L^i U_R \tilde{\phi} + H.C. \quad . \qquad (3)$$

Therefore, after spontaneous symmetry breaking the 4 × 4 mass matrices for the up- and down-type quarks, which are not diagonal in general, are transformed into diagonal forms by unitary transformations from weak to mass eigenstates:

$$u_{L,R}^\alpha = A_{L,R}^{u\ \alpha\beta} u'^\beta_{L,R}, \quad d_{L,R}^\alpha = A_{L,R}^{d\ \alpha\beta} d'^\beta_{L,R}, \qquad (4)$$

where we use prime to denote the mass eigenstates. The interesting property of the vectorlike quark model (VQM) is that the above transformations result in the intergenerational mixing among quarks not only in the charged current sector but also in the neutral current interactions as follows:

$$J_{CC}^{W\,\mu} = \sum_{\alpha,\beta=1}^{4} I \frac{g}{\sqrt{2}} \bar{u}'^{\alpha}_L V^{\alpha\beta} \gamma^{\mu} d'^{\beta}_L W^+_{\mu} + H.C. \,, \tag{5}$$

and

$$J_{NC}^{Z\,\mu} = I \frac{g}{\cos\theta_w} \sum_{\alpha,\beta=1}^{4} \left(I_w^q U^{\alpha\beta} \bar{q}'^{\alpha}_L \gamma^{\mu} q'^{\beta}_L - Q_q \sin^2\theta_w \delta^{\alpha\beta} \bar{q}'^{\alpha} \gamma^{\mu} q'^{\beta} \right) \,, \tag{6}$$

where the 4×4 quark mixing matrix V

$$V^{\alpha\beta} = \sum_{i=1}^{3} (A_L^{u\dagger})^{\alpha i} (A_L^d)^{i\beta} \,, \tag{7}$$

is nonunitary

$$\begin{aligned}(V^{\dagger}V)^{\alpha\beta} &= \delta^{\alpha\beta} - (A_L^{d\,4\alpha})^* A_L^{d\,4\beta} \,, \\ (VV^{\dagger})^{\alpha\beta} &= \delta^{\alpha\beta} - (A_L^{u\,4\alpha})^* A_L^{u\,4\beta} \,. \end{aligned} \tag{8}$$

and

$$U^{\alpha\beta} = \sum_{i=1}^{3} (A_L^{qi\alpha})^* A_L^{qi\beta} = \delta^{\alpha\beta} - (A_L^{q\,4\alpha})^* A_L^{q\,4\beta} = \begin{cases} (V^{\dagger}V)^{\alpha\beta}, & q \equiv \text{down-type} \\ (VV^{\dagger})^{\alpha\beta}, & q \equiv \text{up-type} \end{cases} . \tag{9}$$

Five model parameters appear in our calculations[1]: the U-quark mass m_U, the nonunitarity parameter $U^{sb} = |U^{sb}|e^{i\theta}$, where θ is a weak phase, and $V_{4s}^* V_{4b} = U^{sb} - (V_{CKM}^{\dagger} V_{CKM})^{sb}$ where V_{CKM} is the 3×3 submatrix of the mixing matrix V and consists of the elements representing mixing among the three ordinary generations of quarks. $(V_{CKM}^{\dagger} V_{CKM})^{sb}$ presents the deviation from unitarity of the 3-generation CKM mixing matrix in the VQM context. The one-loop terms with coefficient $V_{4s}^* V_{4b}$ are significant only if $V_{4s}^* V_{4b} \approx -(V_{CKM}^{\dagger} V_{CKM})^{sb} \gg U^{sb}$, and therefore, we parametrize our results in terms of $(V_{CKM}^{\dagger} V_{CKM})^{sb}/|V_{cb}| = \varepsilon e^{i\phi}$ instead of $V_{4s}^* V_{4b}$, where $\varepsilon = |(V_{CKM}^{\dagger} V_{CKM})^{sb}|/|V_{cb}|$ and ϕ is another weak phase of the model. We use $|V_{cs}| \approx 0.97$, $|V_{cb}| \approx 0.04$, $|V_{ts}|/|V_{cb}| \approx 1.1$, and $V_{us}^* V_{ub} \approx 0$ [3]. $V_{ts}^* V_{tb}$, which is not known experimentally, can be expressed in terms of the VQM parameters as:

$$\frac{V_{ts}^* V_{tb}}{|V_{cb}|} \approx \varepsilon e^{i\phi} - |V_{cs}| \,. \tag{10}$$

In Fig. 1, the experimental results on $BR(B \to X_s \mu^+ \mu^-)$ reported in Ref. [4] are used to obtain the acceptable regions in the $|U^{sb}|$ versus m_U plane for various choices of the

---

[1] For more details see [2] and the references therein

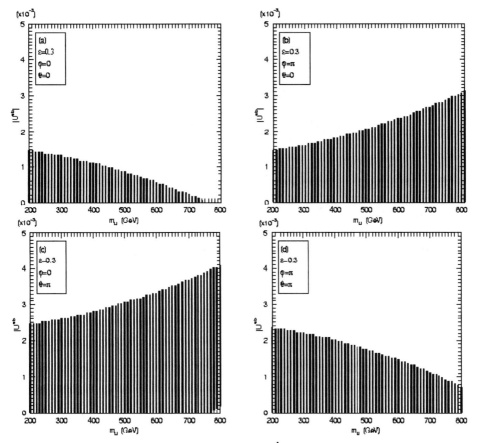

Figure 1: The acceptable region (shaded area) of the $|U^{ub}|$ versus m_U plane for various values of the VQM parameters obtained by using the experimental upper bound on $BR(B \to X_s \mu^+ \mu^-)$.

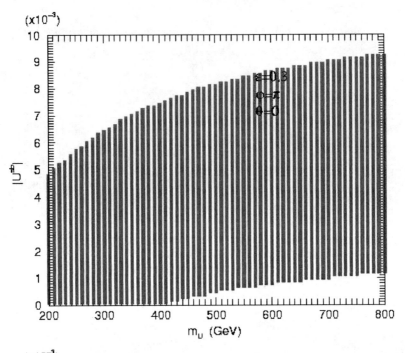
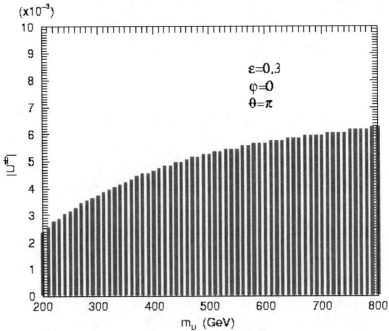

Figure 2: Allowed region of the VQM model parameters extracted from B->Xs gamma

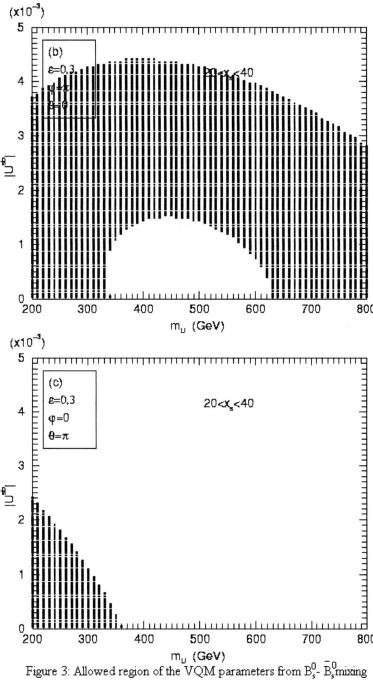

Figure 3: Allowed region of the VQM parameters from B_s^0-$\bar{B}_s^0$ mixing

relative sign of the extra contributions. We Observe that the most stringent constraint is obtained if the relative phases θ and ϕ both vanish. $B \to X_s \gamma$ on the other hand, does not lead to much further constraints on the model parameters, as is evident from Fig. 2. The constraints from the B_s°-$\bar{B}_s^\circ$ mixing are illustrated in Fig. 3. In fact, the most stringent constraint is obtained from this process for the case when the extra tree level and loop contributions have the same sign. Future measurement of the $BR(B \to X_s \mu^+ \mu^-)$ and B_s°-$\bar{B}_s^\circ$ mixing will severely restrict the VQM model parameters.

ACKNOWLEDGMENTS

M.A.'s research is partially supported by an internal summer research grant from Mount Allison University.

REFERENCES

1. L. Bento and G. C. Branco, Phys. Lett. B**245**, 599 (1990);
   L. Bento, G. C. Branco and P. A. Parada, Phys. Lett. B**267**, 95 (1991).
2. M. R. Ahmady, M. Nagashima and A. Sugamoto, Phys. Rev. D**64**, 054011 (2001).
3. Particle Data Group, D. E. Groom *et al.*, Eur. Phys. J. C**15**, 1 (2000).
4. CLEO Collaboration, S. Glenn *et al.*, Phys. Rev. Lett. **80**, 2289 (1998).

Chiral Lagrangian Treatment of the Isosinglet Scalar Mesons in 1-2 GeV Region

Amir H. Fariborz

Department of Mathematics/Science, State University of New York Institute of Technology, Utica, New York 13504-3050

Abstract.
In this article, preliminary results on isosinglet scalar mesons below 2 GeV in the context of a non-linear chiral Lagrangian are presented.

INTRODUCTION

Scalar mesons play important roles in low-energy QCD, and are at the focus of many theoretical and experimental investigations. Scalars are important from the theoretical point of view because they are Higgs bosons of QCD and induce chiral symmetry breaking, and therefore, are probes of the QCD vacuum. Scalars are also important from a phenomenological point of view, as they are very important intermediate states in Goldstone boson interactions away from threshold, where chiral perturbation theory is not applicable. There are 9 candidates for the lowest-lying scalar mesons ($m < 1$ GeV): $f_0(980)$ [$I = 0$] and $a_0(980)$ [$I = 1$] which are well established experimentally [1]; $\sigma(560)$ or $f_0(400-1200)$ [$I = 0$] with uncertain mass and decay width [1]; and $\kappa(900)$ [$I = 1/2$] which is not listed but mentioned in PDG 2000 [1]. The $\kappa(900)$ is only observed in some theoretical models. It is known that a simple $q\bar{q}$ picture does not explain the properties of these mesons. Different theoretical models that go beyond a simple $q\bar{q}$ picture have been developed, including: MIT bag model [2], $K\bar{K}$ molecule [3], unitarized quark model [4], QCD sum-rules [5], and chiral Lagrangians [6, 7, 8, 9, 10, 11].

The next-to-lowest scalars (1 GeV $< m <$ 2 GeV) are: $K_0^*(1430)$ [$I = 1/2$], $a_0(1450)$ [$I = 1$], $f_0(1370)$, $f_0(1500)$, $f_0(1710)$ [$I = 0$], and are all listed in [1]. The $f_0(1500)$ is believed to contain a large glue component and therefore a good candidate for the lowest scalar glueball state. These states, are generally believed to be closer to $q\bar{q}$ objects; however, some of their properties cannot be explained based on a pure $q\bar{q}$ structure.

Chiral Lagrangians, provide a powerful framework for studying the lowest and the next-to-lowest scalar states probed in different Goldstone boson interactions ($\pi\pi$, πK, $\pi\eta$,...) away from threshold [6, 7, 8, 9, 11]. In this approach, a description of scattering amplitudes which are, to a good approximation, both crossing symmetric and unitary is possible. To construct scattering amplitudes, all contributing intermediate resonances up to the energy of interest are considered, and only tree diagrams (motivated by large N_c approximation) are taken into account. In this way, crossing symmetry is satisfied, but the constructed amplitudes should be regularized. Regularization procedure in turn unitarizes the scattering amplitude. By fitting the resulting scattering amplitude to experimental data, the unknown physical properties (mass, decay width, ...) of the light scalar mesons can be extracted. In a non-linear chiral Lagrangian framework in [6, 7, 8, 9, 10], and a linear sigma model framework in [11], scalar mesons in several low energy processes are studied and similar conclusions are made:

- A $\sigma(560)$ and a $\kappa(900)$ are required in order to describe experimental data on $\pi\pi$ and πK scattering amplitudes, respectively, and that these states [as well as $f_0(980)$ and $a_0(980)$] are closer to four-quark objects [6, 7, 8, 9, 11].
- Some unexpected properties of the isodoublet $K_0^*(1430)$ and the isovector $a_0(1450)$ can be explained based on a mixing mechanism between underlying $q\bar{q}$ and $qq\bar{q}\bar{q}$ nonets [10, 11].

In this article, the possibility of extending the mixing mechanism of ref. [10] for $K_0^*(1430)$ and $a_0(1450)$, to the isosinglet states is investigated. We briefly review this mechanism in section II, and extend it to the $I = 0$ case in section III. The results are given in section IV followed by a summary in section V.

MIXING MECHANISM FOR I=1/2 AND I=1 SCALAR MESONS

The $a_0(1450)$ and $K_0^*(1430)$ are expected to belong to the same $q\bar{q}$ nonet [1]. However, there are properties of these two states that do not quite follow this scenario:

- In a $q\bar{q}$ nonet, isotriplet is expected to be lighter than the isodoublet, but for these two states:

$$m[a_0(1450)] = 1474 \pm 19 > m[K_0^*(1430)] = 1429 \pm 6 \tag{1}$$

- The decay ratios do not follow a pattern expected from a $q\bar{q}$ nonet:

$$\begin{cases} \frac{\Gamma[a_0^{total}]}{\Gamma[K_0^* \to \pi K]} = \begin{cases} 0.92 \pm 0.12 & \text{PDG} \\ 1.51 & \text{SU(3)} \end{cases} \\ \frac{\Gamma[a_0 \to K\bar{K}]}{\Gamma[a_0 \to \pi\eta]} = \begin{cases} 0.88 \pm 0.23 & \text{PDG} \\ 0.55 & \text{SU(3)} \end{cases} \\ \frac{\Gamma[a_0 \to \pi\eta']}{\Gamma[a_0 \to \pi\eta]} = \begin{cases} 0.35 \pm 0.16 & \text{PDG} \\ 0.16 & \text{SU(3)} \end{cases} \end{cases} \tag{2}$$

where we see that there is a discrepancy between the experimental data and the SU(3) prediction.

In ref. [10] a mechanism for understanding these two issues was proposed. In this mechanism, a $qq\bar{q}\bar{q}$ scalar nonet N mixes with a $q\bar{q}$ scalar nonet N', and as a result, the physical $I = 1$ and $I = 1/2$ states acquire two and four quark components. Specifically, it was shown in [10] that a simple "external" mixing term

$$\mathscr{L}_{mix} = -\gamma \text{Tr}(NN') \tag{3}$$

with
$$0.26 < \gamma^2 < 0.38 \text{ GeV}^4 \tag{4}$$

explains the mass spectrum – the isotriplet states split more than isodoublets, and consequently, the physical isovector state $a_0(1450)$ becomes heavier than the isodoublet state $K_0^*(1430)$. The light isovector and isodoublet states in this mechanism are identified with $a_0(980)$ and $\kappa(900)$.

The decay ratios were also investigated in [10] by considering a scalar-pseudoscalar-pseudoscalar interaction term

$$\mathscr{L}_{int.} = A\varepsilon^{abc}\varepsilon_{def}N_a^d \partial_\mu \phi_b^e \partial_\mu \phi_c^f + 2A' \text{Tr}(N' \partial_\mu \phi \partial_\mu \phi) \tag{5}$$

where the first and second terms are natural interaction terms for a four quark nonet, and a two quark nonet, respectively. It was shown in [10] that these interaction terms result in decay ratios that are consistent with PDG data given in Eq. (2).

ISOSINGLET STATES

The case of $I = 0$ states is of course more involved. In this case, in addition to the external mixing (3), there are "internal" mixings, and mixings with glueballs:

$$\begin{aligned}\mathscr{L}_{mass} &= -a\text{Tr}(NN) - b\text{Tr}(NN\mathscr{M}) - c\text{Tr}(N)\text{Tr}(N) - d\text{Tr}(N)\text{Tr}(N\mathscr{M}) + (N \leftrightarrow N') \\ &\quad -\gamma\text{Tr}(NN') - eG\text{Tr}(N) - fG\text{Tr}(N')\end{aligned} \tag{6}$$

where $\mathscr{M} = \text{diag}(1,1,x)$ with x being the ratio of strange to non-strange quark masses; a,b,c and d are unknown parameters describing mass terms for N, with c and d terms inducing the internal mixing between the two $I = 0$ states. There are similar terms for nonet N' (but with different couplings a', b', c' and d'). The last two terms, with unknown couplings e and f, describe mixing with a glueball G. As a result, the five isosinglets below 2 GeV, are a mixture of five different flavor combinations:

$$\begin{pmatrix} \sigma(550) \\ f_0(980) \\ f_0(1370) \\ f_0(1500) \\ f_0(1710) \end{pmatrix} = \mathscr{O} \begin{pmatrix} f_0^{NS} \\ f_0^{S} \\ f_0'^{S} \\ f_0'^{NS} \\ G \end{pmatrix} \tag{7}$$

where f_0^{NS} and f_0^{S} are non-strange and strange $I = 0$ combinations in nonet N, respectively. Similarly, $f_0'^{NS}$ and $f_0'^{S}$ are non-strange and strange $I = 0$ combinations in N' nonet.

The scalar-pseudoscalar-pseudoscalar interaction takes the general form:

$$\begin{aligned}\mathscr{L}_{int} &= A\varepsilon^{abc}\varepsilon_{def}N_a^d \partial_\mu \phi_b^e \partial_\mu \phi_c^f + B\text{Tr}(N)\text{Tr}(\partial_\mu \phi \partial_\mu \phi) + C\text{Tr}(N\partial_\mu \phi)\text{Tr}(\partial_\mu \phi) \\ &\quad + D\text{Tr}(N)\text{Tr}(\partial_\mu \phi)\text{Tr}(\partial_\mu \phi) + (N \leftrightarrow N') \\ &\quad + EG\text{Tr}(\partial_\mu \phi \partial_\mu \phi) + FG\text{Tr}(\partial_\mu \phi)\text{Tr}(\partial_\mu \phi)\end{aligned} \tag{8}$$

where A, B, C and D are unknown coupling constants describing the coupling of the four-quark nonet N to the pseudoscalars. Similar terms (but with different couplings A', B', C' and D') are considered for the coupling of N' to the pseudoscalars [represented by $(N \leftrightarrow N')$]. E and F describe the coupling of a scalar glueball to the pseudoscalar mesons. The pseudoscalar part of the Lagrangian can be found in refs. [7]. These unknown couplings should be determined by fitting to different experimental data. The mass and decay widths of $f_0(1370), f_0(1500)$ and $f_0(1710)$ are [1]:

$$
\begin{aligned}
m[f_0(1370)] &= 1200 \to 1500 \text{ MeV} & \Gamma^{total}[f_0(1370)] &= 200 \to 500 \text{ MeV} \\
m[f_0(1500)] &= 1500 \pm 10 \text{ MeV} & \Gamma^{total}[f_0(1500)] &= 112 \pm 10 \text{ MeV} \\
m[f_0(1710)] &= 1712 \pm 5 \text{ MeV} & \Gamma^{total}[f_0(1710)] &= 133 \pm 14 \text{ MeV}
\end{aligned}
\tag{9}
$$

We see that the largest uncertainty is on the mass and decay width of $f_0(1370)$. Different decay ratios for $I = 0$ states, are given by WA102 collaboration [12]:

$$
\begin{aligned}
\frac{f_0(1370) \to \pi\pi}{f_0(1370) \to K\bar{K}} &= 2.17 \pm 0.90 & \frac{f_0(1370) \to \eta\eta}{f_0(1370) \to K\bar{K}} &= 0.35 \pm 0.21 \\
\frac{f_0(1500) \to \pi\pi}{f_0(1500) \to \eta\eta} &= 5.5 \pm 0.84 & \frac{f_0(1500) \to K\bar{K}}{f_0(1500) \to \pi\pi} &= 0.32 \pm 0.07 \\
\frac{f_0(1710) \to \pi\pi}{f_0(1710) \to K\bar{K}} &= 0.20 \pm 0.03 & \frac{f_0(1710) \to \eta\eta}{f_0(1710) \to K\bar{K}} &= 0.48 \pm 0.14
\end{aligned}
\tag{10}
$$

RESULTS

We first examine the simple case of external mixing only. In this case, c, d, c', d', e and f in (6), as well as E and F in (8) are zero, and there are four isosinglets with masses [10]:

$$
\begin{aligned}
m[f_1] &= 690 \text{ MeV} & m[f_2] &= 980 \text{ MeV} \\
m[f_3] &= 1450 \text{ MeV} & m[f_4] &= 1460 \text{ MeV}
\end{aligned}
\tag{11}
$$

The first and second states are $\sigma(550)$ and $\kappa(900)$. The last two states are almost degenerate, and none of them can be identified with $f_0(1500)$. However, f_3 is within the mass range of $f_0(1370)$, and its decay ratios are consistent with the data given in (10). Specifically, the ratios of the decay amplitude squared are estimated in Table 1, and are consistent with another theoretical estimate of these quantities by Close in [13]. Also the branching ratios of f_3 are consistent with those of $f_0(1370)$. These are shown in Table 2. The results therefore suggest that the mass and decay properties of $f_0(1370)$ can be understood in a quark mixing scenario without glueballs. However, a more accurate experimental data on mass and decay width of $f_0(1370)$ may change this identification. Also the results suggest that the $f_0(1500)$ may have significant glueball component.

TABLE 1. Ratios of the decay amplitude squared for $f_0(1370)$.

	$\eta\eta/\pi\pi$	$\eta\eta/K\bar{K}$
Close	0.74 ± 0.51	1.64 ± 0.96
This model	0.49 ± 0.16	0.70 ± 0.24

TABLE 2. Branching ratios of $f_0(1370)$.

	$\Gamma_{\pi\pi}/\Gamma^{total}$	$\Gamma_{K\bar{K}}/\Gamma^{total}$
PDG(2000)	0.26 ± 0.09	0.35 ± 0.13
This model	0.35 ± 0.01	0.54 ± 0.02

Next, we allow internal mixings (i.e. the parameters c, d, c' and d' in (6) are non zero). In this case, we identify the four isosinglets of the model, with four physical states that can be selected out of five candidates $\sigma(550), f_0(980), f_0(1370), f_0(1500)$ and $f_0(1710)$. Three different ways of this identification are given below, together with the χ^2 of the fits to masses and decay widths:

- Fit to $\sigma, f_0(980), f_0(1500), f_0(1710)$:

$$
\begin{array}{ll}
m[f_1] = 552 \text{ MeV} & m[f_2] = 984 \text{ MeV} \\
m[f_3] = 1548 \text{ MeV} & m[f_4] = 1678 \text{ MeV}
\end{array}
\quad \left(\begin{array}{l} \chi^2_{mass} = 0.417 \\ \chi^2_{decay} = 12.66 \end{array} \right) \quad (12)
$$

- Fit to $\sigma, f_0(980), f_0(1370), f_0(1500)$:

$$
\begin{array}{ll}
m[f_1] = 554 \text{ MeV} & m[f_2] = 980 \text{ MeV} \\
m[f_3] = 1471 \text{ MeV} & m[f_4] = 1551 \text{ MeV}
\end{array}
\quad \left(\begin{array}{l} \chi^2_{mass} = 0.041 \\ \chi^2_{decay} = 3.47 \end{array} \right) \quad (13)
$$

- Fit to $\sigma, f_0(980), f_0(1370), f_0(1710)$:

$$
\begin{array}{ll}
m[f_1] = 550 \text{ MeV} & m[f_2] = 980 \text{ MeV} \\
m[f_3] = 1432 \text{ MeV} & m[f_4] = 1680 \text{ MeV}
\end{array}
\quad \left(\begin{array}{l} \chi^2_{mass} = 0.020 \\ \chi^2_{decay} = 1.65 \end{array} \right) \quad (14)
$$

We see that excluding the $f_0(1500)$ from the fit [i.e. third fit (14)] gives the best agreement. For this case, however, the mass of $f_0(1710)$ is not within the expected experimental range. Therefore, qualitatively at least, we see that $f_0(1500)$ and $f_0(1710)$ are not exactly described by the model, and this may suggest that they both have a substantial glueball component. For the fit (14), individual decay widths are:

$$
\begin{array}{rl}
\Gamma[\sigma(550) \to \pi\pi] &= 51 \text{ MeV} \\
\Gamma[f_0(980) \to \pi\pi] &= 70 \text{ MeV} \\
\Gamma[f_0(1370) \to \pi\pi] &= 341 \text{ MeV} \\
\Gamma[f_0(1370) \to K\bar{K}] &= 139 \text{ MeV} \\
\frac{f_0(1370) \to \pi\pi}{f_0(1370) \to K\bar{K}} &= 2.45 \ (2.17 \pm 0.90) \\
\frac{f_0(1710) \to \pi\pi}{f_0(1710) \to K\bar{K}} &= 0.18 \ (0.20 \pm 0.03)
\end{array} \quad (15)
$$

Decay properties of $\sigma(550), f_0(980)$ and $f_0(1370)$ are within the expected ranges. The decay ratios of $f_0(1370)$ and $f_0(1710)$ are consistent with the WA102 data given in parentheses. However, the decay widths of $f_0(1710)$ to $\pi\pi$ and to $K\bar{K}$ are not consistent with its total decay width given in (9).

Finally, we include mixing with a scalar glueball. In this case the isosinglet masses are:

$$m[f_1] = 523 \text{ MeV}$$
$$m[f_2] = 982 \text{ MeV}$$
$$m[f_3] = 1288 \text{ MeV} \quad (16)$$
$$m[f_4] = 1551 \text{ MeV}$$
$$m[f_5] = 1712 \text{ MeV}$$

This result shows the importance of the mixing with glueballs; all masses, except the $f_0(1500)$' mass, are within the expected ranges. The model sets a lower bound of 600 MeV for the scalar glueball mass, and indicates that the $\sigma(550)$ is mainly a four quark non-strange combination.

CONCLUSION

Chiral Lagrangians are efficient frameworks for probing scalar mesons. Properties of the lowest and the next-to-lowest $I = 1/2$ and $I = 1$ states can be understood in this framework. The case of $I = 0$ is much more involved due to different types of mixing. Preliminary results for $I = 0$ states suggest a significant mixing with glueballs.

ACKNOWLEDGMENTS

I would like to thank J. Schechter for helpful discussions. I would also like to thank the organizers of the MRST 2002 for a very enjoyable conference. This work has been supported in part by the 2002 State of New York/UUP Professional Development Committee, and the 2002 Summer Grant from the School of Arts and Sciences, SUNY Institute of Technology.

REFERENCES

1. Review of Particle Physics, Euro. Phys. J. C **15**, 1 (2000).
2. R.L. Jaffe, Phys. Rev. D **15**, 267 (1977).
3. J. Weinstein and N. Isgur, Phys. Rev. D **41**, 2236 (1990).
4. N.A. Törnqvist, Z. Phys. C **68**, 647 (1995); E. van Beveren et al, Z. Phys. C **30**, 615 (1986).
5. V. Elias, A.H. Fariborz, Fang Shi and T.G. Steele, Nucl. Phys. A **633**, 279 (1998); Fang Shi, T.G. Steele, V. Elias, K.B. Sprague, Ying Xue and A.H. Fariborz, Nucl. Phys. A **671**, 416 (2000).
6. F.Sannino and J. Schechter, Phys. Rev. D **52**, 96 (1995); M. Harada, F. Sannino and J. Schechter, Phys. Rev. D **54**, 1991 (1996); Phys. Rev. Lett. **78**, 1603 (1997).
7. D. Black, A.H. Fariborz, F. Sannino and J. Schechter, Phys. Rev. D **58**, 054012 (1998); Phys. Rev. D **59**, 074026 (1999).
8. A.H. Fariborz and J. Schechter, Phys. Rev. D **60**, 034002 (1999).
9. D. Black, A.H. Fariborz and J. Schechter, Phys. Rev. D **61**, 074030 (2000).
10. D. Black, A.H. Fariborz and J. Schechter, Phys. Rev. D **61**, 074001 (2000).
11. D. Black, A.H. Fariborz, S. Moussa, S. Nasri and J. Schechter, Phys. Rev. D **64**, 014031 (2001).
12. D. Barberis et al, Phys. Lett. B **479**, 59 (2000).
13. F.E. Close and A. Kirk, Phys. Lett. B **483**, 345 (2000).

BRANE WORLDS

Quest for a Self-Tuning Brane-World Solution to the Cosmological Constant Problem

James M. Cline[*][†] and Hassan Firouzjahi[*]

[*]*Physics Department, McGill University, Montréal, Québec, Canada H2A 2T8*
[†]*presenter*

Abstract. It has been proposed that the geometry of an extra dimension could automatically adjust itself to compensate for an arbitrary energy density on the 3-D brane which we are presumed to inhabit, such that a static solution to Einstein's equation results. This would solve the long-standing cosmological constant problem, of why our universe is not overwhelmed by the enormous energy of the quantum vacuum fluctuations predicted by quantum field theory. I will review some of the attempts along these lines, and present a no-go theorem showing that these attempts are doomed, at least within one of the most promising classes of models.

THE COSMOLOGICAL CONSTANT PROBLEM

One of the most vexing problems in theoretical particle physics is the magnitude of the vacuum energy density, Λ. Ironically, Einstein introduced it into general relativity through the field equation

$$G_{\mu\nu} \equiv R_{\mu\nu} - \frac{1}{2} g_{\mu\nu} R = 8\pi G (T_{\mu\nu} + \Lambda g_{\mu\nu}) \tag{1}$$

in order to find static cosmological solutions. He called it his "biggest blunder" when the universe was subsequently found to be expanding. Generically the presence of a positive cosmological constant leads to a universe which is accelerating, not static, but from the Friedmann equation we see that the scale factor can be static for a fine-tuned value of Λ,

$$\left(\frac{\dot{a}}{a}\right)^2 = \frac{8\pi G}{3}(\rho + \Lambda) - \frac{k}{a^2} \tag{2}$$

(in fact one must also tune $\ddot{a}$ to be zero).

The irony of Einstein's supposed blunder is three-fold. First, of course, the universe is not static; second, particle theorists came to realize that Λ should be present due to quantum fluctuations of the vacuum, which can be visualized as spontaneous creation and annihilation of particle/antiparticle pairs. Third, there is now strong evidence from cosmology that Λ is nonzero,

$$\Lambda \cong (2.4 \times 10^{-3} \text{ eV})^4 \tag{3}$$

since the universe appears to be accelerating and at critical density, even though there is not enough dark matter to account for most of the energy density. The big problem is

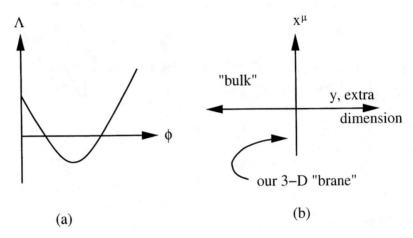

FIGURE 1. (a) A not very imaginative attempt at self-tuning. (b) The braneworld scenario.

that naive computations of the theoretical value of Λ from quantum field theory give a value which is many orders of magnitude greater,

$$\Lambda_{\text{theo.}} \sim (10^{28} \text{ eV})^4 \qquad (4)$$

There must be some mechanism for explaining the difference between the observed and expected values, but so far no really convincing idea has been proposed.

One might suspect some kind of adjustment mechanism is at work, which somehow nullifies the effect of Λ no matter what its underlying value might be. However Weinberg has given a no-go theorem against such ideas [1]. For example, one might imagine that the effective physical value of $\Lambda[\Lambda_0, \phi]$ depends on the underlying value Λ_0 and upon a scalar field ϕ which automatically adjusts itself so that $\Lambda[\Lambda_0, \phi] = 0$. To be concrete, consider $\Lambda = \Lambda_0 + f(\phi)$ as shown in fig. 1(a). Although there are values of ϕ where $\Lambda = 0$, they are not generally stationary, so fine tuning is still required. Although this is very obvious in the present example, Weinberg's theorem shows that essentially the same problem will plague even more clever attempts to implement such an idea.

Weinberg's theorem assumes that the universe is 4-dimensional, so one might hope that recent brane-world ideas might provide a loophole. In the braneworld picture illustrated in fig. 1(b), we have an extra dimension, y, surrounding our 4-D universe which is presumed to be the brane at $y = 0$. One could imagine that the effects of the vacuum energy density on the brane are counteracted by some new physics in the bulk. If the 5-D nature of this setup can't be described by an approximate 4-D picture, then it might be possible to evade the no-go theorem.

SELF-TUNING SOLUTIONS IN 5-D

Some attempts at constructing a self-tuning solution to the cosmological constant problem were presented in references [2, 3]. The idea is to add a scalar field ϕ in the bulk,

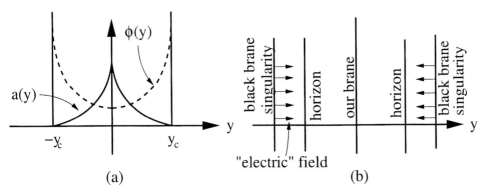

FIGURE 2. (a) Self-tuning brane solution. (b) Brane-like singularities from the visible brane by horizons.

which is described by the metric

$$ds^2 = a^2(y)dx_\mu dx^\mu + dy^2 \qquad (5)$$

The scalar is presumed to couple to the bare energy density of the brane, Λ_0 through a potential $\Lambda_0 e^{-\kappa\phi}$. The coupled field equations for the scalar and the metric give rise to a solution which is singular at some position y_c in the bulk, as illustrated in fig. 2. The value of y_c depends on Λ_0, which is what constitutes the self tuning. In other words, for any value of Λ_0, a static solution can be found with some value of y_c. It is precisely because the solution is static rather than exponentially expanding that an observer would infer that the physical value of Λ is zero.

There are a number of problems with this idea [4]. For one, the static solution is not unique: one can also find expanding or contracting ones [5], which shows that fine tuning of the initial conditions are needed to get the static solution. Moreover we don't like the presence of naked singularities. The cosmic censorship hypothesis asserts that singularities should be hidden behind an event horizon, as is the case with a black hole. Csaki et al. [6, 7] have proposed a modification to the previous self-tuning model which includes a horizon, as shown in fig. 2(b). The points $y \leftrightarrow -y$ are identified with each other by imposing a Z_2 orbifolding.

In this solution, the metric becomes the AdS-Reissner-Nördstrom solution shown in fig. 3,

$$ds^2 = -h(y)dt^2 + a^2(y)d\vec{x}^2 + h^{-1}(y)dy^2 \qquad (6)$$

with $h(y) = y^2/l^2 - \mu/y^2 + Q^2/y^6$ and $a^2(y) = y^2$; l is the radius of curvature of the 5-D anti-deSitter space, μ is mass of the black brane and Q is its charge under a U(1) gauge symmetry. The self-tuning is now accomplished by μ and Q taking on the appropriate values to cancel the effect of the energy density ρ which is on the brane. Note that these quantities appear not as inputs to the Lagrangian, but rather as integration constants in the bulk solution. However, this ρ cannot be quite the same thing as a cosmological constant because the self-tuning solution works only for a rather bizarre equation of

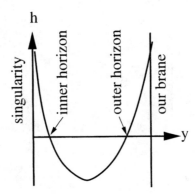

FIGURE 3. Metric function $h(r)$ for the 5-D AdS-Reissner-Nördstrom spacetime.

state for the energy density on the brane:

$$p < -\rho \tag{7}$$

This is in contrast to a vacuum energy density which obeys $p = -\rho = -\Lambda$, and is in violation of the weak energy condition. Such violations may not always be bad, but at least in the case of a classical scalar field theory they are problematic. There we have $\rho = \frac{1}{2}\dot\phi^2 + V(\phi)$ and $p = \frac{1}{2}\dot\phi^2 - V(\phi)$ so that $p = -\rho$ implies $\dot\phi^2 < 0$: negative kinetic energy. This would correspond to a ghost which leads to nonconservation of probability in quantum field theory. Our goal in the present work was to try to find a self-tuning solution with a horizon and without violating the weak energy condition. Unfortunately, we found instead another no-go theorem: one must have $p < -\rho$ either on the brane or in the bulk, as we will now describe [8].

No-go theorem

Before discovering our no-go result, we searched for a generalization of the original solution which would hopefully give us more leeway. A natural possibility is to try adding a scalar field in the bulk, so the Lagrangian becomes

$$\mathscr{L} = \sqrt{|g|}\left(\frac{1}{2}g^{\mu\nu}\partial_\mu\phi\partial_\nu\phi - V(\phi) - V_0(\phi)\frac{\delta(y)}{\sqrt{g_{55}}}\right) \tag{8}$$

We obtained numerical solutions using various potentials and found empirically that a horizon could be obtained only when the black hole charge Q^2 became negative. Interestingly, $Q^2 < 0$ implies that $p < -\rho$ for the bulk stress energy tensor contributed by the gauge field.

It is not difficult to show that the violation of the weak energy condition is actually necessary for obtaining a horizon. By adding together the (00) and one of the spatial (ii) components of the Einstein equations, and integrating from the outer horizon ($y = y_h$) to the brane ($y = y_b$), we obtain

$$h'(y_h) = -\frac{\kappa^2}{a^3}\left(2\int_{y_h}^{y_b}(T_0^0 - T_1^1)a^3 dy + \sqrt{h}(\rho + p)|_{y_b}\right) \tag{9}$$

where κ^2 is the 5-D gravitational constant and T_ν^μ is the bulk stress energy tensor. In particular, $(T_0^0 - T_1^1)$ is the bulk contribution to $(\rho + p)$. Notice the minus sign in front of everything, and the fact that $h'(y_h)$ must be positive at the outer horizon (see fig. 3). These two facts conspire to need $\rho + p$ to be negative somewhere, in the bulk, the brane or both.

This conclusion is independent of any details of the theory like the choice of scalar potentials. We have also found it to be robust against other generalizations: arbitrary field content in the bulk, addition of higher derivative terms to the gravitational action, such as Gauss-Bonnet, and also relaxation of the Z_2 symmetry we assumed around the brane. For example, the GB action is

$$S = \frac{1}{2\kappa^2}\int d^5x\sqrt{|g|}\left(R + \lambda(R^2 - 4R_{ab}R^{ab} + R_{abcd}R^{abcd})\right) \tag{10}$$

We find that the no-go result (9) gets modified by the factor

$$\left(1 - \lambda(a'/a)^2 h\right)h'\Big|_{y_h} = \text{same as in (9)} \tag{11}$$

But since $h = 0$ at the horizon by definition, the left-hand-side is in effect unmodified.

Relaxing the Z_2 symmetry amounts to having two different black holes with different values of μ and Q on the two sides of the branes, with horizons at different distances. This does not change the conclusion.

There is one apparent loophole, which is to allow for spatial curvature on the brane. The Friedmann equation becomes $H^2 = (8\pi G/3)(\rho + \Lambda) - k/a^2$ in 4-D, where $k = 1, 0, -1$ for positive, zero or negative spatial curvature, respectively. In our no-go theorem we assumed $k = 0$, which is consistent with observations of the cosmic microwave background and expectations from inflation. With this modification we obtain

$$h'(y_h) = \text{same as in (9)} + \frac{4k}{a^3}\int_{y_h}^{y_b}a\,dy \tag{12}$$

This term allows us to obtain a positive value for $h'(y_h)$ even if $\rho + p \geq 0$ everywhere, provided that $k > 0$. However, this does not solve the cosmological constant problem because it requires that Λ_0 already be of order $k/Ga^2 \sim (10^{-4}\text{ eV})^4$, which is the original fine-tuning problem all over again. Larger values of Λ_0 would require larger curvatures, which are observationally excluded.

As a final attempt, one might try the same trick except with a very large curvature in some hidden extra dimensions rather than the large ones. However, the new terms in the Einstein equations which depend on the extra-dimensional curvature exactly cancel out of the combination $G_{00} + G_{ii}$ which gave us the no-go result, so this does not help.

CONCLUSION

It seems that Nature abhors the self-tuning idea. Each time we attempt to cure one problem, it introduces a new pathology. Ref.[6] tried to cure the problem of the naked singularity by hiding it behind a horizon, but had to violate the weak energy condition to do so. We find that allowing for positive spatial curvature can cure this problem, but only at the price of reintroducing the original fine-tuning problem. It is a good illustration of why the cosmological constant problem is one of the most daunting in theoretical physics. Some progress with self-tuning seems to have been made recently in ref.[9]. However, it remains a challenge to understand how constants of integration which correspond to physical sources of stress energy can dynamically adjust in a situation where the vacuum energy is changing, as in a phase transition, or alternatively, how they can "know" which value is the ultimate one that must be canceled. It would seem to be a very intelligent mechanism that could fully allow for the effects of vacuum energy during inflation but not in the present.

ACKNOWLEDGMENTS

Our work is supported by the Natural Sciences and Engineering Research Council of Canada and FCAR of Québec.

REFERENCES

1. S. Weinberg, Rev. Mod. Phys. **61**, 1 (1989).
2. N. Arkani-Hamed, S. Dimopoulos, N. Kaloper and R. Sundrum, Phys. Lett. B **480**, 193 (2000) [hep-th/0001197].
3. S. Kachru, M. Schulz and E. Silverstein, Phys. Rev. D **62**, 045021 (2000) [hep-th/0001206].
4. S. Forste, Z. Lalak, S. Lavignac and H. P. Nilles, JHEP **0009**, 034 (2000) [hep-th/0006139].
5. P. Binetruy, J. M. Cline and C. Grojean, Phys. Lett. B **489**, 403 (2000) [hep-th/0007029].
6. C. Csaki, J. Erlich and C. Grojean, Nucl. Phys. B **604**, 312 (2001) [hep-th/0012143].
7. C. Csaki, J. Erlich and C. Grojean, "The cosmological constant problem in brane-worlds and gravitational Lorentz violations," gr-qc/0105114.
8. J. M. Cline and H. Firouzjahi, Phys. Rev. D **65**, 043501 (2002) [arXiv:hep-th/0107198].
9. P. Binetruy, C. Charmousis, S. C. Davis and J. F. Dufaux, arXiv:hep-th/0206089.

Dynamical Stability of the AdS Soliton in Randall-Sundrum Model

C. P. Burgess *, James M. Cline *, N. R. Constable[†] and H. Firouzjahi[**]

*Physics Department, McGill University, Montréal, Québec, Canada H2A 2T8
[†]Center for Theoretical Physics and Laboratory for Nuclear Science Massachusetts Institute of Technology Cambridge, MA 02139, USA
[**]presenter

Abstract. We study a generalization of the Randall-Sundrum mechanism for generating the weak/Planck hierarchy, which uses two rather one warped extra dimension, and which requires no negative tension brane. The dynamical stability of the Radion mode will be studied. It is shown to have a tachyonic instability for certain models of the 4-brane stress energy, while stable or massless in other models. If stable, its mass is in the milli-eV range. The issue of the Radion stabilization by a bulk scalar field is investigated.

INTRODUCTION

Brane-world scenarios have undergone considerable theoretical scrutiny in recent years, largely due to the novel solutions which they provide for the gauge-hierarchy problem. Three kinds of such proposals have been made: the ADD scenario [1], in which two extra dimensions are very much larger than ordinary microphysical scales; the intermediate-scale scenario [2], in which the scale of gravity and the compactification scale are very close to one another; and the RS scenario [3], in which the geometry of the extra dimensions is exponentially warped.

There have been numerous proposals for higher-dimensional generalizations of the RS idea. A particularly attractive six-dimensional warped model has been considered in various contexts by several authors [4], [5]–[6]. This model is related to the AdS soliton [8], a double analytic continuation of a planar AdS Black hole metric, and involves two compact dimensions having the topology of a disc with a conical singularity at its center. The boundary of the disc occurs at a (Planck) 4-brane and a (TeV) 3-brane is placed at the conical defect. The stress energy of the 4-brane requires an anisotropic form which could arise from the smearing of 3-branes around the 4-brane, as suggested in ref. [4], or from Casimir energy of light particles confined to the 4-brane [5].

Since all observable consequences of this (or any other) geometry only involve the theory's low-energy degrees of freedom, essential to understanding its physical implications is a determination of its low energy spectrums. A complete study of the metric modes has been presented in [7]. A review of the radion mode will be presented here.

ADS SOLITON

In this section we review the AdS soliton [8] and its key properties relevant for braneworld applications. we will be interested in the six dimensional AdS soliton for which the line element may be written,

$$ds^2 = a(r)\eta_{\mu\nu}dx^\mu dx^\nu + b(r)d\theta^2 + dr^2 \tag{1}$$

where the metric coefficients are given by

$$a(r) = \cosh^{4/5}(kr); \qquad b(r) = b_0 \frac{\sinh^2(kr)}{\cosh^{6/5}(kr)} \tag{2}$$

and $b_0 = k^{-2}$ if the point $r = 0$ is regular, and $\theta \in [0, 2\pi]$. This is a solution to six dimensional Einstein gravity with negative cosmological constant

$$\Lambda = -\frac{10}{L^2} \equiv -\frac{8}{5}k^2, \tag{3}$$

In general we will suppose the Standard Model particles are living on the 3-brane which has no nvanishing tension. Then the conical singularity at $r = 0$ introduces the deficit angle given by $2\pi(k\sqrt{b_0} - 1)$. There is no curvature singularity at $r = 0$ and the spacetime is asymptotically locally AdS as $r \to \infty$; below we will cut off the radial extent by inserting a 4-brane at a finite value of r. For convenience we have normalized $a(0) = 1$.

To solve the gauge hierarchy problem, let us imagine that all the fundamental scales M_6, k, and $1/R$ are of order TeV. Then the standard reduction of the gravitational action from 6 to 4 dimensions gives the 4-D Planck mass as

$$\begin{aligned} M_p^2 &= M_6^4 \int dr d\theta\, a(r)\sqrt{b(r)} \\ &\sim \frac{M_6^4}{k^2} a^{3/2}(r) = \frac{M_6^4}{k^2} e^{6kR/5} \end{aligned} \tag{4}$$

Thus by taking $e^{3kR/5} \sim 10^{16}$, corresponding to $kR \cong 60$, we can explain the largeness of the Planck scale.

Notice that the relation $M_p^2 \sim a^{3/2}(R)(\text{TeV})^2$ differs from the analogous relation in the 5-D RS1 model, $M_p^2 \sim a(R)(\text{TeV})^2$. The additional factor of $a^{1/2}$ is coming from the $b^{1/2}$ part of the measure, which gives the size of the extra compact dimension that was not present in the 5-D model. This shows that the present model is a hybrid of the RS and ADD scenarios, in that the hierarchy is due to a combination of warping and having a large extra dimension.

The 4-brane we need for compactifying the AdS soliton can be constructed by the standard cutting and pasting procedure. Here, the metric in eq. (1) will be cut along the surface $r = R$ and then pasted onto a mirror image of itself. The resulting space time is then a solution of the Einstein equation

$$\kappa^{-2}\left(R_{MN} - \frac{1}{2}RG_{MN} + \Lambda G_{MN}\right) = S_{ab}\delta^a_M\delta^b_N\delta(r-R), \tag{5}$$

where $\kappa^2 = 8\pi G_6$, and the induced metric on the 4-brane is $g_{ab} = G_{MN}(R)\delta_a^M \delta_b^N$.

S_{ab} is the stress tensor of an infinitely thin brane located at the cutting surface. The form of the stress energy tensor in general is [7])

$$S_{\mu\nu} = \left(T_4 + \frac{T_3}{L_\theta^\alpha}\right) g_{\mu\nu} \equiv V_0 g_{\mu\nu}$$
$$S_{\theta\theta} = \left(T_4 - \frac{T_3'}{L_\theta^\alpha}\right) g_{\theta\theta} \equiv V_\theta g_{\theta\theta} \qquad (6)$$

where $L_\theta = \int d\theta \sqrt{b}|_{r=R}$ is the circumference of the compact 4-brane dimension. Model with $\alpha = 1$, $T_3' = 0$ corresponds to smearing three-brane over internal circle at $r = R$ [4], while the model with $\alpha = 5$, $T_3' = 4T_3$ is due to Casimir energy of any massless fields which are confined to the 4-brane [5].

THE PERTURBATION ANALYSES

We start by writing the perturbed metric as

$$ds^2 = a(r,t)\eta_{\mu\nu} dx^\mu dx^\nu + b(r,t) d\theta^2 + c(r,t) dr^2 \qquad (7)$$

where the perturbation around the static background corresponds to

$$\begin{aligned}
a(r,t) &= e^{-A_0(r)-A_1(r,t)} \equiv a_0(r) e^{-A_1(r,t)} \\
b(r,t) &= e^{-B_0(r)-B_1(r,t)} \equiv b_0(r) e^{-B_1(r,t)} \\
c(r,t) &= 1 + C_1(r,t)
\end{aligned} \qquad (8)$$

and we have now included a subscript '0' to denote the metric functions of the background around which we are expanding. The next step is to linearize the field equations in the dynamical perturbations. We use the relation $\ddot{A}_1 = -m_r^2 A_1$, and similarly for the other perturbations, where m_r is the sought-for radion mass. Expanding the $(tt) + (ii)$, (tr) and (rr) Einstein equations, respectively, to first order gives

$$m_r^2 (2A_1 + B_1 - C_1) = 0 \qquad (9)$$

$$4(A_1' - \frac{1}{2}A_0'C_1) + \left[B_1' - A_1' - \frac{1}{2}(B_0' - A_0')(B_1 + C_1)\right] - 2\kappa^2 \phi_0' \phi_1 = 0 \qquad (10)$$

$$(3A_0' + B_0')A_1' + A_0'B_1' - A_0'\left(\frac{3}{2}A_0' + B_0'\right)C_1 - \frac{m_r^2}{2a_0}(3A_1 + B_1)$$
$$+ \frac{\kappa^2}{2}\left[\phi_0'^2 C_1 + 2\frac{\partial V}{\partial \phi}\phi_1 - 2\phi_0' \phi_1'\right] = 0 \qquad (11)$$

where we also added a bulk scalar field for the later necessity. The appropriate boundary condition at $r = R$ is(see [7] for a detail discussion)

$$A'_1 = \frac{1}{2}A'_0 C_1 - \frac{(\alpha-1)}{8}(B'_0 - A'_0)B_1. \tag{12}$$

We solved the above system of equations for the radion numerically, for the case of $\alpha = 1$ (the general result for arbitrary values of α will be given below) and we find that it has a negative value of m_r^2—it is a tachyon

$$m_r^2 \cong -20L^{-2}a(R)^{-3/2} \qquad (\alpha = 1 \text{ case only}), \tag{13}$$

where we recall that L is the AdS curvature radius. Since $L \sim 1/\text{TeV}$ and $a(R)^{3/2} \cong 10^{32}$ to solve the hierarchy problem, we obtain

$$\sqrt{-m_r^2} \sim 10^{-3} \text{eV}. \tag{14}$$

This is well above the present Hubble scale, so we would have noticed the expansion or contraction of the extra dimension due to the change in the strength of gravity, if eq. (14) were true.

In comparing the radion in this model to that of the 5-D Randall-Sundrum model, we can notice several similarities and differences. Similarly to the 5-D model, in 6-D the radion is an admixture of the radial and brane metric components, such that oscillations of the radial size are accompanied by fluctuations in the scale factor of the 4-D universe which are 180° out of phase. But in 5-D, the radion was exactly massless in the absence of stabilization, whereas in 6-D it is a tunable parameter. Another difference is that, whereas in 5-D the radion has no tower of KK excitations, in 6-D it does. The mass gap is of order $1/L$, i.e., the TeV scale. The first few eigenvalues are $m_r^2 L^2 = 1.012, 6.15, 12.75, 21.31$. One notices that these masses are systematically smaller than those of the graviton and vector modes [7]. Thus the radion excitations would be the first signs of new physics from this model to appear in accelerator experiments, once we have made it viable by stabilizing the radion.

Doing the analyses analytically and for the general form of the stress energy tensor, we find

$$m_r^2 = \frac{\alpha - 5}{2} \frac{|\Lambda|}{a_0^{3/2}(R)} + O\left(\frac{\Lambda}{a_0^4(R)}\right) \tag{15}$$

Interestingly, the radion mass vanishes almost exactly in the case where the anisotropy of the 4-brane stress tensor is provided by the Casimir energy and pressure of fields living on the compact extra dimension. In this case the relevant energy density scales like L_θ^{-5}, i.e., $\alpha = 5$, as expected from dimensional analysis. This is the unique case where no dimensionful parameter is introduced in the anisotropic part of the 4-brane stress tensor, which is the part that also controls the position of the 4-brane, and hence how large the extra dimensions are.

STABILIZATION MECHANISM

We have found that the radion can be massive, massless, or tachyonic, depending on the value α which controls the dependence of the 4-brane stress-energy on its circumference. In the latter two cases ($\alpha \leq 5$), it is certainly necessary to increase the radion mass squared so that we have a stable universe, with Einstein gravity rather than scalar-tensor gravity at low energies. In the 5-D RS1 model, this was achieved by Goldberger and Wise [10] by adding a bulk scalar field, whose VEV's at the two branes were constrained by potentials on the branes to take certain values. The bulk scalar then acts like a spring between the branes, whose gradient energy becomes repulsive if the branes get too close, and whose potential energy (from $m^2\phi^2$) causes attraction if the branes separate too much. We expect that the same mechanism should work in 6-D.

There are three kinds of corrections to consider. First, the scalar field induces a small back-reaction on the static solutions, A_0, B_0, determined by the zeroth order truncation of the Einstein equations. This effect has been analytically computed in [9]. Second, the background scalar configuration couples to the fluctuations of the metric. This arises solely through the term $\kappa^2 \phi_0'^2 C_1$ of the perturbed (rr) component of the Einstein equations, (11). This is the really important effect for stabilizing the radion. The third kind of correction is from fluctuations of the scalar field, which can mix with the radion. These are governed by the perturbed scalar field equation

$$\phi_1'' - \frac{1}{2}(4A_0' + B_0')\phi_1' - \phi_0' H' - \frac{1}{2B_0'}\frac{\partial V}{\partial \phi}\left((B_0' - A_0')H + H' - 2\kappa^2 \phi_0' \phi_1\right)$$

$$-\frac{\partial^2 V}{\partial \phi^2}\phi_1 + \frac{m_r^2}{a_0}\phi_1 = 0 \qquad (16)$$

where $V = \frac{1}{2}m^2\phi^2$ is the bulk potential. In the small back-reaction limit where m^2 is taken to be small, the radion mass is [7]

$$m_r \sim \frac{m^2 \phi(R)}{kM_6^2}e^{-kR/5} \sim \text{MeV} \qquad (17)$$

independents of the details of the 4-brane stress-energy (independent of α).

It is remarkable that the stabilized mass is not of order the TeV scale, as was the case in 5-D RS1. Recalling that the Planck scale hierarchy was set by $a_0^{3/2} \cong 10^{32}$, we see that the stabilized m_r is suppressed by the factor $10^{16/3}$, giving $m_r \sim 10$ MeV. Thus the smallness of the radion mass seems to be associated with the additional dilution of the strength of gravity that comes from the large extra dimension.

Were the couplings of our MeV-scale radion so large, it would easily be observable in low-energy experiments and possibly affect the cooling of supernovae. Computing the 4-D effective Lagrangian for the gravitational fluctuations at quadratic order, we obtain

$$\mathcal{L} \sim M_p^2 \dot{\varphi}_0^2 + \varphi_0 T_\mu^\mu \qquad (18)$$

where T_μ^μ is the trace of the 3-brane stress-energy tensor, and $\varphi_0(t)$ represents the 4-D ground state radion field. This shows that the canonically normalized radion field ground state has Planck-suppressed couplings to TeV-brane matter. This differs from the behavior of the radion in the 5-D RS1 model. There, the wave function of the radion is exponentially peaked at the Planck brane, which overcomes the exponential warp factor in the measure to give $\mathscr{L} \sim (\text{TeV})^2 \dot{\varphi}_0^2 + \varphi_0 T_\mu^\mu$ instead.

CONCLUSION

We have focused on the Ads Soliton as a simple generalization of the RS model to six dimensions. The hierarchy between the Planck and weak scales, while generated mostly (2/3) by warping, is also partly (1/3) due to the exponentially large size of the compact extra dimension, giving it some features in common with the large extra dimension proposal. The radion fluctuations have been investigated and it was found that $m_r^2 \sim (5 - \alpha)\Lambda \text{ TeV}/M_p$. Only for the special case of Casimir energy on the 4-brane ($\alpha = 5$) is it massless. For smaller values of α it is tachyonic, and the spacetime is unstable. Its couplings to the TeV brane are Planck suppressed rather than TeV suppressed, due to the different behavior of its wave function relative to the 5-D case. Once stabilized by a bulk scalar field, the radion mass is not TeV scale, as in 5-D, but rather at the MeV scale. This suppression is related to the presence of the large extra dimension which does not feature in 5-D.

REFERENCES

1. N. Arkani-Hamed, S. Dimopoulos and G. Dvali, Phys. Lett. **B429** (1998) 263 [hep-ph/9803315]; Phys. Rev. **D59** (1999) 086004 [hep-ph/9807344]
2. C. P. Burgess, L. E. Ibanez and F. Quevedo, Phys. Lett. B **447**, 257 (1999) [hep-ph/9810535]; K. Benakli, Phys. Rev. D **60**, 104002 (1999) [hep-ph/9809582]
3. L. Randall and R. Sundrum, Phys. Rev. Lett. **83**, 3370 (1999) [hep-ph/9905221]; L. Randall and R. Sundrum, Phys. Rev. Lett. **83**, 4690 (1999) [hep-th/9906064]
4. F. Leblond, R. C. Myers and D. J. Winters, JHEP **0107**, 031 (2001) [hep-th/0106140]; F. Leblond, R. C. Myers and D. J. Winters, [hep-th/0107034]
5. Z. Chacko and A. E. Nelson, [hep-th/9912186]
6. J. Louko and D. L. Wiltshire, [hep-th/0109099].
7. C. P. Burgess, James M. Cline, Neil Constable, Hassan Firouzjahi, JHEP 0201 0201(2002) 014, [hep-th/0112047].
8. G. T. Horowitz and R.C. Myers, Phys. Rev. **D59** (1999) 026005, [hep-th/9808079].
9. Z. Chacko, P. J. Fox, A. E. Nelson and N. Weiner, [hep-ph/0106343]
10. W. D. Goldberger and M. B. Wise, Phys. Rev. Lett. **83**, 4922 (1999) [hep-ph/9907447]

Order ρ^2 Corrections to Cosmology with Two Branes

Jérémie Vinet[†][*] and James M. Cline[†]

[*]*presenter*
[†]*McGill University, Montréal, Québec, Canada.*

Abstract. We present the results of work in which we derived the $O(\rho^2)$ corrections to the Friedmann equations in the Randall-Sundrum I model. The effects of Golberger-Wise stabilization are taken into account. We surprisingly find that in the cases of inflation and radiation domination, the leading corrections on a given brane come exclusively from the effects of energy density located on the opposite brane.

INTRODUCTION

In recent years, there has been considerable interest in the idea that our world might be a 3-brane embedded in a higher dimensional bulk [1, 2, 3]. The fact that gravity alone is allowed to propagate through the bulk not only accounts for why the extra dimension(s) have so far avoided detection, but also provides an attractive solution to the infamous hierarchy problem.

We will be concerned here mainly with the Randall-Sundrum I model, where two 3-branes bound a slice of AdS(5) space. In this model, the weakness of gravity on our brane, the so-called "TeV brane" which has negative tension, is a consequence of the fact that the extra dimension is warped, *i.e.,* the metric along different slices of the bulk parallel to the 3-branes will depend on their position along the extra dimension.

An interesting feature of models with 3-branes in a 5-dimensional bulk is the fact that the Friedmann equations contain terms of higher order in the energy density than in standard cosmology [4, 5, 6, 7, 8]. More specifically, $H^2 = \frac{8\pi G}{3}\rho + O(\rho^2)$. These new terms could have important implications during inflation [9, 10, 11] and electroweak baryogenesis [12, 13]. Until now however, their exact form had only been worked out in the Randall-Sundrum II model, which doesn't solve the hierarchy problem. (In RS II, the extra dimension is infinitely large, and there is only a single 3-brane present).

We present here the results of work [14] (see also [15]) whose aim was to find the $O(\rho^2)$ corrections to the Hubble rate in the RS I model, taking into account the effect of the Goldberger-Wise (GW) mechanism [16] for stabilizing the extra dimension.

EINSTEIN EQUATIONS

The Einstein equations follow from the action

$$S = \int d^5x\sqrt{g}\left(-\frac{1}{2\kappa^2}R - \Lambda + \tfrac{1}{2}\partial_m\Phi\partial^m\Phi - V(\Phi)\right)$$
$$+ \int d^4x\sqrt{g}\,(\mathscr{L}_{m,0} - V_0(\Phi))|_{y=0} + \int d^4x\sqrt{g}\,(\mathscr{L}_{m,1} - V_1(\Phi))|_{y=1}, \quad (1)$$

where $\kappa^2 = M_5^{-3}$, and the potential $V(\Phi)$ is left unspecified for now. The brane contributions are a sum of matter, represented by

$$\mathscr{L}_{m,0} \sim \rho_*; \qquad \mathscr{L}_{m,1} \sim \rho \quad (2)$$

and tension, which is the value of the brane's scalar potential $V_i(\Phi(y_i))$. The matter Lagrangians cannot be written explicitly for cosmological fluids, but their effect on the Einstein equations is specified through their stress-energy tensors (5). Our ansatz for the 5-D metric has the form

$$ds^2 = n^2(t,y)\,dt^2 - a^2(t,y)\sum_i dx_i^2 - b^2(t,y)\,dy^2$$
$$\equiv e^{-2N(t,y)}dt^2 - a_0(t)^2 e^{-2A(t,y)}\sum_i dx_i^2 - b^2(t,y)\,dy^2. \quad (3)$$

We will make a perturbative expansion in the energy densities ρ, ρ_* of the branes around the static solution, where $\rho = \rho_* = 0$:

$$N(t,y) = A_0(y) + \delta N_1(t,y) + \delta N_2(t,y); \qquad A(t,y) = A_0(y) + \delta A_1(t,y) + \delta A_2(t,y)$$
$$b(t,y) = b_0 + \delta b_1(t,y) + \delta b_2(t,y); \qquad \Phi(t,y) = \Phi_0(y) + \delta\Phi_1(t,y) + \delta\Phi_2(t,y). \quad (4)$$

The subscripts on the perturbations indicate their order in powers of ρ or ρ_*. This ansatz is to be substituted into the scalar field equation and into the Einstein equations, $G_{mn} = \kappa^2 T_{mn}$. (Since the scalar field equation can be derived from a combination of the Einstein equations, we will not worry about it any further). The stress energy tensor is $T_{mn} = g_{mn}(V(\Phi) + \Lambda) + \partial_m\Phi\partial_n\Phi - \tfrac{1}{2}\partial^l\Phi\partial_l\Phi g_{mn}$ in the bulk. On the branes, $T_m{}^n$ is given by

$$T_m{}^n = \frac{\delta(y)}{b(t,0)}\mathrm{diag}(V_0 + \rho_*, V_0 - p_*, V_0 - p_*, V_0 - p_*, 0)$$
$$+ \frac{\delta(y-1)}{b(t,1)}\mathrm{diag}(V_1 + \rho, V_1 - p, V_1 - p, V_1 - p, 0) \quad (5)$$

(Later we will assume that the potentials V_0 and V_1 are very stiff and are vanishing at their minima, so they can be neglected.)

It will be useful to define the following variables, which appear naturally in the Israel junction conditions:

$$\Psi_2 = \delta A_2' - A_0'\frac{\delta b_2}{b_0} - \frac{\kappa^2}{3}\Phi_0'\delta\Phi_2 - \frac{\kappa^2}{6}\left(\delta\Phi_1' + \Phi_0'\frac{\delta b_1}{b_0}\right)\delta\Phi_1;$$

$$\Upsilon_2 = \delta N_2' - \delta A_2'. \tag{6}$$

$$\Psi_2(t,0) = +\frac{\kappa^2}{6}b_0\rho_*\frac{\delta b_1}{b_0}\bigg|_{t;y=0} \quad ; \quad \Psi_2(t,1) = -\frac{\kappa^2}{6}b_0\rho\frac{\delta b_1}{b_0}\bigg|_{t;y=1} \tag{7}$$

$$\Upsilon_2(t,0) = -\frac{\kappa^2}{2}b_0(\rho_*+p_*)\frac{\delta b_1}{b_0}\bigg|_{t;y=0} \quad ; \quad \Upsilon_2(t,1) = +\frac{\kappa^2}{2}b_0(\rho+p)\frac{\delta b_1}{b_0}\bigg|_{t;y=1} \tag{8}$$

The analogous quantities at first order were found in [15] to be the same as (6) with indices 2 replaced by 1 and terms with indices 1 absent. They satisfy the same boundary conditions as in (7,8), but with the replacement $\frac{\delta b_1}{b_0} \to 1$.

In terms of these variables, we can write the second order Einstein equations as

$$\left(\frac{\dot{a}_0}{a_0}\right)^2_{(2)} b_0^2 e^{2A_0} = 4A_0'\Psi_2 - \Psi_2' + \mathscr{F}_\Psi \tag{9}$$

$$2\left(\left(\frac{\dot{a}_0}{a_0}\right)^2_{(2)} - \left(\frac{\ddot{a}_0}{a_0}\right)_{(2)}\right) b_0^2 e^{2A_0} = -4A_0'\Upsilon_2 + \Upsilon_2' + \mathscr{F}_\Upsilon \tag{10}$$

$$0 = -\left(\frac{\dot{a}_0}{a_0}\right)_{(\frac{1}{2})} \Upsilon_2 + \Psi_2 + \mathscr{F}_{05} \tag{11}$$

$$\left(\left(\frac{\dot{a}_0}{a_0}\right)^2_{(2)} + \left(\frac{\ddot{a}_0}{a_0}\right)_{(2)}\right) b_0^2 e^{2A_0} = A_0'(4\Psi_2 + \Upsilon_2) + \mathscr{F}_{55}$$

$$+ \frac{\kappa^2}{3}\left(\Phi_0''\delta\Phi_2 - \Phi_0'\delta\Phi_2' + \Phi_0'^2\frac{\delta b_2}{b_0}\right) \tag{12}$$

where all the dependence on first order quantities squared is contained in the functions $\mathscr{F}_\Psi$, $\mathscr{F}_\Upsilon$, $\mathscr{F}_{05}$ and $\mathscr{F}_{55}$. Here, the equivalent first order equations can be obtained by simply replacing the subscripts 2 for 1, and leaving out the functions $\mathscr{F}$.

We thus have at each order in the perturbative expansion a set of first order differential equations which, when combined with the boundary conditions, allows us to solve for the unknown functions Ψ_n, Υ_n and $\delta b_n/b_0$. (We will work in a gauge where the fluctuations $\delta\Phi_n$ vanish).

One final note regarding the physical value of the Hubble rate. Since the (00) component of the metric receives corrections when carrying out our perturbative expansion, we should use the time coordinate

$$d\tau = \frac{n(t,1)}{n_0(t,1)} dt = e^{-\delta N_1(t,1) - \delta N_2(t,1) - \cdots} dt \tag{13}$$

to define the Hubble rate. This means that physical Hubble rate is given by

$$H \cong (1 + \delta N_1)\frac{\dot{a}}{a} \qquad (14)$$

rather than simply $H = \frac{\dot{a}}{a}$ as we might naively have expected. (See [14] for a more thourough discussion of this issue.)

FRIEDMANN EQUATIONS

Taking into account all that has been said in the previous section, we are now able to give the Friedmann equations to second order in ρ and ρ_*:

$$H^2|_{y=1} = \frac{8\pi G}{3}\left(\bar{\rho} + \rho_* + \frac{2\pi G}{3m_r^2\Omega^2}\left(9(1-3\omega)(1+\omega)\bar{\rho}^2\right.\right.$$
$$\left.\left. + 4(1-3\omega_*)(4+3\omega_*)\Omega^2\rho_*^2 + 4(1-3\omega)(4+3\omega)\bar{\rho}\rho_*)\right)\right) \qquad (15)$$

$$H^2|_{y=0} = \frac{8\pi G}{3}\left(\bar{\rho} + \rho_* + \frac{2\pi G}{3m_r^2\Omega^2}\left(9(1-3\omega_*)(1+\omega_*)\rho_*^2\Omega^4\right.\right.$$
$$- (1-3\omega)(7+3\omega)\bar{\rho}^2$$
$$\left.\left. + 2\Omega^2\bar{\rho}\rho_*[2(1-3\omega)(2+3\omega) - 3(1-3\omega_*)(1+\omega)]\right)\right) \qquad (16)$$

$$\frac{dH}{d\tau}\bigg|_{y=1} = -4\pi G\left(\bar{\rho}(1+\omega) + \rho_*(1+\omega_*)\right.$$
$$+ \frac{4\pi G}{3m_r^2\Omega^2}\left(\Omega^2(1+\omega_*)(1-3\omega_*)(13+9\omega_*)\rho_*^2\right.$$
$$+ 9(1-3\omega)(1+\omega)^2\bar{\rho}^2$$
$$- (1-3\omega)(2(1+\omega_*) + 2(1+\omega) + 6(1+\omega)^2$$
$$\left.\left. + 3(1+\omega)(1+\omega_*))\bar{\rho}\rho_*\right)\right)$$
$$\qquad (17)$$

$$\frac{dH}{d\tau}\bigg|_{y=0} = -4\pi G\left(\bar{\rho}(1+\omega) + \rho_*(1+\omega_*)\right.$$
$$+ \frac{4\pi G}{3m_r^2\Omega^2}\left(-4(1+\omega)(1-3\omega)\bar{\rho}^2 + 9(1+\omega_*)^2(1-3\omega_*)\Omega^4\rho_*^2\right.$$
$$+ \Omega^2\left[6(1-3\omega)(1+\omega)^2 + 2(1-3\omega)(1+\omega) - 2(1-3\omega)(1+\omega_*)\right.$$
$$\left.\left.\left. - 4(1-3\omega_*)(1+\omega) + 3(1+\omega_*)(1-3\omega)(1+\omega)\right]\rho_*\bar{\rho}\right)\right) \qquad (18)$$

where we have used

$$\dot{\rho} = -3H(\rho+p) \equiv -3H(1+\omega)\rho; \quad 8\pi G = \kappa^2 \left(2b_0 \int_0^1 e^{-2A_0} dy\right)^{-1}; \quad (19)$$

$$\dot{\rho}_* = -3H(\rho_*+p_*) \equiv -3H(1+\omega_*)\rho_*; \quad m_r^2 \cong \frac{4}{3}\kappa^2 v_0^2 \varepsilon^2 k^2 \Omega^{2+2\varepsilon}; \quad (20)$$

$$\bar{\rho} = \Omega^4 \rho; \quad \Omega = e^{-A_0(y=1)}. \quad (21)$$

It should also be noted that in order to obtain analytical results, we had to expand our expressions in powers of Ω, and we have only kept the dominant terms.

The careful reader will notice that the second order corrections vanish identically for radiation-like equations of state ($\omega = \omega_* = 1/3$). In that case, we need to look at next to leading order terms to find non-vanishing corrections:

$$H^2|_{y=1} = \frac{8\pi G}{3}\left((\rho_* + \bar{\rho}) + \frac{2\pi G}{3k^2\Omega^4}(\Omega^2 \rho_* - \bar{\rho})\rho_*\right) \quad (22)$$

$$H^2|_{y=0} = \frac{8\pi G}{3}\left((\rho_* + \bar{\rho}) - \frac{2\pi G}{3k^2\Omega^6}(\Omega^2 \rho_* - \bar{\rho})\bar{\rho}\right) \quad (23)$$

and the equations for $dH/d\tau$ have vanishing corrections at this order in ρ and ρ_*.

DISCUSSION

Let's now look at what our results mean. First, we note that the brane on which the hierarchy problem is solved is the one located at $y = 1$. It is thus natural to assume this is the brane we are living on. If we look strictly at the terms linear in ρ, ρ_*, we see that in order for the current Hubble rate on our brane not to be completely dominated by the energy density on the other brane, we must assume that ρ_* is presently very small or vanishing.

If we now look at the second order terms on our brane during inflation ($\omega = -1$), we see that the only corrections come from terms involving ρ_*. So unless we can come up with a model where ρ_* was large in the past and becomes negligible in the present epoch, there can have been no appreciable effect coming from the second order terms during inflation.

The situation during radiation domination is similar. The leading second order corrections on each brane involve the other brane's energy density. Furthermore, even if for the sake of argument we allow ρ_* to have been present during this era, we know that it can't be more than 10% of the value of $\bar{\rho}$ by the time of nucleosynthesis. This means that we can ignore the $\Omega^2 \rho_*$ term, and that the term that is left over suppresses the Hubble rate. But in order to make the sphalerons go out of equilibrium at the electroweak phase transition, we would have needed to make the Hubble rate larger [12]. So once again, we find that the second order corrections do not seem to have any constructive applications in this model.

CONCLUSION

We have derived the second order corrections to the Friedmann equations in the Randall-Sundrum I model, taking into account the effect of the GW stabilization mechanism. We have found that the corrections during radiation domination have the wrong sign to be useful for electroweak baryogenesis. We have also found that any effect on a brane with an inflationary equation of state can only be coming from the opposite brane. Our approach being perturbative in nature, we must however keep in mind that our results should not be expected to hold when the energy density becomes greater than some critical density, which on the TeV brane is the $(TeV)^4$ scale.

Acknowledgements

We wish to thank Hassan Firouzjahi for enlightening discussions. J.V. is supported in part by a grant from Canada's NSERC.

REFERENCES

1. Arkani-Hamed, N.,Dimopoulos, S., and Dvali, G., Phys. Lett. B **429**, 263 (1998) [hep-ph/9803315].
   Antoniadis, I.,Arkani-Hamed, N., Dimopoulos, S. and Dvali, G., Phys. Lett. B **436**, 257 (1998) [hep-ph/9804398].
   Arkani-Hamed, N.,Dimopoulos, S., and Dvali, G., Phys. Rev. D **59**, 086004 (1999) [hep-ph/9807344].
2. Randall, L. and Sundrum, R., Phys. Rev. Lett. **83**, 3370 (1999) [hep-ph/9905221].
3. Randall, L. and Sundrum, R., Phys. Rev. Lett. **83**, 4690 (1999) [hep-th/9906064].
4. Binetruy, P., Deffayet, C. and Langlois, D., Nucl. Phys. B **565**, 269 (2000) [hep-th/9905012].
5. Csaki, C., Graesser, M., Kolda, C. and Terning, J., Phys. Lett. B **462**, 34 (1999) [hep-ph/9906513].
6. Cline, J. M., Grojean, C. and Servant, G., Phys. Rev. Lett. **83**, 4245 (1999) [hep-ph/9906523].
7. Kanti, P., Kogan, I. I., Olive, K. A. and Pospelov, M., Phys. Lett. B **468**, 31 (1999) [hep-ph/9909481].
8. Csaki, C., Graesser, M., Randall, L. and Terning, J., Phys. Rev. D **62**, 045015 (2000) [hep-ph/9911406].
9. Kaloper, N. and Linde, A. D., Phys. Rev. D **59**, 101303 (1999) [hep-th/9811141];
   Arkani-Hamed, N., Dimopoulos, S., Kaloper, N. and March-Russell, J., Nucl. Phys. B **567**, 189 (2000) [hep-ph/9903224]. Cline, J. M., Phys. Rev. D **61**, 023513 (2000) [hep-ph/9904495];
   Nihei, T., Phys. Lett. B **465**, 81 (1999) [hep-ph/9905487];
   Himemoto, Y. and Sasaki, M., Phys. Rev. D **63**, 044015 (2001) [gr-qc/0010035]. Lukas, A. and Skinner, D., hep-th/0106190.
10. Maartens, R., Wands, D., Bassett, B. A. and Heard, I., Phys. Rev. D **62**, 041301 (2000) [hep-ph/9912464].
11. Copeland, E. J., Liddle, A. R. and Lidsey, J. E., Phys. Rev. D **64**, 023509 (2001) [astro-ph/0006421].
12. Servant, G., [hep-ph/0112209].
13. Csaki, C., Graesser, M. L. and Kribs, G. D., Phys. Rev. D **63**, 065002 (2001) [hep-th/0008151].
14. Cline, J.M., and Vinet, J. JHEP **02** (2002) 042 [hep-th/0201041].
15. Cline, J. M. and Firouzjahi, H., Phys. Lett. B **495**, 271 (2000) [hep-th/0008185].
16. Goldberger, W. D. and Wise, M. B., Phys. Rev. Lett. **83**, 4922 (1999) [hep-ph/9907447].

dS/CFT CORRESPONDENCE

Conserved Quantities, Entropy and the dS/CFT Correspondence

R. B. Mann* and A. M. Ghezelbash*

Department of Physics, University of Waterloo, Waterloo, Ontario N2L 3G1, Canada

Abstract. We investigate a recent proposal for defining a conserved mass and entropy in asymptotically de Sitter spacetimes that is based on a conjectured holographic duality between such spacetimes and Euclidean conformal field theory. We show that the class of Schwarzschild-de Sitter black holes up to 9 dimensions has finite action and conserved mass, and construct a definition of entropy outside the cosmological horizon by generalizing the Gibbs-Duhem relation to asymptotically dS spacetimes.

INTRODUCTION

The problem of defining conserved quantities in a generally covariant theory is notoriously subtle and has a long history. The roots of this problem reside in the equivalence principle, which qualitatively states that gravitation is locally indistinguishable from acceleration. A moment's reflection on this empirically testable assertion leads one to the realization that defining a "charge" associated with the gravitational field is quite unlike defining a charge associated with any other natural force. For with other forces space and time play a background role, and it is possible to meaningfully localize their conserved charges within a given closed surface of arbitrarily small volume. For gravity this approach presents considerable conceptual and technical difficulties, and it has been more fruitful to consider global definitions. The idea here is that the entire spacetime can – at least in stationary situations – be meaningfully said to have a conserved energy, momentum, and angular momentum, typically constructed by defining these conserved quantities for a closed surface of some finite mean radius R, and then taking the limit as $R \to \infty$.

It is difficult to underestimate the importance of properly understanding conserved charges in gravitational physics. All of black hole thermodynamics depends upon properly formulating the definitions of and relationships between energy, angular momentum, entropy, electric charge, etc. Indeed there is today only a rough consensus as to how conserved charges can and should be formulated, with a variety of prescriptions available for consideration.

In seeking such global definitions one is led to consider the asymptotic structure of the spacetime. Both asymptotically flat and AdS spacetimes have a timelike Killing vector field at infinity, and so the notion of a conserved quantity can be made meaningful in at least an asymptotic sense for these cases. However asymptotically de Sitter (dS) spacetimes lack both a similar notion of spatial infinity and a global timelike

Killing vector[1]. Consequently the physical meaning of conserved charges outside the cosmological horizon of dS spacetime is not clear.

Here we summarize a proposal for extending the notion of both conserved charges [1] and gravitational thermodynamics [2] to asymptotically dS spacetimes. The proposal is analogous to the Brown-York prescription in asymptotically AdS spacetimes [3, 4, 5], and offers suggestive information about the stress tensor and conserved charges of the hypothetical dual Euclidean conformal field theory (CFT) on the spacelike boundary of an asymptotically dS spacetime. This provides further circumstantial evidence for a holographic dual to dS spacetime that is similar to that conjectured for AdS spacetimes.

QUASILOCAL PRESCRIPTIONS

In $d+1$ dimensions, the Einstein equations of motion with a positive cosmological constant can be derived from the action

$$I = I_B + I_{\partial B} \tag{1}$$

where

$$I_B = \frac{1}{16\pi G} \int_{\mathcal{M}} d^{d+1}x \sqrt{-g} \left(R - 2\Lambda + \mathcal{L}_M(\Psi) \right) \tag{2}$$

$$I_{\partial B} = -\frac{1}{8\pi G} \int_{\partial \mathcal{M}} d^d x \sqrt{\gamma} \Theta \tag{3}$$

and $\mathcal{L}_M$ is the Lagrangian for matter fields Ψ. The first term in (1) is the bulk action over the $d+1$ dimensional Manifold $\mathcal{M}$ with metric g and the second term (3) is a surface term necessary to ensure that the Euler-Lagrange variation is well-defined. The boundary $\partial \mathcal{M}$ (with induced metric γ) of the manifold in general consists of both spacelike and timelike hypersurfaces. For an asymptotically AdS spacetime it will be the Einstein cylinder at infinity; for an asymptotically dS spacetime it will be the union of spatial Euclidean boundaries at early and late times.

Taking the variation of the action (1) and carefully keeping account of all boundary terms, a conserved charge

$$Q_\xi = \oint_\Sigma d^{d-1} S^a \xi^b T_{ab}^{eff} \tag{4}$$

can be associated with a closed surface Σ (with normal n^a), provided the boundary geometry has an isometry generated by a Killing vector ξ^μ. The quantity T_{ab}^{eff} is given by the variation of the action at the boundary with respect to γ^{ab}, and the quantity Q_ξ is conserved between closed surfaces Σ distinguished by some foliation parameter τ. If $\xi = \partial/\partial t$ then Q is the conserved mass/energy M; if $\xi_a = \partial/\partial \phi^a$ then Q is the conserved angular momentum J provided ϕ is a periodic coordinate associated with Σ. Details of this formulation can be found in refs. [3, 4, 6].

---

[1] Indeed, the timelike Killing vector field inside the cosmological horizon becomes spacelike outside the cosmological horizon.

One can proceed further and formulate gravitational thermodynamics by considering the path integral

$$Z = \int D[g] D[\Psi] e^{-I[g,\Psi]} \simeq e^{-I_{cl}} \quad (5)$$

where one integrates over all metrics and matter fields between some given initial and final Euclidean hypersurfaces, taking τ to have some period β_H, determined by demanding regularity of the Euclideanized manifold at degenerate points of the foliation. Semiclassically the result is given by the classical action evaluated on this equations of motion, and application of the Gibbs-Duhem relation to the partition function yields

$$S = \beta_H M_\infty - I_{cl} \quad (6)$$

to this order. The gravitational entropy S is simply the difference between the total energy at infinity M_∞ and the free energy $F = I_{cl}/\beta_H$ multiplied by β_H, which can be interpreted as the inverse temperature. This quantity will be non-zero whenever there is a mismatch between M and F.

Unfortunately neither the action (1) nor the conserved charges (4) are finite when evaluated on a solution of the equations of motion because of the infinite volume of Σ. A natural response to this conundrum is to calculate all quantities with respect to some reference spacetime, interpreted as the ground state of the system. This can be done by subtracting a term $I_{ref}[g_{ref}, \Psi_{ref}]$ from (1) and embedding Σ in the background spacetime; conserved quantities then become differences $\Delta Q_\xi = Q_\xi - Q_\xi^{ref}$ where the latter term is computed from the reference action. For example a computation using the Schwarzschild metric $ds^2 = -\left(1 - \frac{2m}{r}\right) dt^2 + \frac{dr^2}{\left(1-\frac{2m}{r}\right)} + r^2 d\Omega^2$ yields

$$M = -R + 2m + \mathscr{O}\left(\frac{m^2}{R}\right) \qquad F = -R + \frac{3}{2}m + \mathscr{O}\left(\frac{m^2}{R}\right) \quad (7)$$

for a spherical surface Σ of radius R, where $\beta_H = 8\pi m$ to ensure regularity of the Euclideanized manifold at the degenerate point $R = 2m$. It is natural to consider flat space to be the ground state, and we obtain

$$M^{ref} = -R + m + \mathscr{O}\left(\frac{m^2}{R}\right) \qquad F^{ref} = -R + m + \mathscr{O}\left(\frac{m^2}{R}\right) \quad (8)$$

from embedding Σ in a flat reference background. The differences $\Delta M = M - M^{ref} = m$ and $\Delta F = F - F^{ref} = \frac{1}{2}m$ are clearly finite for $R \to \infty$, and the entropy is easily seen to be $S = 4\pi m^2$.

BOUNDARY COUNTERTERMS

Although the preceding prescription has a number of intuitively appealing features, it also has some serious drawbacks. The reference spacetime is not unique, leading to ambiguities in the definition of the charges [7]. Second, it is not always possible to embed a given Σ into a given reference spacetime, limiting the utility of the reference approach even for a spacetime as basic as the Kerr solution [8].

It is here that the conjectured AdS/CFT correspondence [9] provides inspiration. This conjecture states that the partition function of any field theory on AdS_{d+1} defined by

$$Z_{AdS}[\gamma,\Psi_0] = \int_{[\gamma,\Psi_0]} D[g]D[\Psi]e^{-I[g,\Psi]} = \left\langle \exp\left(\int_{(\partial \mathcal{M})_d} d^d x \sqrt{\gamma}\mathcal{O}\Psi_0\right)\right\rangle = Z_{CFT}[\gamma,\Psi_0] \tag{9}$$

where $[\gamma,\Psi_0]$ and the quasi-primary conformal operator $\mathcal{O}$ are defined on the boundary of AdS_{d+1} and the integration is over configurations that approach $[\gamma,\Psi_0]$ when one goes from the bulk of AdS_{d+1} to its boundary. This conjecture has been verified for several important examples wherever explicit calculations can be carried out, encouraging the expectation that an understanding of quantum gravity in a given spacetime (at least an asymptotically AdS one) can be carried out by studying its holographic CFT dual, defined on the boundary of spacetime at infinity.

Following this line of thinking, it is natural to consider adding to the action (1) a term I_{ct} due to the contributions of the counterterms from the boundary CFT [5]. Such a boundary counterterm action is universal, depending only on curvature invariants that are functionals of the intrinsic boundary geometry, leaving the equations of motion unchanged. A straightforward algorithm for generating it involves rewriting the Einstein equations in Gauss-Codacci form, and then solving them in terms of the extrinsic curvature functional of the boundary $\partial \mathcal{M}$ and its normal derivatives to obtain the divergent parts [10]. Since all divergent parts can be expressed in terms of intrinsic boundary data and do not depend on normal derivatives [11], the entire divergent structure can be covariantly isolated for any given boundary dimension d; by varying the boundary metric under a Weyl transformation, it is straightforward to show that the trace of the divergent part is proportional to the divergent boundary counterterm Lagrangian. No background spacetime is required, and computations of the action and conserved charges yield unambiguous finite values intrinsic to the spacetime, as has been verified in numerous examples [12].

Despite that present lack of a clear formulation of a dS/CFT correspondence, this procedure can be carried over to the de Sitter case [2], with $I_{ct} = \int d^d x \sqrt{\gamma} \mathscr{L}_{ct}$ and

$$\mathscr{L}_{ct} = \left\{\left(\frac{1-d}{\ell} + \frac{\ell\Theta(d-3)}{2(d-2)}R\right) - \frac{\ell^3 \Theta(d-5)}{2(d-2)^2(d-4)}\left(R_{ab}R^{ab} - \frac{d}{4(d-1)}R^2\right) + \cdots\right\} \tag{10}$$

where $\Lambda = \frac{d(d-1)}{2\ell^2}$ and R is the curvature associated with the induced metric γ. The series truncates for any fixed dimension, with new terms entering at every new even value of d.

CHARGES AND ENTROPY

As is the case for its AdS counterpart, conserved charges Q_ξ on the spatially infinite boundaries of an asymptotically dS spacetime can be defined using (4). However the interpretation of such charges is somewhat peculiar. As with the AdS case, a collection of observers on the hypersurface whose metric is γ_{ab} would all observe the same value of Q_ξ provided this surface had an isometry generated by ξ^b. However the role of the mean radius R is now played by the mean cosmological time. This means that the surface Σ 'encloses' previous times instead of an interior space. Furthermore, the foliation of surfaces Σ by some parameter τ is now spacelike instead of timelike. A collection of observers defining a surface Σ would find that the value of Q_ξ they measure would not differ from that obtained by another set of observers spatially relocated at a different value of τ. (Of course since they are at spacelike infinity, each set of observers is out of causal contact with the other, and so have no way of comparing their values of Q_ξ).

Despite these peculiarities, the process is well-defined because (as has been shown in asymptotically flat and AdS cases) the quantities Q_ξ depend only on data on Σ; the interior is irrelevant [6]. For these reasons a gravitational entropy can be associated with an asymptotically dS spacetime, as defined by the Gibbs-Duhem relation (6). The inverse temperature is that associated with the cosmological horizon, which is in the past of future spacelike infinity.

To be specific [2], for the set of Schwarzschild-de Sitter metrics

$$ds^2 = -\left(1 - \frac{2m}{r^{d-2}} - \frac{r^2}{\ell^2}\right)dt^2 + \frac{dr^2}{1 - \frac{2m}{r^{d-2}} - \frac{r^2}{\ell^2}} + r^2 d\hat{\Omega}_{d-1}^2 \tag{11}$$

in $(d+1)$ dimensions, where $d\hat{\Omega}_{d-1}^2$ denotes the metric on the unit sphere S^{d-1}, we have a black hole with event horizon at $r = r_H$ and cosmological horizon at $r = r_C > r_H$ provided $m \leq \frac{\ell^{d-2}}{d}\left(\frac{d-2}{d}\right)^{\frac{d-2}{2}} = m_N$. When $m = m_N$, the event horizon coincides with the cosmological horizon and one gets the Nariai solution; demanding the absence of naked singularities yields an upper limit to the mass of the SdS black hole.

In order to compute conserved charges at future spacelike infinity we must work outside of the cosmological horizon, so we rewrite the metric as

$$ds^2 = -f(T)dT^2 + \frac{d\tau^2}{f(T)} + T^2 d\tilde{\Omega}_{d-1}^2 \tag{12}$$

where $f(T) = \left(\frac{T^2}{\ell^2} + \frac{2m}{T^{d-2}} - 1\right)^{-1}$, which diverges at $T = T_+$. The role of radius is now supplanted by that of the cosmological time T. Taking spherical surfaces with unit norm $n^a = (0, \sqrt{f}, \vec{0})$, we find

$$M_{dS}^{(d)} = V^{[d-1]}\frac{(d-1)}{4\pi G_d}\left[-m + \frac{\Gamma\left(\frac{2p-3}{2}\right)\ell^{2p-4}}{2\sqrt{\pi}\Gamma(p)}\delta_{2(p-1),d}\right] \tag{13}$$

where $V^{[d-1]}$ is the volume of the compact $(d-1)$-dimensional space and if p is a positive integer, the second term remains. The entropy is

$$S_d = \frac{(T_+^d - 2(d-2)m\ell^2)\beta_H V^{[d-1]}}{8\pi G \ell^2} \tag{14}$$

with $\beta_H = \left| \frac{-4\pi f'(T)}{f^2} \right|_{T=T_+}$ defining the periodicity of the τ coordinate. We have checked these expressions explicitly up to $d = 8$ using the full counterterm prescription [2].

DISCUSSION

The results we have obtained are tantalizingly similar to those of the AdS case, and provide further evidence for a possible dS/CFT correspondence. The extra term in (13) proportional to ℓ^{d-2} is presumably the analogue of the d-dimensional Casimir energy of the dual Euclidean CFT, and has been shown to be such in restrictive cases [13]. We find the rather strange result that the masses $M_{dS}^{(d)}$ are consistently negative for $m > 0$ in any dimension. Pure de Sitter spacetime has $m = 0$; values of $M_{dS}^{(d)}$ larger than this must have $m < 0$, implying that the spacetime has a cosmological singularity. These results are consistent with the conjecture that any asymptotically dS spacetime with $M > M_{dS}^{(d)}$ has a cosmological singularity [1]. For $m < 0$ the singularity is hidden behind a future horizon from observers located in the "lower patch" of the Penrose diagram. Observers in the "upper patch" will never encounter the singularity, although they can causally receive information from it.

ACKNOWLEDGMENTS

This work was supported by the Natural Sciences and Engineering Research Council of Canada.

REFERENCES

1. Balasubramanian V., de Boer, J. and Minic, D., *Phys. Rev.* **D65**, 123508 (2002).
2. Ghezelbash, A.M., and Mann, R.B., *JHEP* **0201**, 005 (2002).
3. Brown, J.D., and York, J.W., *Phys. Rev.* **D47**, 1407 (1993).
4. Brown, J.D., Creighton, J., and Mann, R.B., *Phys. Rev.* **D50**, 6394 (1994).
5. Balasubramanian, V., and Kraus, P., *Commun. Math. Phys.* **208**, 413 (1999); Henningson, M., and Skenderis, K., *JHEP* **9807**, 023 (1998); Hyun, S.Y., Kim, W.T., and Lee, J., *Phys. Rev.* **D59**, 084020 (1999).
6. Booth, I.S. and Mann, R.B. *Phys. Rev.* **D59** 064021 (1999).
7. Chan, K.C.K., Creighton, J.D.E. and Mann, R.B. *Phys. Rev.* **D54**, 3892 (1996).
8. Martinez, E. *Phys. Rev.* **D50**, 4920 (1994).
9. Maldacena, J., *Adv. Theor. Math. Phys.* **2**, 231 (1998); Witten, E., *Adv. Theor. Math. Phys.* **2**, 253 (1998); Gubser, S.S., Klebanov, I.R., and Polyakov, A.M., *Phys. Lett.* **B428**, 105 (1998).
10. Kraus, P., Larsen, F. and Siebelink, R., *Nucl. Phys.* **B563**, 259 (1999).
11. Fefferman, C., and Graham, C.R., "Conformal Invariants", in asterisque 1995, 95.
12. Mann, R.B., *Phys. Rev.* **D60**, 104047 (1999); Emparan, E., Johnson, C.V., and Myers, R., *Phys. Rev.* **D60**, 104001 (1999); Mann, R.B., *Phys. Rev.* **D61**, 084013 (2000).
13. Klemm, D., *Nucl. Phys.* **B625**, 295 (2002).

Static and Dynamic Vortices in de Sitter spacetimes

A. M. Ghezelbash* and R. B. Mann*

Department of Physics, University of Waterloo, Waterloo, Ontario N2L 3G1, Canada

Abstract. We solve the Abelian Higgs field equations in the background of a four dimensional de Sitter spacetime and find both static (in the static patch of the de Sitter spacetime) and dynamic (in an inflationary patch of the ds Sitter spacetime) solutions with axial symmetry. We find that the effect of these solutions is to create a deficit angle in the spacetime.

INTRODUCTION

The conjecture that the only long-range information associated with the endpoint of gravitational collapse is that of total mass, angular momentum and electric charge is referred to as the no-hair conjecture of black holes. Much work has been carried out over the years on this conjecture, either upholding it in certain instances (e.g. scalar fields [1]) or challenging it in others, such as painting Yang-Mills hair [2] or introducing Nielsen-Olesen vortices [3], [4],[5] and [6] onto black holes.

Virtually all efforts in this area have been concerned with asymptotically flat spacetimes, and it is only recently that extensions to other types of asymptotia have been considered. In [7], [8], and [9], it was shown that the $U(1)$ Higgs field equations have a vortex solution in four dimensional AdS spacetime, AdS-Schwarzschild, Kerr-AdS and Reissner-Nordström-AdS backgrounds. Also it is shown how a gauge vortex can be holographically represented via the AdS/CFT correspondence [7].

We report here on vortex solutions in the background of four-dimensional de Sitter spacetime. A motivation for studying dS spacetimes is connected with the recently proposed holographic duality between quantum gravity on dS spacetime and a quantum field theory living on the past boundary of dS spacetime [10].

Abelian Higgs Field Theory

The Abelian Higgs Field theory is described by the following Lagrangian,

$$\mathscr{L}(\Phi, A_\mu) = \frac{1}{2}(\mathscr{D}_\mu \Phi)^\dagger \mathscr{D}^\mu \Phi - \frac{1}{16\pi}\mathscr{F}_{\mu\nu}\mathscr{F}^{\mu\nu} - \xi(\Phi^\dagger \Phi - \eta^2)^2 \quad (1)$$

where Φ is a complex scalar Klein-Gordon field, $\mathscr{F}_{\mu\nu}$ is the field strength of the electromagnetic field A_μ and $\mathscr{D}_\mu = \nabla_\mu + ieA_\mu$ in which ∇_μ is the covariant derivative.

We employ Planck units $G = \hbar = c = 1$ which implies that the Planck mass is equal to one. Defining the real fields $X(x^\mu), \omega(x^\mu), P_\mu(x^\nu)$ by the following equations

$$\Phi(x^\mu) = \eta X(x^\mu) e^{i\omega(x^\mu)}$$
$$A_\mu(x^\nu) = \frac{1}{e}(P_\mu(x^\nu) - \nabla_\mu \omega(x^\mu)) \quad (2)$$

and employing a suitable choice of gauge, one could rewrite the Lagrangian (1) and the equations of motion in terms of these fields as:

$$\mathcal{L}(X, P_\mu) = -\frac{\eta^2}{2}(\nabla_\mu X \nabla^\mu X + X^2 P_\mu P^\mu) - \frac{1}{16\pi e^2} F_{\mu\nu} F^{\mu\nu} - \xi \eta^4 (X^2 - 1)^2 \quad (3)$$

$$\nabla_\mu \nabla^\mu X - X P_\mu P^\mu - 4\xi \eta^2 X (X^2 - 1) = 0$$
$$\nabla_\mu F^{\mu\nu} + 4\pi e^2 \eta^2 P^\nu X^2 = 0 \quad (4)$$

where $F^{\mu\nu} = \nabla^\mu P^\nu - \nabla^\nu P^\mu$ is the field strength of the corresponding gauge field P^μ. Note that the real field ω is not itself a physical quantity. Superficially it appears not to contain any physical information. However if ω is not single valued this is no longer the case, and the resultant solutions are referred to as vortex solutions [11]. In this case the requirement that Φ field be single-valued implies that the line integral of ω over any closed loop is $\pm 2\pi n$ where n is an integer. In this case the flux of electromagnetic field passing through such a closed loop is quantized with quanta $2\pi/e$.

Static Patch of de Sitter Spacetimes

We use the following four dimensional dS spacetime background:

$$ds^2 = -(1 - \frac{r^2}{l^2})dt^2 + \frac{1}{(1 - \frac{r^2}{l^2})} dr^2 + r^2(d\theta^2 + \sin^2\theta \, d\varphi^2) \quad (5)$$

where the cosmological constant Λ is equal to $\frac{3}{l^2}$. The horizon is at $r = l$ and so the range of the coordinate r is bounded to $0 \leq r \leq l$. In this coordinate system $\partial/\partial t$ is a future-pointing timelike Killing vector in only one diamond of the Penrose diagram, which generates the symmetry $t \to t + t_0$ for any constant t_0. In other diamonds of the Penrose diagram, this Killing vector is spacelike or else past-pointing timelike. After using the gauge $P^\mu(r, \theta) = (0; 0, 0, P(r, \theta))$, the equations of a vortex with winding number N are:

$$(1 - \frac{\rho^2}{l^2})\frac{d^2 X}{d\rho^2} + (\frac{1}{\rho} - \frac{4\rho}{l^2})\frac{dX}{d\rho} - \frac{1}{2}X(X^2 - 1) - \frac{N^2}{\rho^2}XP^2 = 0 \quad (6)$$

$$(1 - \frac{\rho^2}{l^2})\frac{d^2 P}{d\rho^2} - \frac{dP}{d\rho}(\frac{1}{\rho} + \frac{2\rho}{l^2}) - \alpha P X^2 = 0 \quad (7)$$

where $\rho = r\sin\theta$ and $\alpha = \frac{4\pi e^2}{\xi}$. We have solved the above equations numerically for dS spacetimes with different values of cosmological parameters l and unit winding number [12]. Some features of our solutions are:

- For a fixed ρ, the X field of the dS spacetime decreases with decreasing l, in contrast to the AdS spacetime which X field increases with decreasing l. In other words, by increasing the cosmological constant Λ from $-\infty$ to $+\infty$, the X field decreases for fixed ρ. Over this same range of Λ the value of the P field increases.
- In general, at a fixed ρ, X field for AdS spacetime is always greater than X field for any dS spacetime.
- The vortex is confined totally behind the cosmological horizon, i.e. the numerical solution of the equations outside the cosmological horizon $r \geq l$, shows that X and P fields remain at their respective constant values of 1 and 0.

Big Bang Patch of de Sitter Spacetimes

We consider the abelian Higgs Lagrangian (1) in the following 'big bang' coordinate system:

$$ds^2 = -d\tau^2 + e^{2\tau/l}(dx^2 + dy^2 + dz^2) \tag{8}$$

where the coordinate $\tau \in (-\infty, +\infty)$. In this case, the equations of a vortex with winding number N become:

$$e^{-2\tau/l}\frac{\partial^2 X}{\partial R^2} + \frac{e^{-2\tau/l}}{R}\frac{\partial X}{\partial R} - \frac{\partial^2 X}{\partial \tau^2} - \frac{3}{l}\frac{\partial X}{\partial \tau} - \frac{N^2}{R^2}XP^2 - \frac{1}{2}X(X^2 - 1) = 0 \tag{9}$$

$$e^{-2\tau/l}\frac{\partial^2 P}{\partial R^2} - \frac{e^{-2\tau/l}}{R}\frac{\partial P}{\partial R} - \frac{\partial^2 P}{\partial \tau^2} - \frac{1}{l}\frac{\partial P}{\partial \tau} - \alpha PX^2 = 03 \tag{10}$$

where $R = \sqrt{x^2 + y^2} = r\sin\theta$. After selection of appropriate boundary conditions and using the finite difference method, we obtained numerical solutions of equations (9,10), which establish the existence of vortex solution on a dS spacetime. Some features of our solutions are:

- By increasing the time from $-\infty$ to 0, the vortex core increases rapidly to a well defined value at time $\tau = 0$. Then increasing the time to more positive values, the vortex core increases more and more such that at time $+\infty$, the X and P fields approach the constant values 0 and 1 respectively.
- The vortex energy density spreads to bigger and bigger distances as time increases due to the larger size of the constant time slices.
- For time slices near $\tau = 0$, the behaviour of the vortex fields do not change by changing the cosmological parameter l. The physical reason is that at this special time the spacetime is flat and independent of the cosmological constant.
- For a constant positive time slice, by increasing the cosmological parameter l, the string thickness decreases; for a constant negative time slice, by increasing the cosmological parameter l, it increases. The asymmetric behaviour of string

thickness for positive and negative times is due to the inflation function $e^{2\tau/l}$ in (8). Hence by increasing l, the density of vortex field time-constant contours concentrates mainly near the time constant $\tau = 0$ slice.
- Increasing the winding number yields a greater vortex thickness in each constant time slice relative to winding number $N = 1$. This tendency runs counter to that of increasing l, for which the vortex thickness decreases on a constant positive time slice. As $l \to \infty$ we can increase the winding number to larger and larger values, and we find that the resultant thickness increases. Hence the effect of increasing winding number to thicken the vortex dominates over the thinning of the vortex due to decreasing cosmological constant.
- What we have found about the behaviour of the vortex in the big bang patch of dS spacetime can be straightforwardly generalized to the big crunch patch of the dS spacetime. The big crunch patch is given by replacing $\tau \to -\tau$ in the metric (8). Increasing the time from $-\infty$ to $+\infty$, for which the constant time slices of the big crunch spacetime become smaller and smaller, the string thickness becomes narrower and narrower. In fact, increasing the time, the energy density of the string concentrates to smaller and smaller distances due to the smaller size of the constant time slices.

Back-reaction of Vortex on Static and Big Bang Patches

To find the back-reaction of the vortex on the static patch of dS$_4$ spacetime, we must find the solutions of the coupled Einstein-Abelian Higgs differential equations in the background of (5). This is a formidable problem – even for flat spacetime no exact solutions have yet been found.

To obtain physical results, we make some approximations. First, we again assume the thin-core approximation, namely that the thickness of the vortex is much smaller that all other relevant length scales. Second, we assume that the gravitational effects of the string are weak enough so that the linearized Einstein-Abelian Higgs differential equations are applicable. For convenience, in this section we use the following form of the metric of dS$_4$:

$$ds^2 = -\widetilde{\mathscr{A}}(r,\theta)^2 dt^2 + \widetilde{\mathscr{B}}(r,\theta)^2 d\varphi^2 + \widetilde{\mathscr{C}}(r,\theta)(\frac{dr^2}{1-\frac{r^2}{l^2}} + r^2 d\theta^2) \tag{11}$$

which in the absence of the vortex, we must have $\mathscr{A}(r,\theta) = \sqrt{1-\frac{r^2}{l^2}} = \mathscr{A}_0(r,\theta)$, $\mathscr{B}(r,\theta) = r\sin\theta = \mathscr{B}_0(r,\theta)$, $\mathscr{C}(r,\theta) = 1 = \mathscr{C}_0(r,\theta)$, yielding the well known metric (5) of pure dS$_4$. Employing the two assumptions concerning the thickness of the vortex core and its weak gravitational field, we solve numerically the Einstein field equations:

$$G_{\mu\nu} + \frac{3}{l^2} g_{\mu\nu} = -8\pi G \mathscr{T}_{\mu\nu} \tag{12}$$

to first order in $\varepsilon = -8\pi G$, where $\mathcal{T}_{\mu\nu}$ is the energy-momentum tensor of the Abelian Higgs field in the dS background. The results emphasize that the functions $\widetilde{\mathcal{A}}(r,\theta) = \mathcal{A}_0(r,\theta)(1+\varepsilon)$, $\widetilde{\mathcal{B}}(r,\theta) = \mathcal{B}_0(r,\theta)(1+2\varepsilon)$ and $\widetilde{\mathcal{C}}(r,\theta) = \mathcal{C}_0(r,\theta)$ to first order in ε. Hence by a redefinition of the time coordinate in (11), the metric can be rewritten as:

$$ds^2 = -(1-\frac{r^2}{l^2})dt^2 + \frac{1}{(1-\frac{r^2}{l^2})}dr^2 + r^2(d\theta^2 + \alpha^2 \sin^2\theta\, d\varphi^2) \tag{13}$$

which is the metric of dS space with a deficit angle. So, the effect of the vortex on dS$_4$ spacetime is to create a deficit angle in the metric (5) by replacing $\varphi \to \alpha\varphi$, where $\alpha \simeq 1+2\varepsilon$ is a constant, since $\varepsilon < 0$.

To obtain the effect of the vortex on the big bang patch of dS$_4$ spacetime, we use the following transformations:

$$\tau = -t + \tfrac{l}{2}\ln\left|1-\tfrac{r^2}{l^2}\right|$$
$$X = \frac{r}{\sqrt{|1-\frac{r^2}{l^2}|}}\exp(t/l)\sin\theta\cos(\alpha\varphi)$$
$$Y = \frac{-r}{\sqrt{|1-\frac{r^2}{l^2}|}}\exp(t/l)\sin\theta\sin(\alpha\varphi) \tag{14}$$
$$Z = \frac{r}{\sqrt{|1-\frac{r^2}{l^2}|}}\exp(t/l)\cos\theta$$

which transform (13) to the following metric:

$$ds^2 = -d\tau^2 + e^{2\tau/l}(dX^2 + dY^2 + dZ^2) \tag{15}$$

Hence the effect of the vortex on a big bang patch of dS spacetime is to create a deficit angle in the X-Y plane that is constant as the (locally) flat spatial slice evolves in time.

De Sitter c-function

Obtaining evidence in support of (or against) a conjectured dS/CFT correspondence is somewhat harder to come by than its AdS/CFT counterpart. Although it is tempting to think of the former as a 'Wick-rotation' of the latter, a number of subtleties arise whose physical interpretation is not always straightforward [13]. One way of making progress in this area is via consideration of the UV/IR correspondence. In both the AdS and dS cases there is a natural correspondence between phenomena occurring near the boundary (or in the deep interior) of either spacetime and UV (IR) physics in the dual CFT. Solutions that are asymptotically dS lead to an interpretation in terms of renormalization group flows and an associated generalized dS c-theorem. This theorem states that in a contracting patch of dS spacetime, the renormalization group flows toward the infrared and in an expanding spacetime, it flows toward the ultraviolet [14]. More precisely, the c-function in $(n+1)$-dimensional inflationary dS is given by:

$$c = \left(G_{\mu\nu}n^\mu n^\nu\right)^{-(n-1)/2} = \frac{1}{\rho^{\frac{n-1}{2}}} \tag{16}$$

where n^μ is the unit normal vector to a constant time slice and ρ is the energy density on a constant time hypersurface. It is natural to consider the time-dependent solution of the previous section in this context. Although we have not solved the Einstein-Abelian-Higgs equations exactly, we have shown to leading order that the vortex solution we obtained induces a deficit angle into de Sitter spacetime, and so to this order our solution is (locally) asymptotically de Sitter. Furthermore, the gauge field values of our solution asymptotically approach the constant values expected in a de Sitter vacuum, and so at large distances (near the boundary) we expect that our solutions will in general be asymptotically de Sitter, at least locally. So we can examine the behaviour of the (16) in the big bang patch (8) in the presence of a vortex. The energy density of the vortex is given by:

$$\rho = -\frac{1}{2e^{2\tau}}\left(\frac{\partial X}{\partial R}\right)^2 - \frac{1}{2R^2 e^{4\tau}}\left(\frac{\partial P}{\partial R}\right)^2 - (X^2-1)^2 - \frac{X^2 P^2}{2R^2 e^{2\tau}} \qquad (17)$$

Using the curves of the vortex fields we can obtain the c-function in terms of time at some fixed point R. We find that by increasing the time from $-\infty$ to $+\infty$, the c-function monotonically increases as the universe expands. Similar calculations show that in the big crunch patch of dS spacetime, the c-function decreases monotonically as time evolves from $-\infty$ to $+\infty$.

ACKNOWLEDGMENTS

This work was supported by the Natural Sciences and Engineering Research Council of Canada.

REFERENCES

1. Sudarsky, D., *Class. Quant. Grav.* **12**, 579 (1995).
2. Winstanley, E., *Class. Quant. Grav.* **16**, 1963 (1999).
3. Achucarro, A., Gregory, R., and Kuijken, K., *Phys. Rev.* **D52**, 5729 (1995).
4. Chamblin, A., Ashbourn-Chamblin, J.M.A., Emparan, R., and Sornborger, A., *Phys. Rev.* **D58**, 124014 (1998).
5. Bonjour, F., and Gregory, R., *Phys. Rev. Lett.* **81**, 5034 (1998).
6. Bonjour, F., Emparan, R., and Gregory, R., *Phys. Rev.* **D59**, 84022 (1999).
7. Dehghani, M.H., Ghezelbash, A.M., and Mann, R.B., *Nucl. Phys.* **B625**, 389-406 (2002).
8. Dehghani, M.H., Ghezelbash, A.M., and Mann, R.B., *Phys. Rev.* **D65**, 044010 (2002).
9. Ghezelbash, A.M., and Mann, R.B., *Phys.Rev.* **D65**, 124022 (2002).
10. Strominger, A., *JHEP* **0110**, 034 (2001); *JHEP* **0111** 049 (2001).
11. Nielsen, H.B., and Olesen, P., *Nucl. Phys.* **B61**, 45 (1973).
12. Ghezelbash, A.M., and Mann, R.B., *Phys. Lett.* **B537**, 329-339 (2002).
13. Ghezelbash, A.M., and Mann, R.B., *JHEP* **0201**, 005 (2002).
14. Leblond, F., Marolf, D., and Myers, R.C., *JHEP* **0206**, 052 (2002).

A short tale from de Sitter space

Frédéric Leblond[1]

Department of Physics, McGill University, Montréal, Québec H3A 2T8[2]

Abstract. In the context of Einstein gravity coupled to matter in d dimensions, we investigate solutions that are dynamically evolving between de Sitter and anti-de Sitter regions. We demonstrate the existence of a *no go theorem* when the matter content of the theory is 'reasonable', *i.e.*, such that the weak energy condition holds. We find solutions when the energy conditions are violated in a finite region of the spacetime. We speculate on the holographic interpretation of these gravitational backgrounds by combining ideas from the AdS/CFT and the dS/CFT dualities.

An extremely important relation between anti-de Sitter (AdS) space and black holes was discovered through string theory. There is at this time no equivalent connection involving de Sitter (dS) space. Not unlike black holes, the latter spacetime has a Bekenstein-Hawking entropy associated with the area of its horizon. A legitimate question to ask is then: Where are, but perhaps more importantly, what are the microscopic degrees of freedom[3] in terms of which this entropy should be computed ? A reason why these and other questions have been made more pressing is the recent experimental data supporting the fact that our universe is evolving toward a phase which is dominated by a small positive cosmological constant[1].

In this note we investigate a scheme that could help in providing a better understanding of dS space by using gauge/gravity dualities, *i.e.*, the AdS/CFT[2] and the dS/CFT correspondences[3, 4, 5, 6]. Using Einstein gravity coupled to matter, we study spacetimes classically interpolating between a phase with a positive cosmological constant and a region where it is negative. To conclude we speculate on some rather peculiar characteristics a field theory dual to these gravitational backgrounds would have to possess.

SPACETIMES WITH DS AND ADS ASYMPTOTIA

Our starting point is a general metric ansatz compatible with a spacetime dynamically evolving through a transitional region where the sign of the cosmological constant

---

[1] This work was done in collaboration with Hassan Firouzjahi.
[2] The author wishes to thank the Perimeter Institute for its ongoing hospitality during this project.
[3] The horizon of de Sitter space is an observer dependent quantity. Therefore, whatever understanding we will come to have of it in terms of miscroscopic degrees of freedom will most certainly teach us much about backgrounds such as flat space written in Rindler coordinates.

changes,
$$ds^2 = -h(r)dt^2 + \frac{1}{h(r)}dr^2 + a(r)dx^a dx_a, \tag{1}$$

where $a,b = 1, \ldots, (d-2)$ and d is the total number of dimensions. The coordinates $x^\mu = (t, r, x^a)$ all have the same range, i.e., $-\infty < x^\mu < +\infty$. The coordinate r is required to be timelike (spacelike) in the asymptotically dS (AdS) region. As $r \to -\infty$ we chose the boundary condition

$$\lim_{r \to -\infty} ds^2 = \frac{r^2}{l_{dS}^2}dt^2 - \frac{l_{dS}^2}{r^2}dr^2 + \frac{r^2}{l_{dS}^2}d\mathbf{x}^2, \tag{2}$$

which corresponds to a spacetime with a positive cosmological constant, $\Lambda_{dS} = (d-1)(d-2)/(2l_{dS}^2)$. In particular, by using the change of variables

$$r = e^{r'/l_{AdS}} - e^{-r'/l_{dS}}, \tag{3}$$

the asymptotic metric (2) takes the form

$$ds^2 = -dr'^2 + e^{-2r'/l_{dS}}\left[dt^2 + d\mathbf{x}^2\right], \tag{4}$$

which corresponds to dS space in inflationary coordinates (see ref. [7] for details) where $r = -\infty$ corresponds to the spacelike boundary I^-. For $r \to +\infty$, we require the boundary condition to be

$$\lim_{r \to +\infty} ds^2 = -\frac{r^2}{l_{AdS}^2}dt^2 + \frac{l_{AdS}^2}{r^2}dr^2 + \frac{r^2}{l_{AdS}^2}d\mathbf{x}^2, \tag{5}$$

which is the metric of a spacetime with a negative cosmological constant, $\Lambda_{AdS} = -(d-1)(d-2)/(2l_{AdS}^2)$. Using the change of coordinates (3) brings expression (5) to the simple form

$$ds^2 = dr'^2 + e^{2r'/l_{AdS}}\left[-dt^2 + d\mathbf{x}^2\right] \tag{6}$$

corresponding to the Poincaré patch of AdS space with the boundary located at $r = +\infty$. In short, the boundary conditions we impose are

$$\lim_{r \to \pm\infty} a(r) = \frac{r^2}{l_{(A)dS}^2}, \qquad \lim_{r \to \pm\infty} h(r) = \pm a(r). \tag{7}$$

The Einstein-Hilbert action is

$$S_g = \frac{1}{2}\int d^d x \sqrt{-g} R, \tag{8}$$

where we have used the convention $8\pi G_N = 1$ for simplicity. The equations of motion are then simply $G_{ij} = T_{ij}$, where T_{ij} is the stress-energy tensor computed from a matter action S_M. We first consider a scalar field for the matter content of the theory[4]

$$T_{ij} = (d-1)(d-2)\left[\partial_i\phi\partial_j\phi - \frac{1}{2}g_{ij}\left(g^{kl}\partial_k\phi\partial_l\phi + V(\phi)\right)\right], \qquad (9)$$

where $V(\phi)$ is a potential function. The Einstein equations for a metric of the form (1) are then found to be

$$(rr)\quad \frac{d-2}{4}h'\frac{a'}{a} + \frac{(d-2)(d-3)}{8}h\left(\frac{a'}{a}\right)^2 = \frac{(d-1)(d-2)}{2}h\left[\phi'^2 - \frac{V}{h}\right], \qquad (10)$$

$$(tt)\quad \frac{d-2}{4}h'\frac{a'}{a} + \frac{(d-2)(d-5)}{8}h\left(\frac{a'}{a}\right)^2 + \frac{d-2}{2}h\frac{a''}{a} = \frac{(1-d)(d-2)}{2}h\left[\phi'^2 + \frac{V}{h}\right], \qquad (11)$$

$$(x^a x^a)\quad \frac{h''}{2} + \frac{d-3}{2}\left[h'\frac{a'}{a} + h\frac{a''}{a}\right] + \frac{(d-3)(d-6)}{8}h\left(\frac{a'}{a}\right)^2 = \frac{(1-d)(d-2)}{2h^{-1}}\left[\phi'^2 + \frac{V}{h}\right], \qquad (12)$$

where a 'prime' denotes a derivative with respect to r. In general, the RHS of equations (10), (11) and (12) is replaced by T_r^r, T_t^t and $T_{x^a}^{x^a}$ respectively.

Our strategy consists in using the equations of motion in order to determine wether there exist or not consistent solutions with boundary conditions of the form (7). Following ref. [8], we substract eqs. (11) and (12) to obtain

$$\left[a^{\frac{d-2}{2}}\left(h' - h\frac{a'}{a}\right)\right]' = 0, \qquad (13)$$

which has a solution of the form

$$h(r) = a(r)\left(k_1\int_0^r \frac{dr'}{a(r')^{d/2}} + k_2\right), \qquad (14)$$

where k_1 and k_2 are constants of integration. Imposing a dS$\to$AdS boundary condition of the form (7), we obtain

$$k_1 = \left(\int_{-\infty}^{+\infty}\frac{dr'}{a(r')^{d/2}}\right)^{-1},$$

$$k_2 = \frac{k_1}{2}\left(\int_{-\infty}^0\frac{dr'}{a(r')^{d/2}} - \int_0^\infty\frac{dr'}{a(r')^{d/2}}\right),$$

where k_1 is positive because of the physical requirement $a(r) > 0$ for all values of r. Finally, we substract eqs. (10) and (11) to get

$$\frac{a^{1/2''}}{a^{1/2}} = -(d-1)(\phi')^2, \qquad (15)$$

which we now use to prove a *no go theorem*.

---

[4] See ref. [7] for conventions.

No go theorem

We assume for the moment that the dS→AdS transitions are classically induced by a rolling scalar field with a stress-energy tensor of the form (9). The solutions of interest are interpolating between two extrema of the potential $V(\phi)$ which correspond to fixed points of the scalar field evolution ($\phi' = 0$). Of course, the boundary conditions (7) guarantee that the solution evolves between extrema for which $V(\phi)$ changes sign. The key observation is that the condition (15) derived from the Einstein equations corresponds to the statement that for a solution to exist we must have

$$\frac{a^{1/2''}}{a^{1/2}} \leq 0 \qquad (16)$$

for all values of r. The problem is that there does not exist any function, with the boundary conditions (7), that can satisfy the condition (16). One might think that this pathology is attributed to the scalar field and that considering more general matter fields will lead to a different result. For a generic matter content, eq. (15) is replaced by

$$\frac{a^{1/2''}}{a^{1/2}} = \frac{T_t^t - T_r^r}{(d-2)h}. \qquad (17)$$

In the region containing the boundary of AdS ($r > 0$) we have $h(r) > 0$. The variable t is then timelike and $T_t^t - T_r^r$ can be written like $-(\rho + p)$, where ρ and p are respectively the energy density and the pressure. In the region with a dS boundary we have $h(r) < 0$ and r has become a timelike coordinate. This means that the combination $T_t^t - T_r^r$ is $(\rho + p)$. Combining these observations leads to the realisation that for condition (17) to be satisfied, one needs to impose $(\rho + p) < 0$ in some finite region of the spacetime. This violates the *weak energy condition* and supports the validity of the following *no go theorem*:

> Einstein gravity coupled to matter obeying the *weak energy condition* has no solutions corresponding to a dynamical evolution from an asymptotic region which is de Sitter (anti-de Sitter) to another region which is asymptotically anti-de Sitter (de Sitter).

We refer the reader to ref. [11] for a more detailed analysis.

Solutions violating the weak energy condition

We have shown that for a solution with boundary conditions (7) to exist the following relation must hold:

$$\frac{a^{1/2''}}{a^{1/2}} = (d-2)\frac{\rho + p}{h}. \qquad (18)$$

For a dS→AdS solution, $h(r) > 0$ for $r < 0$ and $h(r) < 0$ for $r > 0$. At $r = 0$ the function $h(r)$ is degenerate which can be shown to imply the existence of an horizon. The weak energy condition, $T_{\mu\nu}\xi^\mu\xi^\nu \geq 0$ (where ξ^μ is a unit timelike vector), means that

$\rho + p \geq 0$. Solutions $a(r)$ to eq. (18), with the boundary conditions (7), and for which the weak energy condition is satisfied do not exist. A class of possible solutions consists in those for which the weak energy condition is violated in a finite region around $r = 0$. An example of solution is

$$a(r) = \left[\ln\left(e^{r/l_{AdS}} + e^{-r/l_{dS}}\right)\right]^2, \tag{19}$$

which corresponds to a spacetime interpolating between dS and AdS as the coordinate r varies from $r = -\infty$ to $r = +\infty$. One can verify that the solution has $\rho + p = 0$ as $r \to \pm\infty$ but $\rho + p < 0$ in a finite region around $r = 0$. It is interesting to note that this spacetime does not have a curvature singularity at $r = 0$. Examples of such energy violating condition spacetimes are found in theories containing scalar fields with a kinetic term of the wrong sign. The Type II* string theories introduced in ref. [9, 10] generically have such fields.

COMBINING HOLOGRAPHY IN ADS AND DS SPACE

In the AdS/CFT correspondence, quantum gravity in pure AdS space is conjectured to be dual to a conformal field theory in one less dimension that can be regarded as being defined on the boundary of the bulk spacetime. For AdS in Poincaré coordinates,

$$ds^2 = dr^2 + e^{2r/l_{AdS}}\left(-dt^2 + d\mathbf{x}^2\right), \tag{20}$$

the boundary is located at $r = +\infty$. Quite analogously, the dS/CFT correspondence proposes that quantum gravity in dS space has a Euclidean conformal field theory dual defined on its boundary(ies). When this correspondence is formulated using inflationary coordinates,

$$ds^2 = -dr^2 + e^{-2r/l_{dS}}\left(dr^2 + d\mathbf{x}^2\right), \tag{21}$$

the spacelike boundary, I^-, is located at $r = -\infty$. In this note we considered solutions interpolating between asymptotic regions of the form (20) and (21).

A more general formulation of the AdS/CFT and the dS/CFT dualities considers an equivalence between asymptotically (A)dS spacetimes and non-conformal field theories. In the context of the AdS/CFT correspondence, this leads to the so-called UV/IR correspondence[12, 13] stating that asymptotic regions (near the boundary) of AdS space are associated with short distance (UV) physics in the field theory dual, while regions deeper in the AdS are related to long distance (IR) physics. It is then natural to associate evolution along the r-coordinate with a field theory renormalization group (RG) flow. An interesting result then is the derivation of a gravitational c-theorem[14]. The corresponding c-function is a local geometric quantity which appears to give a measure of the number of degrees of freedom relevant for physics in the dual field theory at different energy scales. Requiring that the energy conditions be satisfied on the gravitational side dictates that this c-function must decrease in evolving from the UV into the IR regions. Ideas along the same lines can be used in the context of the dS/CFT correspondence. Again demanding energy conditions to hold leads to a gravitational c-theorem for asymptotically dS spacetimes[4, 7]. It states that the c-function must decrease (increase) in a period of contraction (expansion). For example, when considering spacetimes with both asymptotia (at $r = \pm\infty$) of the form (21), the boundary I^- corresponds to the UV while the horizon at $r = +\infty$ is interpreted as an IR limit of the field theory.

The introduction of a c-function is relevant when the gravitational background under consideration has a field theory dual. In the case of spacetimes interpolating between dS and AdS regions, it would be interesting to find out what kind of field theory could be the dual? While we cannot address this question to a satisfactory level at this point, one thing we can do is describe roughly the rather strange field theory RG flow such a peculiar gravitational background would be dual to. The existence of such a mixed (dS-AdS)/CFT correspondence would imply that the bulk coordinate which is reconstructed by the field theory RG flow is timelike in the region containing the dS boundary and spacelike in the region containing the AdS boundary. Let us for the moment assume that the weak energy condition, which was used in deriving the c-theorems associated to AdS/CFT and dS/CFT correspondences, holds everywhere in the gravitational dual. Then, the boundary at $r = +\infty$ (AdS) would correspond to physics in the conformal UV limit of the field theory. The weak energy condition being satisfied for $r > 0$, we expect that translations along decreasing r, *i.e.,* moving deeper inside of the spacetime, correspond to flowing toward an IR region close to $r = 0$. Past the horizon, the variable r becomes timelike which means that the RG flow should now describe a limit of the field theory which is Euclidean! Since the weak energy condition still holds for $r < 0$, the c-theorem of ref. [4, 7] guarantees that evolving toward increasingly negative values of r corresponds to 'integrating in' degrees of freedom, *i.e.,* moving toward another UV fixed point, this time defined on the spacelike boundary I^-. Since both these UV fixed points appear to be flowing toward the same IR limit, it is tempting to postulate a new self-duality for the field theory. This duality would involve a conformal fixed point of the field theory on a flat metric with Minkowski signature and a conformal fixed point where the theory is Euclidean.

REFERENCES

1. S. Perlmutter *et al.* Astrophys. J. **483**, 565 (1997); S. Perlmutter *et al.* Nature **391**, 51 (1998); A.G. Riess *et al.* Astron. J. **116**, 1009 (1998); N.A. Bahcall, J.P. Ostriker, S. Perlmutter and P.J. Steinhardt, Science **284**, 1481 (1999).
2. O. Aharony, S.S. Gubser, J. Maldacena, H. Ooguri and Y. Oz, Phys. Rept. **323**,183 (2000).
3. A. Strominger, JHEP **0110**, 034 (2001).
4. A. Strominger, JHEP **0111**, 049 (2001).
5. V. Balasubramanian, J. de Boer and D. Minic, Phys. Rev. D **65**, 123508 (2002).
6. E. Witten, "Quantum gravity in de Sitter space", hep-th/0106109.
7. F. Leblond, D. Marolf and R. C. Myers, JHEP **0206**, 052 (2002).
8. J. M. Cline and H. Firouzjahi, Phys. Rev. D **65**, 043501 (2002).
9. C. M. Hull, JHEP **9807**, 021 (1998).
10. C. M. Hull, JHEP **0111**, 012 (2001).
11. H. Firouzjahi and F. Leblond, "The clash between de Sitter and anti-de Sitter space", to appear.
12. L. Susskind and E. Witten, "The holographic bound in anti-de Sitter space," hep-th/9805114.
13. A.W. Peet and J. Polchinski, Phys. Rev. D **59**, 065011 (1999).
14. D.Z. Freedman, S.S. Gubser, K. Pilch and N.P. Warner, Adv. Theor. Math. Phys. **3**, 363 (1999).

Non-Abelian monopoles and dyons in a modified Einstein-Yang-Mills-Higgs system

Joel S. Rozowsky

Department of Physics, Syracuse University, Syracuse NY 13244.

Abstract. We study a modified Yang-Mills-Higgs system coupled to Einstein gravity, with a direct coupling of the Higgs field to the scalar curvature. We are able to write a Bogomol'nyi type condition in curved space and show that the positive static energy functional is bounded. We then investigate non-Abelian spherically symmetric static solutions in a similar fashion to the 't Hooft-Polyakov monopole. After reviewing monopole solutions of this type, we also present dyon solutions.

In these proceedings we review the contents of a recent paper [1]. In a classic paper [2], Dirac proposed the possible existence of a magnetic monopole, the analogue of an isolated electrically charged particle. Motivated principally to restore the symmetries between electric and magnetic forces, Dirac found that the existence of a monopole provided a natural explanation for the quantization of electric charge. A description of such a monopole consistent with quantum mechanics would lead to the famous Dirac charge quantization condition, $eg = n\hbar c/2$, where $e^2/\hbar c$ is the fine structure constant, g is the monopole charge, and n an integer. Later Dirac's theory was reformulated by Wu and Yang [3] within the framework of fiber bundles. The singular Dirac string is avoided at the expense of introducing coordinate patches on a sphere surrounding the monopole. Consequently, during the past decades, extensive effort has gone into experimental search for monopoles, but unfortunately has as yet had no success.

In spite of the lack of experimental success, the monopole continues to thrive in the theoretical laboratory. With the pioneering work of 't Hooft and Polyakov [4], the monopole was reinvented in a new form as a finite energy, particle-like soliton in non-Abelian gauge theories with spontaneous symmetry breaking. Moreover, such objects are generic in any spontaneously broken non-Abelian gauge theory which has an unbroken $U(1)$ gauge symmetry. Such monopoles are expected to be produced in abundance in phase transitions of grand unified theories, which has implications for early universe cosmology.

More recently, a great deal of activity has centered around monopoles in curved spacetime in order to study the effects of gravity. New insights have emerged from a study of a coupled Einstein-Yang-Mills-Higgs [EYMH] system with solutions describing black holes with magnetic charge, black holes within magnetic monopoles and magnetic monopoles within black holes [5, 6]. Consequently, gravity cannot be dispensed with, arguing that the strength of its interaction is weak.

The paper [1] is an extension of the work by Nguyen and Wali [7] to include electric charge and to study dyons coupled to gravity. The starting point is a modified EYMH system with a specific coupling of the Higgs field to the Einstein term in the action.

We begin by defining the action,

$$S = \int d^4x \sqrt{-g} \left(\mathscr{L}_E + \mathscr{L}_M \right), \qquad (1)$$

where

$$\mathscr{L}_E = -\frac{R - 2\Lambda}{16\pi G v^2} \Phi^2, \qquad (2)$$

with R, the Ricci scalar, Λ, the cosmological constant and Φ^a, the Higgs scalar field. Our metric $g_{\mu\nu}$ has signature $(+---)$ with indicies $\mu,\nu,\ldots = 0\ldots3$ and indicies $i,j,\ldots = 1\ldots3$, also $g = \det|g_{\mu\nu}|$. The matter content is given by

$$\mathscr{L}_M = -\frac{1}{4} g^{\mu\rho} g^{\nu\lambda} F^a_{\mu\nu} F^a_{\rho\lambda} + \frac{1}{2} g^{\mu\nu} \left(\mathscr{D}_\mu \Phi^a \right) \left(\mathscr{D}_\nu \Phi^a \right) - \frac{\lambda}{4} \left(\Phi^2 - v^2 \right)^2, \qquad (3)$$

where $\mathbf{F}_{\mu\nu}$ is the field strength associated with the gauge field $\mathbf{A}_\mu$ and the gauge coupling is α. The gauge group for our purposes will be $SU(2)$. We note that in the broken phase of the gauge symmetry, when the Higgs field Φ^a assumes its vacuum expectation value, $\Phi^2 = v^2$, $\mathscr{L}_E$ in (2) is the conventional Einstein-Hilbert Lagrangian. $\mathscr{L}_M$ in (3) represents the standard Yang-Mills-Higgs Lagrangian in curved space-time.

By varying the action S with respect to $\mathbf{A}_\mu$, Φ^a and $g_{\mu\nu}$, we obtain the coupled Yang-Mills, Higgs and Einstein equations of motion.

$$\frac{1}{\sqrt{-g}} \mathscr{D}_\mu \left(\sqrt{-g} F^{a\mu\nu} \right) = \alpha f^{abc} \Phi^b \mathscr{D}^\nu \Phi^c, \qquad (4)$$

$$\frac{1}{\sqrt{-g}} \mathscr{D}_\mu \left(\sqrt{-g} \mathscr{D}^\mu \Phi^a \right) = \left(\frac{R - 2\Lambda}{8\pi G v^2} + \lambda \left(\Phi^2 - v^2 \right) \right) \Phi^a, \qquad (5)$$

$$G_{\mu\nu} = R_{\mu\nu} - \frac{R - 2\Lambda}{2} g_{\mu\nu} = \frac{8\pi G v^2}{\Phi^2} T_{\mu\nu}, \qquad (6)$$

where the energy-momentum tensor $T_{\mu\nu}$ is given by

$$T_{\mu\nu} = -\left(\mathscr{L}_M + \frac{1}{2} \Box \Phi^2 \right) g_{\mu\nu} - F^a_{\mu\rho} \cdot F^{a\rho}_{\nu} + \mathscr{D}_\mu \Phi^a \cdot \mathscr{D}_\nu \Phi^a + \nabla_\mu \nabla_\nu \Phi^2. \qquad (7)$$

We are interested in static solutions to equations (4) - (6). Setting the time derivatives of all fields equal to zero slightly simplifies the equations of motion. In order to find a Bogomol'nyi-type first-order equation [8] to our problem, we make the ansatz [9],

$$F^a_{ij} = \sqrt{-\tilde{g}} \varepsilon_{ijk} \left(\mathbf{D}^k + u^k \right) \Phi^a, \qquad (8)$$

where $u^k = \partial^k f$ is an arbitrary time-independent function, and $\tilde{g} = \det|g_{ij}|$. With this ansatz, we find that that if we make the following restrictions on the metric,

$$g_{00} > 0, \quad g_{\alpha\beta} = 0 \quad \alpha \neq \beta, \quad g_{ii} < 0, \qquad (9)$$

then with some algebra it can be shown that the energy functional is bounded below by,

$$\mathcal{E} = \sin\theta \int d^3x \mathbf{D}_i \left(\frac{\sqrt[4]{-g}}{\sqrt{g_{00}}} \mathbf{E}^i \cdot \chi \right) + \cos\theta \int d^3x \mathbf{D}_i \left(\frac{\sqrt[4]{-g}}{\sqrt{g_{00}}} \mathbf{B}_i \cdot \chi \right)$$
$$+ \int d^3x \left[\frac{\sqrt{-g}}{g_{00}} \mathbf{D}_0 \chi \cdot \mathbf{D}^0 \chi + \frac{1}{2} \partial_i \left(\frac{\sqrt{-g}}{g_{00}} \chi^2 \partial^i f \right) \right]. \tag{10}$$

where $\chi = \sqrt{g_{00}} \Phi$, and

$$\mathbf{E}_i = \frac{\sqrt[4]{-g}}{\sqrt{g_{00}}} \mathbf{F}_{0i}, \quad \mathbf{B}_i = \frac{1}{2} \sqrt[4]{-g} \sqrt{-\tilde{g}} \varepsilon_{ijk} \mathbf{F}^{jk}. \tag{11}$$

We define a spherically symmetric static metric

$$ds^2 = A^2(r) dt^2 - B^2(r) dx^2 - r^2 \left(d\theta^2 + \sin^2\theta d\varphi^2 \right), \tag{12}$$

and assume the spherically symmetric 't Hooft-Polyakov ansatz [4] for the gauge and Higgs fields,

$$A_0^a = \frac{\hat{x}^a}{\alpha} J(r), \quad \eta_{ab} A_i^b = \varepsilon_{aij} \hat{x}^j \frac{(1-W(r))}{\alpha r}, \quad \eta_{ab} = (-1,-1,-1), \tag{13}$$

and

$$\Phi^a = \hat{x}^a v H(r). \tag{14}$$

Expressed in spherical polar coordinates, our ansatz (8) takes the form

$$\mathbf{D}_r \chi + \left(f'(r) - \frac{A'(r)}{A(r)} \right) \chi = -\frac{A(r) B(r)}{r^2 \sin\theta} \mathbf{F}_{\theta\varphi}, \tag{15}$$

$$\mathbf{D}_\theta \chi = -\frac{A(r)}{B(r)} \frac{1}{\sin\theta} \mathbf{F}_{\varphi r}, \quad \mathbf{D}_\varphi \chi = -\frac{A(r)}{B(r)} \sin\theta \mathbf{F}_{r\theta}. \tag{16}$$

We seek solutions to the set of coupled non-linear equations in the Higgs vacuum, $H = 1$. In order to simplify the equations we introduce the dimensionless co-ordinate $x = \alpha v r$ and rescale $J/\alpha v \to J$. We shall also work in units of the dimensionless coupling $4\pi G v^2$ set to unity and henceforth all the field variables are functions of x and 'prime' will denote derivatives with respect to x. Following these conventions the Bogomol'nyi equations (15) and (16) can be simply expressed as

$$\frac{A'}{AB} = \frac{J^2}{A^2} + \frac{B(1-y)}{x^2}, \quad B = -\frac{1}{2} \frac{y'}{y}, \tag{17}$$

where we have defined $y = W^2$. The Yang-Mills equation is given by

$$J'' + \left(\frac{2}{x} - \frac{A'}{A} - \frac{B'}{B} \right) J' - \frac{2}{x^2} B^2 J y = 0, \tag{18}$$

and the Higgs and Einstein field equations are similar and can be seen explicitly in [1].

MONOPOLE SOLUTIONS

Equations for the monopole are obtained by setting $J = 0$ in the above set of equations. It can readily be shown that monopole solution can only exist for a vanishing cosmological constant. We are left with three independent equations,

$$y' = -\frac{2xy}{x+y-1}, \qquad B = \frac{x}{x+y-1}, \qquad \frac{A'}{A} = \frac{1-y}{x(x+y-1)}. \tag{19}$$

The first equation in (19) is the well known Abel's differential equation of the second type. The two equations on the right of (19) are determined in terms of its' solutions. Abel's equation has no known analytical solution other than $y = 0$, in which case, we have an Abelian magnetic monopole which is a Reissner-Nördstrom black hole with mass $M = 4\pi G v/\alpha$ and charge $Q = M/\alpha v$.

In the general case, with non-vanishing y, we can solve equations (19) numerically. We observe that we have a family of solutions, distinguished by $y(1) = y_{\text{init}}$, that

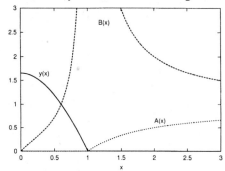

FIGURE 1. Plot of $A(x)$, $B(x)$ and $y(x)$ against x for an extremal non-Abelian monopole ($y(1) = 0$). This solution is a black hole since we see the event horizon at $x = 1$.

have exponentially vanishing non-Abelian components $y = W^2$. These solutions are characterized by a non-Abelian core. Once one fixes the asymptotic behavior of A ($\lim_{x \to \infty} A \to 0$) it is uniquely determined. The metric component B automatically has the correct asymptotic behavior. We see that as the initial value of $y(1)$ approaches zero, the minimum of A also approaches zero at $x = 1$ (and the maximum B at $x = 1$ gets larger). The solution for $y(1) = 0$ (see Fig. 1) represents the extremal case where an event horizon has formed, i.e. a black hole with a non-Abelian magnetic monopole.

In Fig. 1 we see that the metric coefficient for A vanishes inside the horizon, $x \leq 1$. This is not a problem since A has been normalized at $x = \infty$. An observer at infinity can not observe the interior of the black hole. If one chose to normalize A at the origin one would have a perfectly well defined metric inside the horizon which is infinite outside. This does not affect the determination of the non-Abelian magnetic field y which is independent of the normalization of A. This behavior is similar to the monopole solutions of Lue and Weinberg [6].

DYONIC SOLUTIONS

Due to the complexity of the equations in the dyonic case (see [1]), we are forced to look for numerical solutions. Examination of the asymptotic behavior of these equations reveals that the cosmological constant must vanish. With some algebra the equations can be reduced to the following two equations,

$$y'' - y'^3 \frac{(x^2 - 2y)}{2yx^3} - y'^2 \frac{(1 - 7y + 2x^3)}{2yx^3} + y'\frac{2}{x} + \frac{2y}{x}(x+y')\left(\frac{y'}{y} + \frac{J'}{J}\right)\frac{J'}{J} = 0, \quad (20)$$

$$\left(\frac{J'}{J}\right)^4 + \frac{y'}{y}\left(\frac{J'}{J}\right)^3 - \left(\frac{1}{x} + \frac{y'}{2x^2}\right)\left(\frac{y'}{y}\right)^2\left(\frac{J'}{J}\right)$$
$$+ \frac{1}{x^2}\left(\frac{y'}{y}\right)^2\left[\left(\frac{y'}{y}\right)^2 \frac{(x^2 - 5y^2 + 2y - 1)}{16x^2} + \frac{y'(1-5y)}{y \; 4x} - \frac{5}{4}\right] = 0. \quad (21)$$

The solutions to y and J yield the metric functions A and B via:

$$\frac{A'}{A} = -\frac{1}{2}\frac{y'}{y}\left(1 + \frac{1-3y}{x^2}\right) + 2\frac{y}{y'}\left(\frac{J'}{J} + \frac{1}{2}\frac{y'}{y}\right)^2 + \frac{2}{x}, \quad B = -\frac{1}{2}\frac{y'}{y}. \quad (22)$$

Equation (21) is a quartic polynomial in J'/J in terms of y and y'. Since J is positive and the requirement that J vanish asymptotically implies that J'/J should be negative everywhere (as long as J is montonic). Thus we select the negative real root of the quartic polynomial. Equation (20) is a non-linear second order differential equation in y which can be readily integrated numerically using a 4th order Runga-Kutta with initial conditions set at $x = 1$.

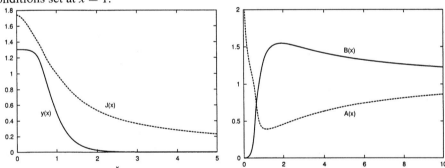

FIGURE 2. Plots of $y(x)$, $J(x)$ and $A(x)$, $B(x)$ against x for a non-Abelian dyon with initial conditions $y(1) = 0.5, y'(1) = -1.3, A(1) = 0.4, J(1) = 1$.

Inspection of equations (20) and (21) reveals that y can only vanish at $x = 1$. In figure 2 we see a solution for a non-Abelian dyon. As the initial conditions approach critical values numerical solutions develop singularities, this is discussed in [1].

CONCLUSIONS

We have shown that with an ansatz, eqn. (8), we are led to Bogomol'nyi type equations and an energy functional that is bounded from below in the case of a dyon. In flat space-time, solutions that saturate the Bogomol'nyi bound are known to exist [10]. Demonstrating this in the case of a non-Abelian dyon in curved space-time is a new result. These results follow from the form of the action in eqn. (1), where the Higgs scalar is directly coupled to the Ricci scalar. In the monopole case, this unconventional coupling led to a first order classic Abel's equation, the solution of which yielded solutions to all the equations in the problem [7]. In the dyon case, the problem turns out to be more complicated; nonetheless, contains similar simplifications in contrast to the more standard approaches in the literature. We also present numerical evidence for non-Abelian dyonic solutions. For a range of initial conditions the defining equations yield stable numerical results which are well-defined to the center of the dyon. However, as the initial conditions are tuned closer to critical values, the numerical results have singularities that are likely to correspond to the formation of horizons.

ACKNOWLEDGEMENTS:

This work was done in collaboration with A.C. Cornell, G.C. Joshi and K.C. Wali. This work was supported by the DOE contract no. DE-FG02-85ER40237. We would like to thank R. F. Sawyer and P. Silva for helpful discussions.

REFERENCES

1. A. S. Cornell, G. C. Joshi, J. S. Rozowsky, K. C. Wali, arXiv:hep-th/0207043.
2. P. A. Dirac, Proc. Roy. Soc. Lond. **A133**, 60 (1931).
3. T. T. Wu, C. N. Yang, Phys. Rev. **D12**, 3845 (1975).
4. G. 't Hooft, Nucl. Phys. **B79**, 276 (1974); A. M. Polyakov, JETP Lett. **20**, 194 (1974).
5. K. M. Lee, V. P. Nair, E. J. Weinberg, Phys. Rev. **D45**, 2751 (1992) [arXiv:hep-th/9112008]; P. Breitenlohner, P. Forgacs, D. Maison, Nucl. Phys. **B383**, 357 (1992); P. C. Aichelburg, P. Bizon, Phys. Rev. **D48**, 607 (1993) [arXiv:gr-qc/9212009]; M. S. Volkov, D. V. Gal'tsov, Phys. Rept. **319**, 1 (1999) [arXiv:hep-th/9810070]; Y. Brihaye, B. Hartmann, J. Kunz, N. Tell, Phys. Rev. **D60**, 104016 (1999) [arXiv:hep-th/9904065]; A. R. Lugo, F. A. Schaposnik, Phys. Lett. **B467**, 43 (1999) [arXiv:hep-th/9909226].
6. A. Lue, E. J. Weinberg, Phys. Rev. **D60**, 084025 (1999) [arXiv:hep-th/9905223]; A. Lue, E. J. Weinberg, Phys. Rev. **D61**, 124003 (2000) [arXiv:hep-th/0001140].
7. A. V. Nguyen, K. C. Wali, Phys. Rev. **D51**, 1664 (1995) [arXiv:hep-ph/9310370].
8. E. B. Bogomol'nyi, Sov. J. Nucl. Phys. **24**, 449 (1976); M. K. Prasad, C. M. Sommerfield, Phys. Rev. Lett. **35**, 760 (1975).
9. A. Comtet, P. Forgacs, P. A. Horvathy, Phys. Rev. **D30**, 468 (1984).
10. B. Julia, A. Zee, Phys. Rev. **D11**, 2227 (1975); S. R. Coleman, S. Parke, A. Neveu, C. M. Sommerfield, Phys. Rev. **D15**, 544 (1977).

DEVELOPMENTS IN QUANTUM THEORY

Realism in the Realized Popper's Experiment

Geoffrey Hunter

Department of Chemistry, York University, Toronto, Canada M3J 1P3. ghunter@yorku.ca

Abstract. The realization of Karl Popper's EPR-like experiment by Shih and Kim (published 1999) produced the result that Popper hoped for: no "action at a distance" on one photon of an entangled pair when a measurement is made on the other photon. This experimental result is interpretable in local realistic terms: each photon has a definite position and transverse momentum most of the time; the position measurement on one photon (localization within a slit) disturbs the transverse momentum of that photon in a non-predictable way in accordance with the uncertainty principle; however, there is no effect on the other photon (the photon that is not in a slit) no action at a distance. The position measurement (localization within a slit) of the one photon destroys the coherence (entanglement) between the photons; i.e. decoherence occurs.

This realistic (albeit retrodictive) interpretation of the Shih-Kim realization of what Popper called his "crucial experiment" is in accord with Bohr's original concept of the nature of the uncertainty principle, as being an inevitable effect of the disturbance of the measured system by the measuring apparatus. In this experiment the impact parameter of an incident photon with the centerline of the slit is an uncontrollable parameter of each individual photon scattering event; this impact parameter is variable for every incident photon, the variations being a statistical aspect of the beam of photons produced by the experimental arrangement.

These experimental results are also in accord with the proposition of Einstein, Podolski and Rosen's 1935 paper: that quantum mechanics provides only a statistical, physically incomplete, theory of microscopic physical processes, for the quantum mechanical description of the experiment does not describe or explain the individual photon scattering events that are actually observed; the angle by which an individual photon is scattered is not predictable, because the photon's impact parameter with the centerline of the slit is not observable, and because the electromagnetic interaction between the photon and the matter forming the walls of the slit is not calculable.

POPPER'S CONCEPT OF THE CRUCIAL EXPERIMENT

Karl Popper was a philosopher who was deeply concerned with the interpretation of quantum mechanics since its inception in 1925. In a book originally published in 1956 he proposed an experiment which he described as "an extension of the Einstein-Podolsky-Rosen argument" [1, pp.ix,27-30]. In summary: two photons are emitted simultaneously from a source that is mid-way between (and colinear with) two slits A and B; the coincident diffraction pattern is observed beyond each slit twice: once with both slits present, and again with one slit (slit B) "wide open".[1]

Popper (in common with Einstein, Podolsky and Rosen and with the proponents of the Copenhagen interpretation of quantum mechanics) believed that *quantum theory predicts* that localization of a photon within slit A will not only cause it to diffract in accordance with Heisenberg's uncertainty relation, but will also cause the other photon

---

[1] T. Angelidis has restated this as "slit B being absent".

at the location of slit B to diffract by the same angle - regardless of whether slit B is present or not. Such spatially separated correlations would violate the causality principle of Special Relativity; i.e. that physical interactions cannot travel faster than the speed of light; thus Einstein called them "spooky actions at a distance".

Popper was "inclined to predict" [1, p.29] that in an actual experimental test, the photon at the location of the absent slit B would not diffract ("scatter") when its partner photon (with which it is in an entangled state) is localized within slit A. He emphasized that "this does not mean that quantum mechanics (say, Schrödinger's formalism) is undermined"; only that "Heisenberg's claim is undermined that his formulæ are applicable to all kinds of indirect measurements".

Popper [1, p.62] notes that Heisenberg agreed that *retrodictive* values of the position *and* the momentum can be known by knowing the position of a particle (e.g. as it passes through a small slit) followed by a measurement of the momentum of the particle after it has passed through the slit - from the position in the detection plane where it is located. Feynman has also emphasized that the uncertainty principle does not exclude *retrodictive* inferences about the simultaneous position and momentum of a particle [5, Ch.2, pp.2-3], but the uncertainty principle does exclude precise *prediction of both* position or momentum.

Popper also notes that momentum values are usually inferred from two sequential position measurements; in this regard he concurs with Heisenberg, who wrote [1, p.62]:

> "The ... most fundamental method of measuring velocity [or momentum] depends on the determination of position at two different times ... it is possible to determine *with any desired degree of accuracy* what the velocity [or momentum] of the particle *was* before the second measurement was made"

Indeed precise inferences of particle positions and momenta are widely used in the analysis of observations in high-energy particle accelerators, by which the modern plethora of "fundamental particles" have been discovered. This is emphasized by another quote from Popper [1, p.39]:

> "my assertion [is] that most physicists who honestly believe in the
> Copenhagen interpretation do not pay any attention to it in actual practice".

THE REAL EXPERIMENT OF SHIH AND KIM

Yoon-Ho Kim and Yanhua Shih carried out a modern realization of Popper's experiment; their report was published three times [2, 3, 4]. Other experiments that don't confirm the widely held interpretation of quantum mechanics have also been reported [7].

Kim and Shih note that Popper's thought experiment was designed to show:

> "that a particle can have both precise position
> and momentum at the same time" – [2, p.1849, 2nd paragraph], and that

> "it is astonishing to see that the experimental
> results agree with Popper's prediction" – [2, p.1850, 2nd paragraph].

Experimental Parameters

- the source is a CW (continuous wave) argon-ion laser producing highly monochromatic ultra-violet radiation of wavelength, $\lambda = 351.1$ nm;
- a pair of entangled visible photons ($\lambda = 702.2$ nm) are produced by Spontaneous Parametric Down-Conversion (SPDC) in a BBO (β barium borate) crystal;
- the width of slit A and of slit B (when present) $= 0.16$ mm;
- the diameter of the avalanche photodiode detectors (D1 and D2) $= 0.18$ mm;
- during the measurements, D1 was in a fixed position (<u>not</u> scanned along the y-axis) close to a collection lens (of 25 mm focal length) with the lens close to slit A; this was designed to direct *every* photon passing through slit A into detector D1; the photons thus detected were used simply to trigger the coincidence circuit to look for a photon with detector D2 (500 mm beyond slit B).

Transverse Momentum: Slit B Present

The classical (wave-optics) theory of diffraction [8, pp.214-216] based upon Huygens principle[2] predicts that the diffraction pattern produced by a beam of light incident upon a single slit in a screen perpendicular to the incident direction of the beam, will have a first minimum intensity at:

$$y = \frac{\lambda D}{s} \qquad (1)$$

where y is the transverse displacement of the diffraction minimum from the incident direction of the beam in the plane of the detecting screen (at distance D from the slit of width s); λ is the wavelength of the light and y is naturally proportional to D.

Feynman has shown [5] that the classical formula (1) can be re-interpreted as an effect of the Heisenberg Uncertainty Relation, for substitution of the de Broglie relation, $\lambda = h/P$ between a photon's wavelength, λ and its momentum, P, into eqn.(1) produces:

$$yP = \frac{hD}{s} \qquad \text{or} \qquad \Delta y \Delta P_y = h \qquad (2)$$

because the photon's *transverse* momentum, ΔP_y, is $P \times y/D$. Thus diffraction towards the first diffraction minimum is interpreted as being the measure of the "uncertainty" in the transverse (i.e. in the y direction) momentum of the photon, ΔP_y, caused by location of its y-coordinate within the width of the slit, $\Delta y = s$.

I have carried out calculations showing that (1) and (2) are in accord with the first diffraction minimum in the experimental curve in Figure 5 of [2] when slit B is present.

---

[2] that each point in a beam is a source, the resulting beam being the result of interference between the radiation from all the sources

It is noteworthy that Kim and Shih report [2, top of p.1856]:

> "the *single* detector counting rate of D2 is basically the same as that of the coincidence counts except for a higher counting rate".

In other words: the outer (wider) diffraction peak (the curve for slit B present) of Figure 5 of [2] is obtained regardless of whether the coincidence detection circuit is active or not. This observation indicates that this peak is produced by the diffraction of the photons incident upon slit B *regardless* of their entanglement with the photons incident upon slit A; i.e it is a single-photon phenomenon; indeed it is nothing more than the central diffraction maximum predicted by the classical theory of light [8, p.214].

The higher counting rate is explained by the effective size of the source only being about 0.16 mm diameter for coincidence counting, whereas for non-coincidence counting it is the full diameter of the laser beam of 3 mm.

Transverse Momentum: Slit B Absent

The inner curve of Figure 5 of [2] is drawn from the experimental coincidence counts obtained when slit B is "wide open". This shows that the photons are deflected up to 0.9 mm, and that from 0.9 mm to 1.45 mm the detection rate is constant and close to zero.[3] This range (0.9 to 1.45 mm) over which the counting rate is constant precludes discernment of a diffraction minimum (unlike the curve for slit B present, which has one point above zero as the onset of the second diffraction maximum); there are in fact 5 data points ($y = 1.0 - 1.45$ mm) all of which have the same (very small) value; this suggests that the origin of this peak is *not* diffraction through a slit.

It is noteworthy that Kim and Shih report [2, p.1856, 2nd *paragraph*]:

> "the single detector counting rate of D2 keeps constant in the entire scanning range"

In other words: when the coincidence circuit is switched off D2 detects the same count rate at all values of y at which it was placed ($y = 0$ to $y = 1.45$ mm); this would be a horizontal straight line if added to Figure 5 of [2]. This measurement was simply seeing the beam emanating from the source towards D2 with a uniform intensity over the scanning range of ≈ 3 mm ($y = \pm 1.5$ mm).

Kim and Shih also report [2, p.1856, 2nd *paragraph*]:

> "the width of the pattern is found to be much narrower than the actual size of the diverging SPDC beam at D2".

The probable cause of this narrow peak is a convolution of the finite size of the source,[4] with the geometry of possible coincidences; if it were interpreted as a diffraction pat-

---

[3] These observed counting rates are only zero within the plotting precision of Figure 5 of [2]; the observed counting rates were not recorded as actual, recorded values in any of the published records of the experiments.

[4] a cylinder 3 mm diameter and 3 mm long

tern, then in terms of eqn.(2), there would be an observed violation of the Heisenberg Uncertainty Principle by a factor of about 3.

A RETRODICTIVE REALIST ACCOUNT

- A pair of photons produced by SPDC is in an entangled state from the moment of generation until one of them enters a slit; entanglement means that their positions and momenta are correlated; knowledge of the position of one photon allows one to infer the position of the other photon; likewise for their momenta.
- When one photon enters a slit it interacts with that slit and this destroys the coherence (entanglement) between them; i.e. decoherence occurs. The interaction can be attributed to the photon being a localized electromagnetic wave [9],[5] which interacts with the electrons in the surface of the solid that forms the slit. That photons are localized waves is supported experimentally by the production of laser pulses as short (in time) as two optical periods [10]; thus the photon cannot be longer than two wavelengths along its direction of propagation.
- Measurement of the diffracted position (y coordinate) of a photon (coincidence detection by D2) with slit B absent, allows one to calculate not only the momentum vector of this photon as it travels from the source to D2, but also the momentum of the other photon as it travels from the source to slit A; however when this latter photon enters slit A its interaction with the walls of the slit causes it to diffract at an angle which is predictable only statistically - in accord with the uncertainty principle.
- Thus coincidence measurements <u>with slit B absent</u> provide the positions of both photons (from the detection of a photon having passed through slit A) with a precision equal to the width of slit A. Likewise the measurement of the deflected position (y coordinate) of a photon by D2 allows one to calculate the momentum vectors of *both* photons of the entangled pair – during their trajectories from the source to the plane of slit A (for one photon), and from the source to the scanning plane of D2 for the other photon. These *in principle*, precise, retrodictive calculations of the trajectories of both photons are unfortunately limited in precision by the actual experimental results because of the relatively large size of the non-point source.[6]
- It is especially noteworthy that <u>individual events</u>[7] are not limited by the uncertainty principle: any diffraction of a photon to a position of D2 smaller than the y-coordinate of the first diffraction minimum will yield a position-momentum product that is smaller than Planck's constant – even when slit B is present. In particular, the most probable diffraction angle (to the top of the central peak) yields a transverse momentum of zero, which when multiplied by the uncertainty in its position (the slit width of 0.16 mm) yields an uncertainty product of zero !

---

[5] In [9] the photon is an ellipsoidal soliton of length λ, and diameter λ/π.
[6] a cylinder 3 mm diameter and 3 mm long
[7] the generation and detection of a particular entangled photon pair

Concluding Remarks

Karl Popper regarded his experiment as a "crucial" test of the inconsistency between the inferred non-locality of quantum mechanics and the causality principle of Special Relativity. While Shih and Kim conceded [2, p.1858,last paragraph] that: "Popper and EPR were correct in the prediction of the physical outcomes of their experiments.", their subsequent sentences (in the same paragraph), ("...Popper and EPR made the same error ...") are incongruous; Popper and EPR made no error - they agreed with Bohr, Heisenberg and other proponents of the Copenhagen interpretation [11] that quantum theory apparently predicts an instantaneous action at a distance on one particle of an entangled pair, when a measurement is made on the other particle of the pair; Popper and EPR's crucial point was that if such actions are a distance are not in fact observed (as in the Shih-Kim experiment), then quantum theory must be an incomplete (only statistical) theory of the physical world, and as a statistical theory that does not describe individual events, it is entirely consistent with the causality principle of Special Relativity.

The above interpretation of the Shih-Kim experiment in locally realistic terms, is not easily extended to the interpretation of some other experiments: Bohm's EPR experiment involving the Bell inequalities [12, Ch.11], and the various single-particle, double-path experiments conducted with particles as large as C_{60} molecules [13].

ACKNOWLEDGMENTS

Thomas Angelidis brought the Shih-Kim experiment to the author's attention at conferences in Baltimore (1999) and Berkeley (2000), and Yanhua Shih provided helpful elaborations of the published accounts of the experimental work. The Natural Sciences and Engineering Research Council of Canada provided financial support.

REFERENCES

1. Popper, K.R. *Quantum Theory and the Schism in Physics*, Rowan and Littlefield, Totowa, New Jersey, 1982.
2. Kim, Y-H. and Shih, Y. *Foundations of Physics*, **29**, 1849-1961 (1999).
3. Shih, Y. and Kim, Y-H. *Fortschritte der Physik*, **48**, 463-471 (2000).
4. Shih, Y. and Kim, Y-H. *Optics Communications*, **179**, 357-369 (2000).
5. Feynman, R.P. *Lectures on Physics*, Addison-Wesley, Reading, Massachusetts, 1965.
6. Einstein, A., Podolsky, B., and Rosen, N. *Physical Review*, **47**, 777-780 (1935).
7. Bell, J.S. *Science*, **177**, 880 (1972).
8. Longhurst, R.S. *Geometrical and Physical Optics*, Longmans Green & Co., London (2nd Edition, 1967).
9. Hunter, G. and Wadlinger, R.L.P. *Physics Essays*, **2**, 158-172 (1989).
10. Dietrich, P., Krausz, F. and Corkum, P. B. *Optics Letters*, **25**, 16-18 (2000).
11. Cushing, James T. *Quantum Mechanics: Historical Contingency and the Copenhagen Hegemony*, University of Chicago Press, 1994.
12. Holland, P.R. *The Quantum Theory of Motion*, Cambridge University Press, 1993.
13. Arndt, M., Nairz, O., Vos-Andreae, J., Keller, C., van der Zouw, G., & Zeilinger, A. *Nature*, **401**, 680-682, 1999.

The Dirac Equation in Classical Statistical Mechanics

G. N. Ord

M.P.C.S.
Ryerson University
Toronto Ont.

Abstract. The Dirac equation, usually obtained by 'quantizing' a classical stochastic model is here obtained directly within classical statistical mechanics. The special underlying space-time geometry of the random walk replaces the missing analytic continuation, making the model 'self-quantizing'. This provides a new context for the Dirac equation, distinct from its usual context in relativistic quantum mechanics.

INTRODUCTION

The title of this talk requires some explanation. Statistical mechanics is in some respects the simplest branch of physics; all you ever have to do is count. The trick is of course to find the right class of objects to count.

The implication of the title is that the Dirac equation is simply related to the operation of counting objects. If we were talking about the diffusion or heat equation then there would be no mystery. It is well known that the diffusion equation may be obtained by counting random walks in an appropriate limit.

However, the likelihood of being able to make a similar claim for the Dirac equation seems small. We all know that the Dirac equation has elements of quantum mechanics, special relativity and half-integral spin. None-the-less the claim in the title is true for the Dirac equation in one dimension [1], and may well be true in three dimensions for a free particle. Among other things, this means that the Dirac equation has a context as a phenomenology which may be distinct from its role as a fundamental equation of quantum mechanics. Although this may be surprising from the perspective of quantum mechanics, it is not so extraordinary in the context of general partial differential equations. For example we are used to calling the same PDE either the Diffusion equation or the heat equation, depending on context. In the case of Dirac, we have only one name for the equation, but there are still multiple contexts.

Today I am going to talk about the Dirac equation as a classical phenomenology. However since we all think 'quantum mechanics' at the mention of Dirac's name, let's recall where Quantum mechanics begins and where it ends. Typically we pass from classical physics to quantum mechanics through a formal analytic continuation (FAC). The canonical FAC is to replace the momentum by the operator $-i\hbar\nabla$ and the energy by $i\hbar\frac{\partial}{\partial t}$. In a sense this is where quantum mechanics begins and classical physics ends.

The quantum equations describe the evolution of the initial conditions in time, and

TABLE 1. The relation between Classical PDE's based on stochastic models, and their 'Quantum' cousins. FAC accomplishes the trick, but then the stochastic basis for the equations becomes formal.

	Classical	Quantum
Microscopic basis	Kac (Poisson)	Chessboard
First Order	$\frac{\partial \mathbf{U}}{\partial t} = c\boldsymbol{\sigma}_z \frac{\partial \mathbf{U}}{\partial z} + a\boldsymbol{\sigma}_x \mathbf{U}$	$\frac{\partial \boldsymbol{\Psi}}{\partial t} = c\boldsymbol{\sigma}_z \frac{\partial}{\partial z}\boldsymbol{\Psi} + im\boldsymbol{\sigma}_x \boldsymbol{\Psi}$
Second Order	$\frac{\partial^2 U}{\partial t^2} = c^2 \frac{\partial^2 U}{\partial z^2} + a^2 U$	$\frac{\partial^2 \psi}{\partial t^2} = c^2 \frac{\partial^2 \psi}{\partial z^2} - m^2 \psi$
'Non-relativistic'	$\frac{\partial U}{\partial t} = D \frac{\partial^2 U}{\partial x^2}$	$\frac{\partial \psi}{\partial t} = iD \frac{\partial^2 \psi}{\partial x^2}$

we relate the results of this evolution to our macroscopic world by the measurement postulates. These postulates are the weakest link in the theory, and it is here that the (many) interpretations of quantum mechanics vie for supremacy. At this point we do not really know where quantum mechanics ends and classical physics begins. So let us return to the point where it begins, at the FAC.

The the canonical FAC just mentioned specifically relates the Schrödinger equation to Hamiltonian mechanics. We will not be so specific. Table(1) compares classical PDE's with their corresponding 'quantum' counterparts. In the left 'Classical' column we start out with a two component form of the Telegraph equations due to Marc Kac. Here U is related to a two component probability density, c is a mean free speed and a is an inverse mean free path. Note that if we replace the real positive constant a by im we get a form of the Dirac equation.

Similarly the second order form of the Telegraph equations continues to the Klein-Gordon or Relativistic-Schrödinger equation using the same FAC. The 'non-relativistic limit' of the Telegraph equation gives the Diffusion equation, with the usual FAC to the Schrödinger equation. The equations on the left are phenomenologies which are interpreted through the underlying statistical mechanical models (Poisson or Brownian motion). The equations on the right are regarded as fundamental equations which have no realistic microscopic basis, and are interpreted through postulates. The analog of Poisson paths for the Telegraph equations are the Chessboard paths of Feynman.

What we will do today is to show that the Dirac equation in 1-dimension is also a phenomenological equation with a microscopic basis, and is accessible directly through classical statistical mechanics ... all we have to do is to find the right objects to count. But first let us see how Kac[2] obtained the Telegraph equations from a microscopic model.

THE KAC MODEL

Imagine a particle on a discrete space-time lattice with spacings Δz and Δt. The speed of particles are fixed at c and occasionally they scatter backwards with probability $\alpha \Delta t$. Fig. (1.A) shows a typical Kac path. Considering the density of particles moving in the plus and minus directions we can write difference equations for their conservation.

$$F_n^+(z) = (1 - a\Delta t)F_{n-1}^+(z - c\Delta t) + a\Delta t F_{n-1}^-(z) \qquad (1)$$

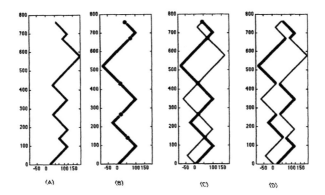

FIGURE 1. (A) Paths in the Kac process are broken-line segments with slope $\pm c$. The average distance between corners is $1/a$. (B) The Kac Stochastic process with a stutter. (C) With the return Path (indicated by the thinner line). (D) The outer envelopes of the Entwined paths.

Similarly,
$$F_n^-(z) = (1 - a\Delta t)F_{n-1}^-(z + c\Delta t) + a\Delta t F_{n-1}^+(z) \quad (2)$$
In the continuum limit these give the coupled PDE's:
$$\frac{\partial F^+}{\partial t} = -c\frac{\partial F^+}{\partial z} - aF^+ + aF^- \quad (3)$$

$$\frac{\partial F^-}{\partial t} = c\frac{\partial F^-}{\partial z} + aF^+ - aF^- \quad (4)$$

It is easy to identify the streaming and scattering terms here, but if we remove the exponential decay, and write in 2-component form we get.
$$\frac{\partial \mathbf{U}}{\partial t} = c\sigma_z\frac{\partial \mathbf{U}}{\partial z} + a\sigma_x\mathbf{U} \quad (5)$$
where the σ are the usual Pauli Matrices.

Note the suggestive form of the coupled equations. We are a FAC away from the Dirac equation. However the FAC destroys any coherent interpretation of our microscopic model, so we will strictly avoid it!

ENTWINED PAIR MODEL

As mentioned above, counting Kac Paths gives the telegraph equations. So what do we count if we want to go directly to the Dirac equation? And how is i going to appear if we are not allowed a FAC? To answer the second question first, we do not need i to get the quantum equations, we only need the fact that $i^2 = -1$. This may seem like a trivial point but it is extremely important. To illustrate it, think of the even and odd parts of e^t ... these are $\cosh(t)$ and $\sinh(t)$. Now analytically continue and think of the even and

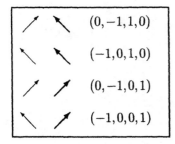 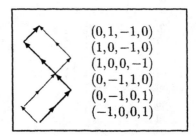

FIGURE 2. The four possible states of Entwined Pairs and two cycles of an Entwined Pair. Note the alternating signs in the envelope.

odd parts of e^{it}. These are $\cos(t)$ and $i\sin(t)$. It is true that here i distinguishes the even and odd parts, but the important effect of the analytic continuation is to produce the alternating signs in the series expansion of the trigonometric functions. It is the same in the quantum equations. We do not actually need i, what we need is the alternating pattern of sign that gives oscillatory behaviour, and we have to get it from the geometry of particle trajectories, not from a FAC.

Consider the Entwined paths of Fig(1. B-D). We will generate these with the same process we used to generate the Kac paths of Fig.(1. A), except here we will employ a periodic 'stutter'. That is, wherever we would change direction in a Kac path, we alternately change direction or leave a marker in the entwined path. At the first marker past some specified time t_R, we reverse our direction in t and follow the markers back to the origin. In Fig(1. C), the return path has a thinner line-width to distinguish it from the forward path. Even in Classical physics, particles which move backwards in time behave like anti-particles to a forward moving observer, so if we record (+1) as a charge carried on blue portions of the trajectory, we shall associate a -1 with the red portions. The task will then be to calculate the expected average charge deposited by an ensemble of these paths. Note that each entwined pair can be regarded as two osculating envelopes with a periodic colouring(Fig.1. D). Each envelope is just a Kac path. Each has the same statistics and geometry as a Kac path, the only difference is the periodic colouring. We can then set up a difference equation as we did for the Telegraph equations. This time there are 4 states instead of Kac's 2 states. Fig.(2. A) shows the four possible states of an entwined pair and Fig.(2. B) shows a path evolving through two loops.

$$\begin{align}
\phi_n^1(z) &= (1-a\Delta t)\phi_{n-1}^1(z-c\Delta t) - a\Delta t \phi_{n-1}^2(z-c\Delta t) \\
\phi_n^2(z) &= (1-a\Delta t)\phi_{n-1}^2(z+c\Delta t) + a\Delta t \phi_{n-1}^1(z+c\Delta t) \\
\phi_n^3(z) &= (1-a\Delta t)\phi_{n-1}^3(z-c\Delta t) - a\Delta t \phi_{n-1}^4(z-c\Delta t) \\
\phi_n^4(z) &= (1-a\Delta t)\phi_{n-1}^4(z+c\Delta t) + a\Delta t \phi_{n-1}^3(z+c\Delta t)
\end{align} \tag{6}$$

Note the alternating minus sign from the crossover of paths (cf. Feynman Chessboard Model). Here the ϕ are real ensemble averages of a net 'charge' density. They are *not* quantum mechanical amplitudes.

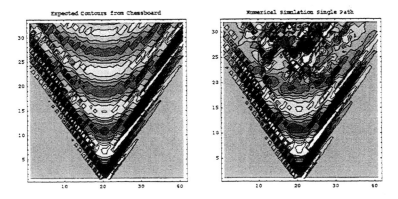

FIGURE 3. Contour plot of the sum of the real and imaginary parts of the Dirac Propagator. Left, discrete Dirac, right, single path.

Removing the exponential decay and writing $p_z = -i\frac{\partial}{\partial z}$, setting $c = 1$ and $a = m$ with
$\alpha_z = \begin{pmatrix} -\sigma_z & 0 \\ 0 & \sigma_z \end{pmatrix} \beta = \begin{pmatrix} \sigma_y & 0 \\ 0 & \sigma_y \end{pmatrix}$ we get the Dirac equation

$$i\frac{\partial \psi}{\partial t} = (\alpha_z . p_z + \beta m)\psi \qquad (7)$$

in the continuum limit. Notice there is no FAC here. The i in Eqn.(7) was introduced simply to show a familiar form of the Dirac equation. The ψ here are real, four component, and oscillatory in character. The oscillation is implemented through the presence of $\sigma_q = i\sigma_y$, which is a real, anti-hermitian matrix. σ_q arises because of the periodic exchange of particle and antiparticle in entwined paths, and it has the important feature that $\sigma_q^2 = -1$. We have not forced an analytic continuation on the system here. The space-time geometry itself has rendered the system 'self-quantizing'!

DISCUSSION

The above suggests that the Dirac equation appears as an ensemble average of a net charge over our entwined pairs. Since all pairs meet at the origin the entire ensemble can be regarded as being generated *by a single particle traversing all entwined paths*. However the above analysis only shows that the Dirac propagator will result if we cover the ensemble exactly. There have been other cases where the quantum equations have been recovered as projections without using a FAC [3, 4, 5, 7, 6, 8], however all of these have required a complete ensemble. What happens if we just watch the stochastic process with its inherent fluctuations? Will the process converge to the propagator or will the fluctuations swamp the signal? The answer appears to be that the signal survives stochastic fluctuations. The Dirac propagator is a stable feature of entwined paths! Figures (3) and (4) show the propagator drawn by a single path in comparison to the discrete Dirac propagator.[1] In Fig.(3) we see that the propagator formed by

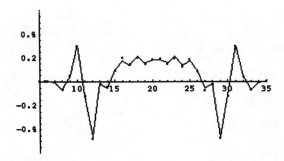

FIGURE 4. A time slice through the above figure at $t = 15$. The curve is the Dirac propagator and the black dots are the single path.

the entwined paths is quite accurate near the origin but gets less accurate the father out we go. This is to be expected since the configuration space grows exponentially as the number of time steps and the farther away we are from the origin, the poorer the ensemble coverage.

Considering the original context of the Dirac Equation, the above demonstration that the equation is easily obtained within classical statistical mechanics is surprising, to say the least. Clearly, the algebraic route followed by Dirac to obtain his equation was, and is, elegant, concise and algebraically compelling. However, for all its comparative inelegance, the heuristic approach via entwined paths strongly suggests that canonical quantization, Dirac's starting point, might well benefit from a re-evaluation. The FAC represented by canonical quantization gives an algebraic connection to Hamiltonian mechanics without any suggestion as to what the connection actually means in terms of the propagation of a classical particle. By comparison, the entwined paths approach replaces a formal algebraic requirement by *a physical constraint on space-time geometry*. What we lose in elegance we may well regain in physical cogency. As Dirac suggested in the preface to his book[9]

> "Mathematics is the tool specially suited for dealing with abstract concepts, ...All the same, the mathematics is only a tool and one should learn to hold the physical ideas in one's mind without reference to the mathematical form."

REFERENCES

1. G. N. Ord and J. A. Gualtieri, quant_ ph 0109092 (2001).
2. M. Kac, Rocky Mountain Journal of Mathematics p. 4 (1974).
3. G. N. Ord, Int. J. Theor. Physics **31**, 1177 (1992).
4. G. N. Ord, Phys. Lett.A **173**, 343 (1993).
5. G. N. Ord, J.Phys. A: Math. Gen. **29**, L123 (1996).
6. G. N. Ord and A.S.Deakin, J.Phys.A **30**, 819 (1997).
7. D. G. C. McKeon and G. N. Ord, Phys. Rev. Lett. **69**, 3 (1992).
8. G. N. Ord and A.S.Deakin, Phys. Rev. A. **54**, 3772 (1996).
9. P. Dirac, *The Principles of Quantum Mechanics* (Oxford at Clarendon Press, 1958).

MRST 2002 Conference Schedule

==== Tuesday, May 14 (Evening) =======================

19:00 – 21:00: Early registration and reception at Perimeter Institute

==== Wednesday, May 15 ===========================

07:30: Registration and welcome

08:50: **Howard Burton** (Executive Director, Perimeter Institute):
Opening remarks

Plenary Session [Gabor Kunstatter, Session Chair]

09:00: **Art McDonald** (Sudbury Neutrino Observatory):
Direct evidence for neutrino flavor transformation from neutral-current interactions in SNO

10:00 – 10:20: Break

Superstrings [Gabor Kunstatter, Session Chair]

10:20: **Pedro Silva** (Syracuse University):
String bits and the Myers effect
10:45: **Garnik Alexanian** (Syracuse University):
Fixed-topology solutions in the Myers effect
11:10: **Konstantin Savvidy** (Perimeter Institute):
A new non-commutative field theory
11:35: **Wenfeng Chen** (University of Guelph):
External and internal superconformal anomalies in gauge/gravity dual

12:00 – 13:25: Lunch

Topics in Field Theory [Laurent Freidel, Session Chair]

13:25: **Sergey Kruglov** (International Educational Centre, Toronto):
Trace anomaly and quantization of Maxwell's theory on non-commutative space
13:50: **Andrei Terekidi** (York University):
Variational two fermion wave equation in QED

14:15: **Sergiy Kuzmin** (University of Western Ontario):
Breaking of supersymmetry in a U(1) model with Stueckelberg fields

14:40 – 15:00: Break

15:00: **John Cannellos** (SUNY Buffalo):
Self energy of chiral leptons in a background hypermagnetic field
15:25: **Richard MacKenzie** (Université de Montréal):
Magnetic properties of SO(5) superconductivity
15:50: **John Moffat** (University of Toronto & Perimeter Institute):
Cosmological constant problem: possible solutions

==== **Thursday, May 16** ==============================

General Relativity [Lee Smolin, Session Chair]

08:35: **Gabor Kunstatter** (University of Winnipeg):
Spherically symmetric scalar field collapse in any dimension
09:00: **Ivan Booth** (University of Alberta):
Dynamic black hole mechanics
09:25: **Kayll Lake** (Queen's University):
Algebraic computing in general relativity: recent developments
09:50: **Edward Schaefer** (Washington Gas Light Company):
Imposing a 4D background on general relativity

10:15 – 10:35: Break

Matrix Models [Konstantin Savvidy, Session Chair]

10:35: **Herbert Lee** (University of Waterloo):
A string bit Hamiltonian approach to two-dimensional quantum gravity
11:00: **Artem Starodubtsev** (University of Waterloo & Perimeter Institute):
On quantization of matrix models
11:25: **Levent Akant** (University of Rochester):
Large N matrix models and noncommutative Fisher information

11:50 – 13:15: Lunch

Particle Phenomenology [Gabriel Karl, Session Chair]

13:15: **Alakabha Datta** (Université de Montréal):
Effect of KK excited W in single top production

13:40: **Mohammad Ahmady** (Mount Allison University):
Constraints on the vector quark model from rare B decays
14:05: **Amir Fariborz** (SUNY Institue of Technology, Utica, New York):
Chiral Lagrangian treatment of the isosinglet scalar mesons in 1-2 GeV region
14:30: **Kim Maltman** (York University):
Model independent determination of the $K \to \pi\pi$ EW penguin matrix element in the chiral limit

14:55 – 15:15: Break

Plenary Session [Jim Cline, Session Chair]

15:15: **Steve Giddings** (University of California at Santa Barbara):
Black holes in high-energy collisions
16:15: **Dick Bond** (Canadian Institute for Theoretical Astrophysics & U of Toronto):
Cosmic structure and the microwave background radiation

19:00 – 21:00: Banquet at Perimeter Institute

==== **Friday, May 17** ================================

Brane Worlds [Rob Myers, Session Chair]

08:00: **Jim Cline** (McGill University):
Quest for a self-tuning brane-world solution to the cosmological constant problem
08:25: **Hassan Firouzjahi** (McGill University):
Dynamical stability of the AdS soliton in Randall-Sundrum model
08:50: **Jérémie Vinet** (McGill University):
Order ρ^2 corrections to Randall-Sundrum 1 cosmology
09:15: **Joel Trudeau** (McGill University):
Inflation in Randall-Sundrum I Cosmology with Radius Stabilization

09:40 – 10:00: Break

Plenary Session [Robert Mann, Session Chair]

10:00: **Juan Maldacena** (Institute for Advanced Study):
Strings on flat space and plane waves from Yang Mills theory
11:00: **Lee Smolin** (Perimeter Institute & University of Waterloo):
Quantum gravity with a positive cosmological constant

12:00 – 13:15: Lunch

dS/CFT Correspondence [Eric Poisson, Session Chair]

13:15: **Robert Mann** (University of Waterloo & Perimeter Institute):
Conserved quantities, entropy and the dS/CFT correspondence
13:40: **Amir Masoud Ghezelbash** (University of Waterloo):
Vortices in dS spacetimes
14:05: **Frederic Leblond** (McGill University & Perimeter Institute):
Two tales from dS space
14:30: **Joel Rozowsky** (Syracuse University):
Non-abelian monopole and dyon solutions in a modified Einstein-Yang-Mills-Higgs system

14:55 – 15:10: Break

Developments in Quantum Theory [Fotini Markopoulou-Kalamara, Session Chair]

15:10: **Wim van Dam** (University of California at Berkeley):
Quantum computation
15:35: **Geoffrey Hunter** (York University):
Interpretation of the realized Popper's experiment
16:00: **Garnet Ord** (Ryerson University):
The Feynman Propagator from a single path

Thanks to all the participants of MRST 2002!

MRST 2002 List of Participants

Name	Affiliation	E-mail Address
Abdel-Rehim, Abdou	Syracuse University	abdou@physics.syr.edu
Ahmady, Mohammad	Mount Allison University	mahmady@mta.ca
Akant, Levent	University of Rochester	akant@pas.rochester.edu
Alexanian, Garnik	Syracuse University	garnik@physics.syr.edu
Allen, Ted	Hobart & William Smith Cols.	tjallen@hws.edu
Berndsen, Aaron	McGill University	aberndsen@physics.mcgill.ca
Bond, Dick	CITA & U. of Toronto	bond@cita.utoronto.ca
Booth, Ivan	University of Alberta	ibooth@phy.ualberta.ca
Cannellos, John	SUNY Buffalo	jc@acsu.buffalo.edu
Chen, Wenfeng	University of Guelph	w3chen@sciborg.uwaterloo.ca
Christensen, Dan	University of Western Ontario	jdc@uwo.ca
Cline, Jim	McGill University	jcline@physics.mcgill.ca
Corlett, Ian	York University	yu214773@yorku.ca
Darewych, Jurij	York University	darewych@yorku.ca
Datta, Alakabha	Université de Montréal	datta@lps.umontreal.ca
Descheneau, Julie	McGill University	jdesch@physics.mcgill.ca
Drozd, John	University of Western Ontario	jdrozdl@uwo.ca
Dyck, Alfred	Queen's University	dyck23@canada.com
Easson, Damien	McGill University	easson@hep.physics.mcgill.ca
Epp, Richard	Perimeter Institute	repp@perimeterinstitute.ca
Fariborz, Amir	SUNY Institute of Technology	fariboa@sunnyit.edu
Fernando, Sharmanthi	Northern Kentucky University	fernando@nku.edu
Firouzjahi, Hassan	McGill University	Firouzh@physics.mcgill.ca
Ghadab, Sofiane	Syracuse University	sghadab@phy.syr.edu
Ghezelbash, Amir M.	University of Waterloo	amasoud@sciborg.uwaterloo.ca
Giddings, Steve	UC Santa Barbara	giddings@physics.ucsb.edu
Hagen, C. R.	University of Rochester	hagen@pas.rochester.edu
Homayouni, Sirous	University of Western Ontario	shomayou@uwo.ca
Hunter, Geoffrey	York University	ghunter@yorku.ca
Janca, Andrew	University of Waterloo	ajjanca@uwaterloo.ca
Karl, Gabriel	University of Guelph	gk@physics.uoguelph.ca
Kiriushcheva, Natalia	University of Western Ontario	nkiriusc@uwo.ca
Kruglov, Sergey	International Educational Centre	krouglov@sprint.ca
Kunstatter, Gabor	University of Winnipeg	g.kunstatter@uwinnipeg.ca
Kuzmin, Sergiy V.	University of Western Ontario	skuzmin@uwo.ca
Lake, Kayll	Queen's University	lake@astro.queensu.ca
Laskin, Nick	University of Toronto	nlaskin@rocketmail.com
Leblond, Frederic	McGill University	fleblond@hep.physics.mcgill.ca
Lee, Herbert	University of Waterloo	h11lee@math.uwaterloo.ca
MacKenzie, Richard	Université de Montréal	rbmack@lps.umontreal.ca
Maldacena, Juan	Institute for Advanced Study	malda@ias.edu
Maltman, Kim	York University	maltman@fewbody.phys.yorku.ca
Mann, Robert	University of Waterloo	mann@avatar.uwaterloo.ca
Martel, Karl	University of Guelph	martelk@physics.uoguelph.ca

McDonald, Art	Queen's University	mcdonald@owl.phy.queensu.ca
McDonald, Mark	University of Waterloo	ml4macda@uwaterloo.ca
Moffat, John	U. of Toronto & Perimeter Inst.	john.moffat@utoronto.ca
Myers, Rob	Perimeter Institute	rmyers@perimeterinstitute.ca
Ord, Garnet	Ryerson Polytechnic University	gord@ryerson.ca
Parent, Mike	University of Guelph	mparent@uoguelph.ca
Phi, Tan-Trao	Mount Allison University	tph@mta.ca
Poisson, Eric	University of Guelph	poisson@uoguelph.ca
Potvin, Geoff	University of Toronto	gpotvin@physics.utoronto.ca
Rozowsky, Joel	Syracuse University	rozowsky@phy.syr.edu
Saremi, Omid	University of Toronto	omidsar@yahoo.com
Savvidy, Konstantin	Perimeter Institute	konstantin@perimeterinstitute.ca
Schaefer, Edward	Washington Gas Light Co/	ems57@erols.com
Silva, Pedro	Syracuse University	psilva@phy.syr.edu
Smolin, Lee	Perimeter Institute	lsmolin@perimeterinstitute.ca
Sofiane, Ghadab	Syracuse University	sghadab@phy.syr.edu
Spector, Donald	Hobart & William Smith Cols.	spector@hws.edu
Squires, Amgad	Mount Allison University	aasqrs@mta.ca
Starodubtsev, Artem	U. of Waterloo & Perimeter Inst.	astarodubtsev@perimeterinstitute.ca
Terekidi, Andrei	York University	terekidi@yorku.ca
Trudeau, Joel	McGill University	trudeau@physics.mcgill.ca
Turnbull, Joseph	University of Western Ontario	jdturnbu@uwo.ca
Valcarcel, Luis	McGill University	luis.valcarcel@mail.mcgill.ca
Valluri, Sreeram	University of Western Ontario	valluri@uwo.ca
van Dam, Wim	UC Berkeley	vandam@cs.berkeley.edu
van Leeuwen, Greg	University of Western Ontario	gvanleeu@uwo.ca
Veerasubramanian, P.	McGill University	pveera@sympatico.ca
Vinet, Jérémie	McGill University	vinetj@physics.mcgill.ca
Winters, David	McGill University	winters@hep.physics.mcgill.ca
Yavin, Tzahi	York University	t_yavin@yorku.ca
Zhiliba, Anatoly	Tver University, Russia	p000262@tversu.ru

Author Index

A

Agarwal, A., 173
Ahmady, M. R., 183
Akant, L., 173
Alexanian, G. G., 83

B

Balachandran, A. P., 83
Berenstein, D., 3
Birukou, M., 139
Bond, J. R., 15
Burgess, C. P., 203
Burton, H., xvii

C

Cannellos, J., 117
Cline, J. M., 197, 203, 209
Constable, N. R., 203
Contaldi, C., 15

D

Darewych, J. W., 105
Datta, A., 179
de la Incera, V., 117
Durhuus, B., 161

F

Fariborz, A. H., 189
Ferrer, E. J., 117
Firouzjahi, H., 197, 203

G

Ghezelbash, A. M., 217, 223
Giddings, S. B., 34

H

Hunter, G., 243
Husain, V., 139

K

Krishnaswami, G. S., 173
Kruglov, S. I., 99
Kunstatter, G., 139
Kuzmin, S. V., 111

L

Lake, K., 147
Leblond, F., 229
Lee, C.-W. H., 161
Leibbrandt, M. C. C., x

M

MacKenzie, R., 123
Maldacena, J., 3
Mann, R. B., 217, 223
Mason, B., 15
McDonald, A. B., 43
McKeon, D. G. C., 111
Moffat, J. W., 130
Myers, S., 15

N

Nagashima, M., 183
Nastase, H., 3

O

Olivier, M., 139
Ord, G. N., 249

P

Pearson, T., 15
Pen, U.-L., 15
Pogosyan, D., 15
Preston, B., 139
Prunet, S., 15

R

Rajeev, S. G., 173
Readhead, T., 15
Rozowsky, J. S., 235

S

Savvidy, K., 89
Schaefer, E. M., 152

Sievers, J., 15
Silva, P. J., 77, 83
Smolin, L., 59
Starodubtsev, A., 167
Sugamoto, A., 183

T

Terekidi, A. G., 105

V

Vaz, E., 139
Vinet, J., 209